T0259959

CAMBRIDGE LIBRARY COLLECTION

Books of enduring scholarly value

Earth Sciences

In the nineteenth century, geology emerged as a distinct academic discipline. It pointed the way towards the theory of evolution, as scientists including Gideon Mantell, Adam Sedgwick, Charles Lyell and Roderick Murchison began to use the evidence of minerals, rock formations and fossils to demonstrate that the earth was older by millions of years than the conventional, Bible-based wisdom had supposed. They argued convincingly that the climate, flora and fauna of the distant past could be deduced from geological evidence. Volcanic activity, the formation of mountains, and the action of glaciers and rivers, tides and ocean currents also became better understood. This series includes landmark publications by pioneers of the modern earth sciences, who advanced the scientific understanding of our planet and the processes by which it is constantly re-shaped.

Travels in the Island of Iceland, During the Summer of the Year MDCCCX

Sir George Steuart Mackenzie (1780–1848) was a Scottish baronet whose interests included chemistry and geology. This work, first published in 1811, is his account of his voyage to Iceland in 1810 for the purposes of mineralogical research. Accompanied by physicians Henry Holland and Richard Bright, Mackenzie surveyed volcanoes, geysers and the other geological features of the island. In addition to reporting the results of the expedition's scientific exploration, this charming and evocative journal describes the history, culture, attire and cuisine of the islanders. Also included are Richard Bright's observations on the zoology and botany of Iceland and a survey of the health of the population by Henry Holland, who introduced smallpox inoculation during his visit. Written in an easy, accessible style, this account brings to life the sights, smells and tastes of the tour and the often rudimentary accommodation and travel conditions.

Cambridge University Press has long been a pioneer in the reissuing of out-of-print titles from its own backlist, producing digital reprints of books that are still sought after by scholars and students but could not be reprinted economically using traditional technology. The Cambridge Library Collection extends this activity to a wider range of books which are still of importance to researchers and professionals, either for the source material they contain, or as landmarks in the history of their academic discipline.

Drawing from the world-renowned collections in the Cambridge University Library, and guided by the advice of experts in each subject area, Cambridge University Press is using state-of-the-art scanning machines in its own Printing House to capture the content of each book selected for inclusion. The files are processed to give a consistently clear, crisp image, and the books finished to the high quality standard for which the Press is recognised around the world. The latest print-on-demand technology ensures that the books will remain available indefinitely, and that orders for single or multiple copies can quickly be supplied.

The Cambridge Library Collection will bring back to life books of enduring scholarly value (including out-of-copyright works originally issued by other publishers) across a wide range of disciplines in the humanities and social sciences and in science and technology.

Travels in the Island of Iceland, During the Summer of the Year MDCCCX

GEORGE STEUART MACKENZIE

CAMBRIDGE UNIVERSITY PRESS

Cambridge, New York, Melbourne, Madrid, Cape Town,
Singapore, São Paolo, Delhi, Tokyo, Mexico City

Published in the United States of America by Cambridge University Press, New York

www.cambridge.org
Information on this title: www.cambridge.org/9781108030212

© in this compilation Cambridge University Press 2011

This edition first published 1811
This digitally printed version 2011

ISBN 978-1-108-03021-2 Paperback

TRAVELS IN ICELAND.

TRAVELS

IN THE

ISLAND OF ICELAND,

DURING THE SUMMER OF THE YEAR

MDCCCX.

By SIR GEORGE STEUART MACKENZIE, Baronet,

FELLOW OF THE ROYAL SOCIETY OF EDINBURGH,

&c. &c. &c.

EDINBURGH:

Printed by Thomas Allan and Company,

FOR ARCHIBALD CONSTABLE AND COMPANY, EDINBURGH;

LONGMAN, HURST, REES, ORME, & BROWN; CADELL & DAVIES; WILLIAM MILLER,

AND JOHN MURRAY, LONDON.

1811.

But where to find that happiest spot below,
Who can direct, when all pretend to know?
The shudd'ring tenant of the frigid zone,
Boldly proclaims that happiest spot his own;
Extols the treasures of his stormy seas,
And his long nights of revelry and ease.

— — — — — — — — —

Such is the Patriot's boast, where'er we roam,
His first best country ever is—at home.

Goldsmith.

TO HIS EXCELLENCY

FREDERIC, COUNT TRAMPE,

LATE GOVERNOR OF ICELAND,

NOW OF DRONTHEIM, IN NORWAY,

&c. &c. &c.

AS A TESTIMONY

OF RESPECT FOR HIS PUBLIC CHARACTER,

AND OF GRATITUDE FOR HIS KIND EXERTIONS

TO RENDER THE JOURNEY THROUGH ICELAND

AGREEABLE AND SUCCESSFUL,

THIS VOLUME

IS INSCRIBED

BY

THE AUTHOR.

PREFACE.

THE Island of ICELAND is but little distant from that of Britain: it has long been known to contain many extraordinary natural phenomena; and yet very few have been induced to visit it, either from private curiosity, or from the more general views of science. The first British travellers who attempted to explore the country, either observed nothing, or thought their observations too uninteresting to be communicated to the public; and even the Letters of Von Troil, who accompanied Sir Joseph Banks about forty years ago, though in many respects valuable, were, perhaps, chiefly so, by awakening the curiosity of science to that neglected, but remarkable country.

In the year 1789, Sir John (then Mr) Stanley conducted an expedition to Iceland, with more extended views, and visited some parts of the country that had not been examined by Sir Joseph Banks and his party. The very interesting account which he has given of the Geysers and the Reikum Springs, in his letters to Dr Black, leaves room for regret that he has not given to the public a complete narrative of his voyage.

The importance which the study of Mineralogy has of late years acquired, and the intimate connection which it is now acknowledged to hold, with all legitimate views of geological science, rendered the examination of Iceland particularly desirable. Impressed with this idea, and several circumstances combining to strengthen the desire I had long entertained of visiting that island, I resolved to gratify my wishes during the summer of the last year. One of my principal inducements for undertaking the voyage at that time, was the good fortune of having accidentally met, when visiting part of my estate in Ross-shire, with a young man, a native of Iceland. He had been on his voyage to Copenhagen, when stormy weather forced the vessel he was in, to take shelter in the harbour of Stornoway, in the island of Lewis. By this time the war with Denmark had broken out, and he was detained a prisoner with the rest of the crew and passengers; but was afterwards liberated. This young man, by name Olave Loptson, having acquired some slight knowledge of medicine, from a residence with the physician at Reikiavik, in his native country, contrived to make himself useful to the people in the Lewis and neighbouring islands, and even extended his practice to the mainland where I found him. The circumstances of his story, and compassion for his situation, induced me to befriend him. I carried him with me to Edinburgh, to give him the advantages of a better medical education; and from the information I received from him, I resolved not to delay my long projected visit to his country.*

* Loptson attended us on the voyage; and from his knowledge of the language, we undoubtedly derived some benefit. He did not choose to return with us; and though I had much reason to be displeased with him, I should have been very glad to have heard, even after he had forfeited all right to my patronage, that he had availed himself of the advantages which I was enabled to afford him during his residence in Scotland.

b

My intentions being known, two gentlemen of the University of Edinburgh, Mr Henry Holland (now Dr Holland), and Mr Richard Bright, expressed a wish to accompany me; and I did not hesitate to meet their wishes, knowing them to be young men of very superior talents and acquirements, in a high degree pleasing in their manners, and promising me the hope of numbering them (as I now have the happiness of doing) among my friends. I was the more gratified by this incident, as particular circumstances had deprived me of the company of my friend Thomas Allan, Esq. a gentleman well known to mineralogists; and of that of Mr Hooker, whose reputation as a botanist is already very great. Mr Hooker, at the desire of Sir Joseph Banks, had gone to Iceland in 1809, and had made considerable collections in natural history, which were unfortunately destroyed, by the burning of the ship in which he was returning to England. He was desirous of repairing his losses, with the comfort of a companion to share his toil. Notwithstanding the destruction of his notes, and of almost every thing he had collected during his stay in Iceland, this gentleman has given to the world a volume replete with valuable information. It was only a short time ago, that Mr Hooker resolved to publish his work, (having first printed it merely for the use of his friends), and before he informèd me of his intentions, I had, by his kind permission, extracted and printed his botanical remarks.*

Though mineralogical research was the principal object of the voyage, yet having enjoyed the opportunity of seeing much more of Iceland than any former British traveller, and having bestowed considerable attention upon every thing that appeared interesting in the island and its inhabitants, I cannot but conceive myself as still under the obligation of submitting to the public, the information that has

---

* Every one who peruses Mr Hooker's work, must value the information it contains; but the account which he has given of the revolution, which happened while he was in Iceland, appears to me to be very partial. A brief account of it is given in a note, page 80, in this volume, but no opinion is offered. After the reader has perused the note above referred to, I request his attention to No. VIII. of the Appendix, where, since Mr Hooker has published his work, I have thought it necessary to state my opinion.

b 2

been acquired, even although Mr Hooker's book may, in some respects, have anticipated me.

The labour of composing the following work has been divided, that it might be sent to the press with as little delay as possible. To some readers, perhaps, the difference of style, which such a division of labour necessarily creates, may be displeasing: yet, I trust, that the advantages of it will be still more apparent; that it will be felt, when each individual of a party devotes himself to those particular objects of research, which are most suitable to his inclination and habits of thought, there is a probability of much more accurate, as well as extensive information being acquired, than if every one embraced the whole field; and that, in a work of this kind, accuracy of observation will be considered of far greater importance than mere uniformity of composition.

Whoever reads the Preliminary Dissertation on the History and Literature of Iceland, the account of the present state of Literature, and the chapter on Government, Laws, and Religion, will find no cause to regret that these parts of the work have been executed by Dr Holland; and medical men cannot but be satisfied with his account of the diseases of the Icelanders, which will be found in the Appendix. The Agricultural Report of Cheshire has given an early celebrity to Dr Holland's name; and I have great pleasure in anticipating, from the labour he has employed in this work, a material addition to the reputation which he has so deservedly acquired.

Mr Bright has made the most of the materials we had time to collect, for an account of the Zoology and Botany of Iceland. To him we are indebted for the preservation of the plants we gathered, and indeed for by far the greatest part of the collection; and I shall ever retain a grateful remembrance of the chearful and ready exertion he always displayed, and the undeviating good humour, with which he submitted to the cross accidents which sometimes befel us. In the midst of professional studies, he has found time to furnish me with many valuable remarks, and much useful information, which, without his assistance, I could not have procured. These, in addition to the materials of Dr Holland's journal, and my own, have enabled me to draw up a narrative of our Voyage and Travels, in which simplicity has been chiefly studied. For what relates to Rural Affairs, Commerce, and Mineralogy, I am alone answerable.

I do not pretend, even with the able assistance I have received, to have accomplished all that might have been done, or to have rendered the future visits of naturalists to Iceland useless or unimportant. The eastern and northern coasts still remain unexplored, and from what I have learned, are well worthy the attention of mineralogists and botanists.

We went to Iceland, believing that we had provided against any difficulties which might occur; but we were so far deceived, that we found ourselves obliged to spend much of the time in Reikiavik, the chief place in the island, which we had expected to employ in travelling through the country. The extreme slowness of the Icelanders, their awkward mode of loading horses, and the badness of the roads, (if any thing in Iceland can be called a road), occasioned much unlooked-for trouble and delay; and will be regarded, I trust, as an apology

for any disappointment in expectations which may
have been excited.

The geography of Iceland is still very imperfect;
but a survey of the coast, in which two Danish
officers have during some years been engaged, will
doubtless, in a great measure, remove this defect.
Ignorant of this undertaking, I took instruments
with me for the purpose of improving the geography
as much as I was able, and of making meteorological observations; but I soon found that it was
impracticable to carry them safely, and to attend at
the same time to the objects I had principally in
view. Having obtained a copy of Minor's charts
of the south western coast of the island, I have
been enabled to delineate those parts through which
we passed, so as to convey a tolerably correct idea
of the face of the country. By means of a sex-
tant, I found the latitudes laid down by Minor to be
very correct; that of Reikiavik differing only 4″
from my observations. From the neighbourhood of
Reikiavik, I took a set of angles with a theodolite,
and from these, and observations taken with a compass as we travelled, our route has been traced.

But, though the map I have drawn certainly gives a better idea of the country than any hitherto published, it must not be considered as by any means accurate; for I had no opportunity of rectifying my observations; and those made with a compass are liable to many errors. The small general map of Iceland is a mere sketch, made for the purpose of giving an idea of the shape of the island, and of the relative positions of different places mentioned in the course of the work.

Our deficiency in meteorological observations arises from causes already noticed. The register of the weather is annexed, as it was made up at the time.

Though it was inconsistent with the plan of this volume, to enter into details respecting the history of Iceland, it has been thought proper to give a very general historical sketch of the island, which may be useful to the reader. A very good account of the state of Iceland in the eighteenth century, has been published by Mr Stephenson, the present President of the Supreme Court of Justice in the island. From that book some valuable information has been extracted ; and, in consulting other works of merit in reference to the same object, we have been actuated by the earnest desire of conveying to the public a distinct view of a country, which must ever be interesting, both to the moral and physical observer.

Edinburgh,
October 1811.

CONTENTS.

ERRATA.

Page   Line

PRELIMINARY DISSERTATION

ON THE

HISTORY AND LITERATURE

OF

ICELAND.

DISSERTATION, &c.

THE History of Iceland, though possessing little importance in its relation to the political events of other nations,
is nevertheless curious and interesting in many of its features.
It is the narrative of a distinct and peculiar race of people ;
of a community which, oppressed by all the severities of soil
and climate, and secluded amidst the desolation and most
destructive operations of nature, has preserved, through the
progress of nearly a thousand years, an enlightened system of
internal policy, an exalted character in all religious and social
duties, liberal methods of education, and the culture of even
the more refined branches of literature and knowledge. Few
of the events which are most prominent on the page of general history, are to be met with in the simple chronicles of
this island. The battle, the siege, the rebellion, and the revolution, are here almost utterly unknown. In place of these
social and self-engendered calamities, we find the record, pathetic in its simplicity, of the various physical evils by which
the Icelanders have been oppressed,—the severities of a winter—the famine of the ensuing year—the virulence of an
epidemic disease—or the horrors of an earthquake and volcanic eruption. With the civil history of the people, that of

their literature is intimately combined ; for, in the government and internal concerns of the country, mental attainments and civil influence have at all periods been closely linked together.   The progressive state of religion forms another principal feature in their annals ; and more appears of individual biography, than is common to historical writing ; a natural effect, as it would seem, of the situation of the inhabitants, and of the condition of society which has existed among them from early times.   The records of Iceland, in short, are not so much those of kings and governments, as of a community of families and a people : and the philosopher or moralist, while they glean over the fields of history for the materials of their study, will find a harvest provided even in the annals of this remote and desolate island.

The native historians of Iceland are exceedingly numerous. Of their particular merits; more will hereafter be said : at present it may be sufficient to state, that they have successfully elucidated even the most remote periods in the history of their country, and that their simplicity and distinctness furnish strong internal evidence of authenticity.   The minuteness of narrative, however, which forms their best recommendation to those who desire to study the character of a people and of a passing age, would deter the majority of readers from an examination of their contents.   In the following sketch of the history of Iceland, the most conspicuous and remarkable features have been selected, to the exclusion of the numerous details with which the native historians have crowded the picture ; while an attempt is made to trace the circumstances which have given to the Icelanders that peculiar and distinctive character, which, as a nation, they have always exhibited.   In reference to the latter object, it was requisite that some notice should be taken of the literature of the country at different periods ; and, accordingly, with the narrative of events, are connected those sketches of the intellectual habits and pursuits of the people, which may best serve to illustrate this singular feature in their history. To a subsequent chapter is reserved the account of the present state of literature and education in the island.

Much argument has taken place on the question, whether Iceland was the Thule of the ancients. Though the affirmative opinion has not been without its authorities, and these of eminent character, yet there appears, on the whole, little that is satisfactory or probable in this hypothesis. Were it true that the Romans had ventured upon the northern ocean as far as the shores of this island, we might expect to find some distinct and well-marked record of the fact. The passages, however, from their poets and historians, in which the Thule is mentioned, do not afford any such testimonies; and some of these would seem to be decidedly opposed to the opinion in question. *

The discovery of Iceland, as first authenticated by history, is due to the adventurous spirit of certain Norwegian and Swedish pirates, about the year 860. Naddodr, the first voyager who visited the island, was accidentally driven upon the eastern coast, while sailing from Norway to the Faroe isles :

* Those who wish to examine further into the merits of this question, may consult the writings of Saxo Grammaticus, Casaubon, Bochart and Mallet ; the Crymogæa of Arngrim Jonas ; the Historia Ecclesiastica Islandiæ, &c. The two first mentioned writers maintain the identity of Iceland with the Thule of the Romans. Arngrim Jonas is the principal opponent of this opinion. Bochart, Mallet, and Bishop Jonson, the learned author of the Ecclesiastical History, hold an intermediate belief; supposing it not impossible that Iceland was known to the ancients ; but contending that they applied the name Thule, without particular discrimination, to several places near the northern boundary of their empire.

he ascended to the summit of a hill near the shore ; and, seeing around him only a solitary desart, the mountains of which were covered with snow, though it was then the autumn of the year, he gave to the country the name of Sniá-land. Led by the report of this casual discovery, a Swede, called Gardar, pursued the same track ; and, in the summer of 864, succeeded in circumnavigating the island, which he named Gardarsholm, or the island of Gardar. A pirate of much celebrity in those times, Floke by name, was the third adventurer in this track. Directed by the flight of ravens, which he had carried out with him from Norway, he reached Iceland ; where he remained two winters ; exploring, during this time, a great extent of the southern and western coasts, and giving to the island the name which it still retains. This name was derived from the observation he made of a large quantity of ice in some of the bays on the northern coast ; probably floated hither from Greenland.

These voyagers carried back with them various and contradictory reports of the island thus recently discovered. Their expeditions, though affording some knowledge of the country, would probably not have led to its colonization, had not other causes contributed to this effect. At the period in question, Harold the Fair-haired, who had been successful in subjugating the several petty states of Norway, obtained the sovereignty over the whole of that kingdom. Many noble Norwegians, dissatisfied with the change, and aggrieved by the attempts of Harold to controul and reduce their privileges, determined to abandon the country which gave them birth. A body of these voluntary exiles, under the conduct of Ingolf, one of the discontented subjects, sailed from Norway, A. D. 874, and directed their course to Iceland, where they hoped to retain unimpaired their former rights and independence. It would appear that this emigration was the result of a mature and well-concerted scheme; since Ingolf, with his kinsman Hiorleif, visited Iceland in 870; and afterwards returned to Norway, to assemble the different families, who were content to resign their fortunes to his guidance.

It is not necessary to the history of Iceland to trace back into ages more remote, the Norwegian people by whom the island was thus colonized. Forming one branch of the great Gothic or Teutonic family, which occupied at this period the northern kingdoms of Europe, it cannot be doubted that they were derived from one common source; and we look to their origin in the Scythian emigration, which, nine centuries before, under the banners of the victorious Odin, carried conquest and usurpation throughout all the vast regions of the north.

It is stated by the Icelandic historians, that the first settlers upon the island, found, on certain parts of the coast, wooden crosses and other instruments; from which it might be inferred that the country had before been visited, either by accident or design. The nature of the relics thus discovered, led to the belief that these earlier visitors were Christians; and it was conceived probable that fishermen from Britain or from Ireland had been accidentally driven upon the coast, and had either perished there, or succeeded in refitting their vessels, so as to return to their own country. That they did not remain long in the island, is rendered probable by there being no vestige of habitations when the Norwegians arrived. Some of the Icelandic historians, however, make a different statement; and assert that there were actual settlements on the island, previously to the period of the Norwegian emigration. The Landnama Book, one of the earliest of the records of Iceland, speaks of the discovery of crosses, bells, and certain writings in the Irish language; and mentions in another place the residence of some of these foreigners at Kirkiubai, on the southern coast of the island. *   Are Frode, an eminent writer at the commencement of the 12th century, who had a part in the composition of the Landnama Book, states in another of his works, that the *Papæ*, as these original inhabitants were termed by the Norwegians, were actually upon the island at the arrival of the colony from Norway, and that they migrated elsewhere, from an aversion to the society of the Heathen strangers. †   It is somewhat singular that a discrepancy of statement should exist on this subject; more especially when we consider the early period at which the Icelanders began to compose history, and the minute attention they bestowed upon historical facts.   The first opinion which has been mentioned, though not free from uncertainty, is probably, on the whole, that which approaches most nearly to the truth. ‡

* Landnama Bok. Proem. et Part IV. Cap. 11.   Havniæ 1774.

† Arii Sched. c. 2.

‡ The testimonies of Alfred and the venerable Bede have been adduced, to prove that Iceland was known to the natives of Britain before the discovery of the Norwegian pirates.   The description, however, given by Bede, of *an island, called Thule, six days' sail from Britain, where there are no nights in the summer solstice, and no days in the winter*, seems to be derived solely from the testimony of Pliny, in describing the northern voyage of the celebrated Pytheas Massiliensis (*Hist. Nat. l. ii. c. 75, & l. iv. c. 16*.); and it appears probable that the Thule of Pytheas is to be found in the Shetland or Faroe isles, rather than in the more remote coasts of Iceland.   In Hackluyt's Voyages (*London, 1559*), *vol. I. p.* 1, it is mentioned, on the authority of Galfridus Monumetensis, and the Αρχαιονομια of M. Lambard, that the ancient British king, Arthur, about the beginning of the 6th century, subdued Iceland and Greenland, as well as many of the northern kingdoms of Europe.   It is further stated, in the same book, likewise from Gal-

However this may be, Ingolf and his associates found no other difficulty in establishing themselves on the island, than what arose from the nature of the country; at that time so much covered with thickets, that it was necessary in some places to open tracks through them. The infant colony which, directed by certain superstitious observances, was first settled in the south-western part of Iceland, received a rapid augmentation of its numbers. * Many other noble Norwegian families, dissatisfied with their condition under the imperious authority of Harold, left their country, attended by large trains of dependants, and followed the course of the earlier emigrants, who had sought liberty in the desolate regions of Iceland. Unassociated in their plans, and arriving at different times, they did not confine themselves to any particular district, but spread their settlements along all the coasts of the island. It appears that Harold at first opposed no hindrance to this emigration of his subjects, but even afforded them in some instances his assistance and advice, in the distribution of the lands upon which they fixed their new abodes.† Afterwards, however, its frequency so much alarmed him, that he issued an edict, imposing a fine of four ounces of silver upon every person who should leave Norway to settle in Iceland.

fridus, that Malgo, a successor of King Arthur, a second time conquered Iceland, and subjected it to his power. These accounts are evidently entitled to very little credit.

* When Ingolf approached the coast of Iceland, he threw into the sea the wooden door of his former habitation in Norway; and some time afterwards, finding it cast upon the shore at Reikiavik, he fixed his abode on this spot. We meet with relations of a similar practice among others of the first settlers in the country; it being regarded as a method of ascertaining the will of the Deities as to their place of settlement. *Landnama Bok*, p. 1, c. 6.

† Landnama Bok, p. 5, c. 1.

The emigrations, however, in despite of restriction, continued to a great extent; and, in little more than half a century, the coasts of Iceland were peopled in a degree fully equivalent to the means of subsistence which the country afforded. The Landnama Book, before mentioned, the object of which is to give a picture of the colonization of the island, describes with singular minuteness the arrival and spreading of the different settlers; and connects this narrative with a profusion of genealogical detail, unexampled perhaps in the annals of any other country. From this record it appears, that together with the Norwegians, many Danes and Swedes came over to the country; and the names also are preserved of several individuals of the Scotch and Irish nations, who at this time chose Iceland as their place of abode.

The period occupied in this progressive colonization of the country, may be considered the first in the history of Iceland. The Norwegians brought with them to the island their language, their religion, their customs, and their historical records. Their method of settling was accordant to the usages which existed at this period among the northern nations of Europe: a sort of feudal arrangement was immediately adopted, by which every leader of a body of emigrants, having occupied a district of country, distributed lands to his followers, under certain implied conditions of vassalage and subservience. Thus all the habitable parts of the coast were speedily parcelled out among the leaders of numerous small communities; and, fresh bands of emigrants still arriving, frequent contests arose between the new comers, and those already dwelling in the country. These contests for possession became in time so common among the petty feudal chiefs in different parts of the island, that a necessity was universally felt for the adoption of some system, which by connecting the yet separate communities, might regulate the concerns, assimilate the interests, and unite the feelings of the whole. The beneficial change, thus required, was effected A. D. 928, fifty-four years after the arrival of the first settlers ; and a constitution was established, well calculated to provide for the emergencies which gave it birth.

The general features of the new government were undoubtedly those of an aristocracy ; though possessing certain peculiarities which may distinguish it perhaps from any other recorded in history. The island was divided into four provinces ; over each of which presided an hereditary governor or judge. The southern and western provinces each comprehended three subdivisions or prefectures ; the northern province four of these prefectures, on account of its greater extent; the eastern province only two; so as to give a distribution of the whole island into twelve secondary districts. Over each of these divisions was placed a prefect ; who, as well as the governor of a province, held his office by an hereditary right, being originally appointed from one of the principal families in the district. The prefectures were further subdivided into smaller districts, called *Hreppar;* in each of which there were five civil officers, elected from among those of the inhabitants, who to a certain value of property, united the reputation of general sense and integrity. It was the province of these magistrates to maintain the peace and good order of their several districts ; and more especially to attend to the condition and management of the poor.

In each of the three civil divisions thus established, were vested certain powers of assembly for public purposes. In the *Hreppar*, or smaller districts, these assemblies were com-

posed of all the inhabitants who possessed more than an assigned value of estate, and were of an unblemished moral
character.   Their proceedings, which related almost entirely
to matters of local police, were under the cognizance of the
court of Prefecture ; formed by the prefect and the depu-
ties from the several included districts.   At these second assemblies of judicature, which were held once in the year, a
great part of the civil business of the country was transact-
ed, and a superintendance exercised also in certain matters
of religion ; an appeal, however, being still reserved, in the
first instance to the States of the whole Province, which met
at particular times, and finally to the Supreme General Assembly of Iceland.

This great assembly, called the *Althing*, * was held annually on the shores of the lake of Thingvalla ; and was attended by all the civil officers, and by a certain number of the
clergy and laity from every part of the island.   In it were
vested the legislative and supreme judicial powers of the
state, and the decision in all weighty and important matters
of national policy.   Every appeal from the lower courts was
determined here ; and every magistrate made responsible to
this tribunal for his public conduct.   The functions of the
executive government were committed by the assembly to a
magistrate, elected by their votes, and retained in power
during their pleasure.   This supreme officer had the title
of *Laugman*, or administrator of the laws.   Chosen, in consequence of merit, from among the chiefs of the state, he
was invested with all the dignities suitable to his office.   He

---

* From *all*, all, and *thing*, a forum or place of justice.   This assembly corre-
sponds in many circumstances to the Wittena-gemot of the Anglo-Saxons.   Similar institutions, indeed, existed among most of the northern nations at this period.

presided at the general assembly, conducted its counsels, and registered its decisions. He interpreted the laws of the nation, and pronounced the sentences which the administration of these laws required. His authority, though dependent for its duration upon the will of the states, was usually continued for many years, and often for life; while at the same time this uncertainty afforded an important check to abuses of his office. As the supreme magistrate of the country, he was wisely gifted with all the externals of dignity and power ; and with equal wisdom restricted in all those points where his influence might prove dangerous to the welfare of the republic. *

In this sketch of the constitution which the Icelanders created for themselves, a distinct relation will be traced to the progressive institutions of several of the European states. One obvious peculiarity, however, offers itself in the present instance. Elsewhere, the progress towards regular government was gradual, and every step made through contest and bloodshed. In Iceland, all was effected by a single and simultaneous effort. The necessity appeared for some bond of union among the several communities of the island : the voice of sage deliberation gave to the people this common bond, in a constitution which was received without tumult, and brought into action without delay. On this subject it would be an injustice to the reader not to quote the words of the elegant and ingenious Mallet.

‘ Le génie de ces peuples, leur bon sens naturel, et leur
‘ amour pour la liberté y paroissent sans aucun nuage. Au-
‘ cune force extérieure ne les croise, ni ne les gêne ; c’est
‘ une nation livrée à elle même, qui s’établit dans un pays

* A minute account of the constitution of the Icelandic commonwealth is given in the Crymogæa of Arngrim Jonas.

' isolé, et comme separé du reste du monde ; dans tous ses
' établissemens, on ne voit que la plus pure expression de ses
' inclinations, et de ses sentimens ; et ils lui sont en effet si
' naturels, que l'on n' apperçoit pas dans les recits aussi
' naïfs, q'étendus des Chroniques Islandois qu'aucune déli-
' beration génerale, aucune irrésolution, aucune expérience
' des états différens, ayent précedé chez eux l'institution de
' cette forme politique.   Tout y nait, et s'y arrange de soi-
' même ; et comme les abeilles forment leurs ruches, les nou-
' veaux Islandois établissent chez eux ce gouvernement, qui
' semble ne devoir être le fruit què d'une longue experience
' et d'une étude réflechie des hommes ; et dont un grand gé-
' nie de ce siecle (*Montesquieu*) remarque avec étonnement,
' qu'il a été trouvé dans les bois. ' *

The code of laws, adopted in connection with this new
form of government, and progressively altered and amended by the decisions of the public assembly, is another striking specimen of the genius and habits of the Icelanders of
this age.   It was constructed with a minute attention to the
usages of the people, and to the various objects in their internal economy.   All the contingencies of society were provided for ; the relations and duties of different classes prescribed ; and other regulations introduced, which had in
view the convenience and utility of the whole.   As instances,
may be mentioned the laws respecting marriage, divorce, and
inheritance ; those which regarded the management of the
poor, the distribution of landed property, and the regulation
of weights and measures.   The criminal laws were equally
reduced to a systematic form, and adapted to the character
of society in this age.   In conformity to the custom of

* Introduction a l'Histoire de Dannemarc, p. 118.

the other Scandinavian nations, corporeal punishment was rarely inflicted ; and the atonement for almost every species of offence was made by the payment of certain fines ; which in the cases of more heinous guilt, amounted sometimes to the confiscation of the whole property of the offend- er. * As in the Spartan and Roman laws, no punishment was provided for parricide ; from the same conviction that the nature of the crime was in itself a sufficient security a- gainst its commission. †

The constitution, thus adopted by the Icelanders, was preserved with little change for more than three centuries ; during which period the records exist of thirty-eight Laug- men, who in succession sustained the executive powers. Were it allowed to apply the term to a desolate island on the con- fines of the Arctic Circle, this might be called the golden age of Iceland. Secured by physical circumstances from the ambition of more powerful states, an efficient government and well directed laws provided for the people all the advan- tages of justice and social order. Education, literature, and even the refinements of poetical fancy, flourished among them. Like the Aurora Borealis of their native sky, the poets and historians of Iceland not only illuminated their own country, but flashed the lights of their genius through the night which then hung over the rest of Europe. Com-

---

* For an account of the *Wiigslode*, or criminal laws of the Icelandic common- wealth, see the Crymogæa of Arngrim Jonas.

† Before the introduction of Christianity into Iceland, a superstitious test of the truth of evidence, frequently resorted to in criminal and other cases, was to set the witness upon an oblong piece of turf, so far separated from the ground as to be elevated about the middle into an arch. If this arch did not break with the weight of the person, he was declared a valid witness ;—if it fell, his testimony was rendered void.

merce was pursued by the inhabitants with ardour and success ; and they partook in the maritime adventures of discovery and colonization, which gave so much merited celebrity to the Norwegians of this period.   Many of their chiefs
and learned men visited the courts of other countries, formed connections with the most eminent personages of the time,
and surveying the habits, institutions and arts of different
communities, returned home, fraught with the treasures of
collected knowledge.   Nor was there among the Icelanders
of this period an extinction of the elevated spirit, common
to their forefathers and to the age.   The Sagas, or tales of
the country, afford many striking pictures of that high feeling
of honour, and of those deeds of personal prowess, which
were cherished by the disposition of the northern nations,
and which refused not to exist even in this remote and desolate region.

Of the several features which distinguish this remarkable
period in the history of Iceland, the literary character of the
people is doubtless the most extraordinary and peculiar. We
require much evidence to convince us of the fact, that a nation, remote from the rest of Europe, dwelling on a soil so
sterile, and beneath such inclement skies, should have sent
forth men whose genius, taste and acquirements did honour
to their country, and to the times in which they lived.   Such
evidence, however, of the most distinct and decisive kind,
we possess in the many writings which have come down from
this period to the present age, and in the testimonies afforded by the contemporary writers of other countries.   The reality of the fact, indeed, can admit of no doubt ; and it is
only left for us to speculate upon the causes which led to this
singular anomaly in the history of literature.

The original settlers in Iceland were men who had posses-sed eminence and hereditary rank in the kingdom of Nor-way. Deserting their country from an abhorrence of des-potic sway, they carried with them to their new abodes the language of their ancestors, (the Gothic or Teutonic root, from which so many branches have sprung;) and numerous records, both of individual family descent, and of the ge-neral history of the northern nations; these annals being preserved for the most part in that poetical form, which dwelt with an equal vivacity of impression upon the memory and feelings. The mythology of the Scandinavians, in its nature propitious to poetic fiction and ornament, was at the same time transplanted into Iceland; the gloomy scenery of which was not wholly unaccordant with the character and usages of this religion. The majesty and the terrors of Odin and of Thor might well be supposed to reside amidst the desolation of na-ture; and the future bliss of the Valhalla was pictured with a simplicity of description, which gave it even to the imagi-nation of the Icelander, who knew but the names of ease, luxury, and splendour. The attributes of the other deities, and the events of the mythology, were equally adapted to the situation and conceptions of the people during this in-fancy of their existence as a civilized community.

Possessing thus the materials for history and the elements of poetical composition, and aided by various remembrances and associations connected with their former country, it is not wonderful that the Icelanders should have been early led to the cultivation of these branches of literature. Other causes also contributed to this effect. The institution of a regular form of government, though it diminished the number of inci-dents which might have been adorned by the language of po-etry, yet afforded a peaceable security of condition, which fa-

c

voured the culture of the mind, and enabled those who had
learning or genius to record the events and atchievements of
ages that were gone by. The peculiar form, too, of this government, afforded a powerful incentive to mental exertions ;
and at the great annual assembly of the people, those who
possessed quickness and promptitude of talent, with powers
of composition and oratory, never failed to obtain the applause of their fellow-citizens, and an influence in the deliberations of the state. Nor were the more domestic occupations of the Icelanders such as to preclude an indulgence of
these dispositions. The summer sun saw them indeed labo-
riously occupied in seeking their provision from a stormy
ocean and a barren soil ; but the long seclusion of the winter gave them the leisure, as well as the desire, to cultivate
talents, which were at once so fertile in occupation and delight. During the darkness of their year, and beneath the
rude covering of wood and turf, they recited to their assembled families the deeds and descent of their forefathers ;
from whom they had received that inheritance of liberty,
which they now dwelt among deserts to preserve.

But it was not solely as reviving the memory of former
times, or as a source of domestic enjoyment, that the Icelanders of this age devoted themselves to the composition of
history and poetry. The ambition of wealth and glory further animated their pursuits. Their bards and historians visited other countries, resided amid the splendours of courts,
were caressed by the greatest monarchs of the time, and returned to their native island, covered with honours, and enriched by the gifts which their genius had won. Thus, interest and emulation preserved the character the people had
acquired, long after some of the causes producing it had
ceased to operate ; and literature became with the Iceland-

ers a species of commerce, in which the fruit of their mental endowments was exchanged for those foreign luxuries or
comforts, which nature had denied to them from their own
soil.

Such appear to have been the circumstances which gave
rise to this singular condition of Iceland during the period from
the 10th to the 14th century; nor will it seem extraordinary, when the nature of the causes is considered, that they
should have had so much effect upon the dispositions and
character of the people. It may be observed, in concur-
rence with the view that has been given, that their attention
was chiefly engaged by the two branches of literature already mentioned, history and poetry; and that the more severe
departments of knowledge, though not entirely disregarded,
were by no means held in equal estimation. To these favourite pursuits they applied their utmost powers; cultivating
them in various forms, yet reducing the whole to a system,
which in its structure displayed great refinement and skill.

The poetry of the ancient Icelanders, though cherished
by them with so much success, was not, however, essential-
ly distinct in its character from that common to the other
northern nations at this period. Before the emigration which
originally peopled Iceland, the Scandinavian kings and chieftains retained in their courts, and about their persons, bards
who might celebrate their greatness, and convey the memory of their deeds to future times. These men were called
Skaldr or Skalds : * they exercised poetry as a profession,

---

* The origin of the word *Skald* has been variously stated. It has been derived from *Skiall, narratio;* from *skall, sonus;* from *gal-a, canere;* and by Torfæus (*Præfat. Hist. Orcad.*) from *Skalla, depilare.* The most probable derivation
seems to be from *Skiael,* signifying wisdom or counsel; whence also the English
word *skill.*

and their exertions were munificently rewarded by those whose praises they sung.   After the Icelanders were established as a people, and when from the causes just enumerated, they had devoted themselves to poetical composition, their native poets assumed the highest rank among these bards of the age.   The style of their composition was nearly the same as that of their predecessors in the art ; but, from their more complete devotion to the pursuit, they appear to have acquired greater skill, and a superior excellence in the qualities which were deemed essential to this kind of poetry. We accordingly find that the Icelandic Skalds obtained a singular celebrity, not only in their own country, but throughout all the north of Europe.   They visited the courts of England, Ireland, Sweden, Denmark and Norway, and were everywhere received with the hospitality and honours due to their talents, and to the exertions they made in the service of their patrons.   From catalogues which are preserved to us of the Skalds who flourished in the three last mentioned kingdoms, it appears that the majority of the whole number were natives of Iceland ; and numerous testimonies exist of the superior reputation and influence which these islanders enjoyed in the exercise of their profession. *

So much has been written concerning the Scandinavian poetry of this age, that it will not be requisite here to enter with much minuteness into the subject.   The character of this poetry exhibited certain peculiarities, derived partly

* See the *Skalda-tal,* or list of Skalds, in the appendix to Wormius's Litt. Runic. Also the accurate catalogue in the Sciograph. Hist. Literar. Island. p. 49 ; an excellent work, written by Halfdan Einarsen, rector of the school at Hoolum, and published at Copenhagen in 1777. Under the form of a *catalogue raisonnée,* with notes and indexes, it affords the most complete account, yet extant, of the literature and learned men of Iceland.

from the mythology with which it was connected, partly from the situation and circumstances of the northern nations. The religion of the Scandinavians, highly metaphorical in its nature, and embracing many strange and exaggerated fictions, gave a corresponding tone to the poetical composition of the people. It abounded in allegory and abstruse images. The events and language of mythology, associated with the various objects of nature, threw over them a shade of mystery, somewhat akin to the sublime. Even where the subject was of a narrative kind, or the praises of a hero were sung, a studied obscurity was still preserved;—every name assumed some fictitious and figurative shape;—the thought was strained to meet the conceptions of the poet;—and the imagination oppressed by the magnitude of the metaphors employed. Owing to these causes, many of the compositions which have come down to us from this age, are either wholly unintelligible, or have little accordance with the rules and observances of modern taste.

The circumstances of the northern nations, and especially of the Icelanders, further modified the character of their poetry. Dwelling in countries where the softer features of nature were but rarely seen, and simple from necessity in all their habits and modes of life, their compositions seldom exhibit much refinement either of imagery or feeling. We do not find any extended descriptions of nature, or of the mental passions and emotions. All the allusions of this kind are short and abrupt; while yet they often involve a degree of hyperbole which would be inadmissible in the poetry of the present day. No resemblance was too vague or fanciful to form the basis of a metaphor; and the imagination being urged to the discovery of such relations, numerous allegorical phrases were thus obtained, which were habitually employed by the writers of the age; though with every license to the novelties which individual genius might suggest. Their phraseology, however, was sometimes much less extravagant; and, in the description of common events, we occasionally find an homeliness and simplicity of style, which are strikingly contrasted with the qualities just described. *

The structure of the verse in the northern poetry of this period, as exemplified by that of the Icelanders, was equally peculiar. Its harmony appears to have depended less upon the arrangement and number of syllables, (though this also was the subject of minute attention,) than upon certain alliterations, and repetitions of sound, which were studiously introduced into all their poetical writings. The assonance, thus sought for, was of a much more various and complicated kind than is admissible in the mechanism of modern verse. The simple artifice of rhyme was indeed rarely employed; but upon the disposition of the consonants and vowels, and the repetition of these according to certain rules, infinite skill and labour were bestowed. Though it is difficult now to appreciate the beauty or propriety of these alliterations, we may presume that it was not merely as a demonstration of skill that they were used; and it is probable they had severally their adaptations to the nature of the Gothic language; which, abounding in consonants, might derive much increase of harmony from this artful disposition of sounds. The varieties of alliteration were exceedingly numerous. Sometimes

---

* Instances of the metaphors employed in the poetry of the Skalds might be indefinitely multiplied. The earth is *the daughter and wife of Odin ;*—hunger, *the knife of Hela or Death ;*—mountains, *the bones of the giant Ymer ;*—giants, *the sons of frost ;*—a warlike mind is *an angry sword ;*—a battle, *a storm of blood ;*—the raven is said to *rejoice over the hard game of war ;*—*a cloud of bloody drops covers the head of the wounded,* &c.

a repetition of the initial letters of verses was required ;— sometimes a correspondence in the initial letters of several words occurring in a distich or a verse ; while in other cases, more complex repetitions of sound were obtained, by using not only the same initial letters in different parts of a distich, but also certain correspondent syllables, with regular intervals between them. These varieties were connected with an almost equal diversity in the metre of the poetry ; of which some have estimated more than a hundred different forms. It has been supposed that certain of these metres have a correspondence with the Sapphic measure of the lyrical poets of antiquity ; but this opinion may probably be considered more fanciful than certain.*

The Scandinavian poetry was thus an art of the most refined kind ; and, as such, exercised with peculiar success by the Icelanders of this age. The skill of the poet being called into action even more than his imagination, contests in va-

---

* The metre most frequently used among the Icelanders was one in which the stanza was composed of four couplets ; each line of the couplet consisting of six syllables ; as in the following example from the Gunnlaugi Saga, the address of Gunnlaug to Helga, at the time of their last parting :

Bramani skein brima	Enn sa geisli syslir
Brims und liosum himni	Sidan menia-fridar
Hristar horvi glœstrar	Hvarma-tungls ok hringa
Hauk-frann a mik lauka.	Hlinar uthurft mina.

' Like that of the falcon, the bright eye of the beautiful maid, shining from ' beneath an eye-brow, which is curved as the horned moon, hath enlightened me ' by its splendour. But the beam from beneath the moon-like eye-brow of the ' maiden adorned with rings, is the cause of evil, both to herself and to me.'

Some valuable critical remarks on the Scandinavian poetry of this age will be found in various parts of Mallet's *Introd. a l'Histoire de Dannemarc* ; in the notes of his excellent translator, Dr Percy ; and also in Dr Van Troil's Letter on Icelandic Poetry.

riety and facility of versification were very frequent, and much credit was attached to impromptus, as a proof of poetical talent. Instances of such extemporary composition are extremely common, and may be found related in most of the Icelandic Sagas.*   From the same works it appears that these short pieces of poetry, (called *Flockr,* to distinguish them from the longer and more finished compositions, which had the name of *Drapa,*†) were frequently the vehicle of queries and enigmas, proposed to the ingenuity of competitors in the art. A striking example of this is recorded in the Hervarar Saga, where, to decide in a contest between a certain king and his vassal, relatively to the payment of tribute, a strict trial was made before judges of the intellectual skill of the two parties, in proposing and solving such poetical enigmas. ‡ Satire, too, was by no means an uncommon subject of these poems ; and it was at one time exercised with such severity against Harald, a king of Denmark, who had offended the Icelanders, that an invasion of the island was threatened ; and it became necessary to pass an edict, making it a capital crime to satirize the Danish, Norwegian, or Swedish kings. Other laws also were enacted in Iceland about the same period, respecting the use of personal allusions in general,

* See the Gunnlaugi ok Skalld-Rafni Saga, Hervarar Saga, Eyrbyggia Saga, Egills Saga, &c.

† See Gunnlaugi Saga, Havniæ, 1775, p. 112, and note p. 113.

‡ Hervarar Saga, Havniæ, 1785, cap. xv, p. 128.   Several other examples occur of the kings and princes of this age assuming the character of poets.   The verses of Regnar Lodbrok, a warlike and celebrated king of Denmark, are still preserved : see Wormius's Litt. Runic. 195 : also the fragment called *Nordymra,* published by Prof. Thorkelin, Lond. 1788.   An instance is elsewhere recorded of a poet, Hiarno by name, who was invested with the royal diadem, on account of the excellence of an epitaph which he composed upon his predecessor.

whether of censure or applause, in consequence of the extreme propensity to such topics which the poets of the country displayed.

Poetry having so entirely the character of an art among the ancient Icelanders, we might expect to find them possessing some common means of education and instruction in this favourite pursuit. The Edda, one of the most valuable remnants of northern antiquity, is a work designed expressly for these purposes. Much controversy has existed respecting this singular and celebrated performance; the period at which it was written, and the writers, being made equally the subjects of question. Though certain points of the discussion have never been completely decided, yet we may now consider ourselves as possessing all those facts respecting the work, which are of any material importance. It seems to be well ascertained, that the Edda is not entirely the composition of one person, or of one age, but that it derives its present form from several distinct sources. The name has been assigned to two different works; one of which is called the ancient Edda, or Edda of Sæmund; the other, supposed to be of more modern date, bears the name of the celebrated Snorro Sturleson, to whom it is ascribed. It must be remarked, however, that these titles were given at a period much later than the composition of either of the works; and that their accuracy has been disputed, inasmuch as regards the names of the authors affixed to them. *

* Different derivations have been given of the name *Edda*: some have derived it from *Edde, a grandmother*, thus making it to signify the *parent of poetry*; or from *atta, a father*, with the same use of the prosopopeia. Others have referred it to *Odde*, the residence of Sæmund Sigfuson. Arnas Magnæus considers the name as a feminine of the old word *Odr*, signifying wisdom, or reason.

The ancient Edda consists of various odes; which, as some allege, are the fragments only of a much larger work, now lost to the world. These writings, suppressed during a long period by the mistaken zeal of the Catholic clergy, were brought to light about the year 1630, by Brynjolfus Suenonius, bishop of Skalholt. The most important of the poems are the Völuspa, and the Hávámal. The Völuspa, or Prophetess of Vola, is a digest of the ancient Scandinavian mythology, short and extremely obscure; the Hávámal, a singular collection of moral precepts, professing to be derived from the god Odin himself. These poems have generally been attributed to Sæmund Sigfuson, an eminent Icelander, born A. D. 1056; who, from his knowledge, writings, and various acquirements, has been called by succeeding authors, Frode, or the learned. This opinion, however, as before mentioned, has had its opponents; and strong reasons have been urged for believing that Sæmund did not compose, perhaps not even compile, the Edda which is ascribed to him. *

The second work, bearing this name, has come to us under a more perfect form, and though itself losing the garb of poetry, is much better adapted to the object of instructing others in the poetic art. It is distributed into two principal parts. The first contains an extensive view of the mythology of Odin, under the form of dialogue; in which are explained the attributes of the deities, their several actions, and the other remarkable events of the mythology. This was a code from which the Skalds, or bards of the age,

---

* The principal opponent of Sæmund's claim to the first Edda is Arnas Magnæus; whose recondite inquiries into the early literature of Iceland have given him much celebrity. See his Life of Sæmund Frode, prefixed to the *Edda Sæmundar, Hafniæ* 1787.

might derive incidents and allusions for the ornament of their verses. The second part of the Edda, which has been called Skalda, is a still more singular instance of the attention which was given at this period to poetry, as an art. It is a collection of synonymes, epithets, and prosodiacal rules, carefully arranged, and admirably adapted to increase the accuracy and facility of composition. The different errors of style are distinctly pointed out, and a minute account is given of the varieties of figure and of metre, which may be admitted into poetical writing. The origin of this extraordinary work, like that of the ancient Edda, is still a matter of dispute. Most authors concur in ascribing it to Snorro Sturleson, admitting, however, that certain additions were afterwards made to the Skalda, either by Gunnlaug, a monk who lived about the beginning of the thirteenth century, or more probably, by a poet called Olaf Huitaskald, the nephew of Sturleson. The learned Arnas Magnæus, and some other writers, have contradicted this opinion, and suppose it more probable that the Edda was greatly altered, if not composed, in the fourteenth century; an idea which is the less probable, since at this period the art of poetry had greatly declined among the Icelanders, and the office and reputation of the Skalds were now become almost wholly extinct.*

* See Vita Sæmund. Mult. Edd. Sæmund. præfix. p. 14; also Sciagraph. Hist. Lit. Island. p. 17. The controversy respecting the origin of the Edda, and the examination of this singular work, have engaged many writers of great eminence. Besides those just referred to, we find connected with this subject the names of Wormius, Bartholin, Rudbeck, Resenius, Mallet, Suhm, Ihre, Thorkelin, &c. from whose several works the curious reader may obtain ample information on the subject. The principal editions of the Edda are those of Resenius, (*Copenhagen,* 1665), and of Mr Goranson, a Swede, who obtained his text from the

It would be exceeding the limits of the present dissertation to detail the names of all the eminent poets who adorned this period of Icelandic history. Besides the two celebrated men, whose names have been mentioned in connection with the Edda, those most conspicuous for their talents were, Egill Skallgrimson, a celebrated warrior as well as bard, who in the court of Norway rescued himself by his verses from impending death; Thorleif Jarlaskald, whose skill was so great, that while every one admired and applauded, no one knew whether his poetry spoke praise, satire, or reproach; Sighvatr Thordson, whose talents raised him to the counsels and friendship of two successive Norwegian kings; and the two brothers, Olaf and Sturla Thordson, who, in the thirteenth century, carried away the prize of fame from most of their competitors, both in Iceland and the kingdoms of the north. The former of these, surnamed *Huitaskald*, or the *White Poet*, from the colour of his hair, has just been noticed, as the supposed author of a part of the Edda. The latter, besides his celebrity as a poet, acquired much eminence in the departments of history and jurisprudence, and in other branches of knowledge. The chronicles of the country recite the names and compositions of many contemporary bards, little inferior in skill, and who exercised with scarcely less reputation the art to which they were attached.

Great, however, as was the poetical celebrity of the Icelanders of this age, they derived a still higher character from

Upsal Manuscript of the work. A French translation of the greater part of it has been given by Mr Mallet in his Introduct. a l'Hist. de Dannemarc; and this has been transferred to our own language by Dr Percy, in his Northern Antiquities. To the pens of Gray, Herbert, and Cottle, we owe poetical translations of several passages in the ancient Edda.

their historical writings. These may with propriety be divided into the genuine, the fictitious, and those of a mixed character. Of the fabulous histories, which were chiefly composed during the earlier periods of the Icelandic commonwealth, some appear to have had a concealed and figurative meaning; others were mere fables, not connected with any such metaphorical allusions.* The Sagas,† or historical narratives, form a much more numerous and valuable class of compositions. These may in general be considered to belong to the last or mixed character of history; in which the fictions of the author are to a certain extent blended with the events he describes. Many of them, however, possess throughout all the features of real and authentic narrative, and afford sketches of the state of society during this period, which are highly interesting and important. In the subject of these Sagas there is considerable variety. Some of them furnish the history of particular events, either of a political or religious nature; others give the more simple narrative of a family, or a community of families; others, again, contain biographical sketches of the eminent individuals of the age, the king, the warrior, the poet, or the priest. The merit of these writings is equally various. In many of them we find a minute and wearisome description of events, neither interesting in themselves, nor adorned by any of the graces of narrative; in others we meet with pictures of manners and feelings, in which simplicity itself is the charm, and where the imagination is insensibly led back to the times, the people, and the scenes, which are so faithfully pourtrayed. Of those

* See Torf. Ser. Reg. Dan. lib. i. cap. 1; and Bartholin's Antiq. Dan. lib. i. cap. 11.

† From *segia*, to say.

which bear the latter character, the finest example, perhaps, is the *Gunnlaugs ok Skald-Rafni Saga*, or the history of Gunnlaug and the poet Rafn; of which interesting story, a short sketch is given in the subjoined note, without any other ornament than the original itself affords, and with the necessary omission of many circumstances which confer grace and beauty upon the tale. The authenticity of the narrative, and the reality of all the personages it includes, are fully established by the evidence of contemporary writers.*

* Thorstein and Illugi, both men of wealth and power, dwelt in the great vale of the Borgar-Fiord, in the western part of Iceland. The former, who was son to the celebrated poet Egill, had a daughter named Helga, the pride of her family, and the loveliest among the women of the island. In the house of Illugi, the most remarkable person was his youngest son, Gunnlaug. Born in 988, he early acquired reputation from his stature, strength, and prowess, both of body and mind; but his temper was turbulent and unyielding, and being opposed by his father in his desire to travel, he abruptly left his home, when only fifteen years of age, and took refuge in the house of Thorstein, by whom he was hospitably received. Here, while his mind was instructed by the father, his heart was subdued by the gentleness and elegance of the daughter. Living with Helga, and partaking in all her occupations and amusements, a mutual affection was quickly formed; and the restless impetuosity of the boy passed into the refinement and delicacy of the youthful lover. His character thus changed, Gunnlaug was reconciled to his father, and, during three years, resided sometimes with him, sometimes at the house of Thorstein. When he had reached the age of eighteen, Illugi consented to his going abroad; but he would not leave Iceland, till he had obtained from the father of his secretly betrothed Helga, a solemn promise that the maiden's hand should be given to him, if, after three years had expired, he returned to claim it. Departing from his native country, Gunnlaug visited the courts of England, Ireland, Norway, and Sweden, and was every where received with the honours to which his person and talents entitled him. His extempore poetry was admired and munificently rewarded: this art he had early cultivated, though with so much tendency to satire, that he was called *Ormstunga*, or the snake-tongue. At the court of the Swedish king, Olave, he found the celebrated poet Rafn, likewise an Icelander, and of noble birth. A friendship formed be-

From the Icelandic Sagas, our knowledge of the history and antiquities of the northern nations has derived many important additions. Still more valuable, however, in this respect have been the regular historical writings of the Icelanders; many of which have come down, in a more or less perfect state, to the present time. The causes which led these islanders

tween them, was speedily broken by a dispute, which took place in the royal presence, respecting the comparative merits of their poetry. Rafn, thinking himself disgraced, declares his determination of revenge; and, in pursuance of this, returns to Iceland, where he seeks to obtain in marriage the maiden betrothed to his rival. The three years being gone by, and no tidings received of Gunnlaug, Thorstein, after some delay, gave to Rafn the unwilling hand of Helga, whose heart meanwhile remained with her former lover. The unfortunate Gunnlaug, hastening home to claim his bride, was accidently detained by a hurt received in wrestling, and reached the abode of his father on the very day on which Helga became a wife. A nuptial feast was prepared, with all the splendour suited to the condition of the families concerned. Gunnlaug shewed himself on a sudden among the assembled guests, eminent above all from the beauty of his person and the richness of his apparel. The eyes of the lovers hung upon each other in mute and melancholy sorrow; and the bitterest pangs went to the heart of the gentle Helga. The nuptial feast was gloomy and without joy. A contest between the rivals was prevented by the interference of their friends, but they parted with increased animosity and hatred.

The revenge of Rafn, though thus accomplished, gave him little satisfaction. Helga, refusing all conjugal endearments, spent her days in unceasing sadness. At the great public assembly at Thingvalla, the ensuing summer, Gunnlaug challenged his rival to single combat; and the challenge being accepted, they met on an island in the river, which flows into the lake of Thingvalla. The combat, however, though severe, was indecisive; and a renewal of it was prevented by an edict of the assembly, passed the following day, prohibiting the practice of duels in Iceland. Gunnlaug here sees his beloved Helga for the last time; and in the empassioned language of poetry laments their mutual affliction and sorrows. Restrained from deciding their quarrel in Iceland, and each pursued by his own unhappiness and resentments, the rivals pass over to the territory of Sweden, and meet, attended by their respective companions, at a place called Dynguines. A combat takes place: the companions of each party fall victims to the bloody fray,

thus early to the composition of history, as well as poetry, have already been mentioned. Originally bringing with them from Norway numerous traditionary records of the Scandinavian people, they derived progressive additions to these, from the residence of their poets and learned men in the courts of the northern kingdoms, the princes and chieftains of which cherished the talents by which their own actions might be conveyed to posterity. Provided by these means with ample materials for history, they became the annalists of all the north of Europe ; and the simplicity and precision which their narratives display, prove that they were well entitled to this pre-eminence. The history of their own country was not neg-

and Gunnlaug and Rafn are left alone to decide their contest. The foot of the latter is severed by the sword of Gunnlaug, who wishes now to discontinue the combat ; but Rafn exclaims that he would persevere in it, could he procure some water to alleviate his thirst. The generous Gunnlaug, trusting to the honour of his adversary, brings him water in his helmet from an adjoining lake. Rafn, seizing the critical moment, when the water was presented to him, strikes with his sword the bare head of Gunnlaug ; crying out at the same time, " that he " cannot endure that his rival should enjoy the embraces of the beautiful Helga." The fight is fiercely renewed, and Gunnlaug slays his perfidious opponent ; but dies soon afterwards of the wound he has himself received, when yet only in the twenty-fifth year of his age.

The remainder of the story is short and melancholy. The sorrowing Helga, her husband and lover both destroyed, is compelled to give her hand to Thorkell, a noble and wealthy Icelander. But these nuptials are equally joyless as the former. Her mind is wholly devoted to misery and gloom ; and she sinks an early victim to the grave, bending her last looks upon a robe she had received from Gunnlaug ; and dwelling with her last thoughts upon the memory of her unhappy lover.

A sketch of this story is given by the elegant pen of Mr Herbert, in the first volume of his poems. Were it less interesting, as a specimen of the manners and literature of the ancient Icelanders, the repetition of what he has so ably done, would not have been attempted.

lected amid the more conspicuous events of other nations; but the most careful record preserved of every circumstance occurring in the little community to which they belonged.

Isleif, the first bishop of Skalholt, who died in 1080, was the earliest of the Icelandic historians, and a man of great general learning; but his works are now unfortunately lost. Sæmund Frode, who has before been mentioned as the reputed author of the ancient Edda, was contemporary with the latter years of Isleif. He composed, among other historical works now extinct, the annals called from his place of residence, Odda, which contain a chronicle of events from the beginning of the world to his own time. For this work he was peculiarly qualified from his studies at Paris and Cologne, where in the earlier part of his life, he spent several years in the most ardent devotion to the pursuits of knowledge.* His friend and fellow-traveller Are Thorgilson, also from his learning surnamed Frode, was still more eminent as an historian of this age; but here, too, we have to regret the loss of what would have been among the most important of the ancient records of Danish, Norwegian, and English history, particularly of his Lives of the Norwegian Kings, from Odin to Magnus the son of Olaus. He has left us, however, valuable testimonies of his talents and industry in the Icelandic Chronicle, usually called the Schedæ, and in the Landnama Book, formerly mentioned, of which he appears to have composed the principal part, He died in 1148, when eighty years of age. Succeeding in the same track, was the celebrated Snorro Sturleson, who, born in 1178, became, when yet young, the

---

* The Annales Oddenses have been denied by some to be the work of Sæmund Frode, and have been assigned to a much later period of Icelandic literature. It is impossible, at the present time, to decide with certainty upon a question belonging to an age so remote.

wealthiest and most powerful man in Iceland. Twice he sustained the office of Laugman, or chief magistrate. His establishment was suited to the dignity of his condition, and in
visiting the general assembly of the island, he was frequently
attended by a splendid retinue of 800 armed men. The reputation which he acquired for learning and accomplish-
ments was equally extensive. He had a minute knowledge
of the Greek and Latin languages, was an excellent poet and
historian, an admirable orator, and profoundly skilled in all
the arts of his time. Besides the Edda, which is usually ascribed to him, and one or two fragments,* he has bequeathed
to posterity his Chronicle of the Kings of Norway, called the
Heimskringla;† a work which, while it strikingly displays the
erudition and industry of its author, is scarcely less distinguished for the excellence of its composition and style. The
latter part of his life was less fortunate than its commencement, being clouded by family feuds, which finally subjected
him to a violent death in the sixty-third year of his age. It
would appear that his character was not without its blemish-
es, as well as its noble and exalted qualities. To his private
ambition he is said occasionally to have sacrificed the interests
of his country, and much dissimulation and political artifice
pervaded the whole of his public career.

These are but a few of the illustrious men who adorned at
this period the literature of Iceland. A complete catalogue
of the native historical writers would include nearly two hundred names, some of them scarcely less eminent than those
which have already been mentioned. ‡ Though the sister

---

* The Hattalykill, or Clavis Metrica, and the Bragarbot.

† So named from the initial words of the book

‡ See Torfæi Ser. Reg. Danic. lib. 1. cap. 1. The other most important historical works of the ancient Icelanders are the Annales Flateyenses, Skalholtenses,

muses, history and poetry, were thus principally cherished, there was not however an entire neglect of other branches of knowledge. The ancient calendars of the country, and the extraordinary skill which was exhibited in the maritime adventures of the people, shew that considerable attention was given to astronomical and physical observations; and many learned men in the island, especially Sturleson and Paul, a bishop of Skalholt, were distinguished by their attainments in mathematics and mechanical science.* The study of jurisprudence was pursued with much ardour and industry: it appears, from passages occurring in the Sagas, to have been made a distinct branch of education among the chiefs of the country, and the excellent code of laws which the Icelanders framed for themselves, is a sufficient proof of the success with which its cultivation was attended. Geography could not fail to engage attention, when their travellers not only visited all the kingdoms of Europe, but penetrated even into the remote regions of Asia and Africa.† Philological studies were pursued by the learned men of the island with much diligence; and, in the course of the 11th and 12th centuries, they became familiar with the most celebrated of the Latin authors, deriving assistance to their own compositions from the classical authorities thus laid open to them. The study of the

Holenses, Vetustissimi, Regii, those called from the possessor of the manuscript, the Annales Resenii, the Sturlunga Saga, &c. Of the Annales Regii, an excellent edition will be found in Langebeck's Script. Rer. Danic. Vol. 3.

* See the Blanda and Rimbeigla books, published at Copenhagen under the patronage of M. Suhm. Also a treatise of the same age *de Algorithmo*, which is noticed in the Sciograph. Hist. Lit. Island. p. 161. The Rimbeigla book presents a singular assemblage of astronomical, chronological, and theological facts, and will be found well worthy the attention of the curious reader.

† Gissurus, a Laugman of Iceland in 1181, composed a work entitled *Flos Peregrinationis*, describing the various countries through which he had travelled.

Greek language, though less general, was not however disregarded; as we find from the testimony of several writers of this age.

It was scarcely possible that all this intellectual culture should exist, without some regular system of education, forming its basis and support. The first establishment of this kind appeared about the middle of the 11th century, when Isleif, the first bishop of Iceland, founded a school at Skalholt. This was shortly after followed by the institution of three other schools in different parts of the island, and by provisions for the education of youth in connection with the monasteries which were at this time established. It appears also to have been a common practice for those who possessed wealth and property in the country, to charge themselves with the instruction and advancement of such young men as gave an early promise of eminence in their talents.* In the schools, besides the knowledge of their own language in reading, writing, and various modes of composition, the youth of the island were initiated in classical and theological studies; to the latter of which especially much attention was given. Poetry was made expressly a branch of common education, and even music, or possibly a form of recitation thus termed, appears in some instances to have been taught in a public manner.† Previously to the reception of Christianity in Iceland at the close of the 10th century, the Runic characters, which were brought over by the original emigrants from Norway, seem to have been generally used, where the memory was not alone trusted with the record of events. These were not, however, committed to the form of regular writings, but rudely inscribed upon the walls and rafters of their habitations, upon their shields, wooden staffs, and other implements of common use. About

---

* Hist. Eccles. Island. T. I. p. 190.   † Hist. Eccl. Island. T. I. p. 190 and 327.

the time of this important change in the religion of the
people, the Roman characters were introduced; the adoption of which was attended with manifest advantage to the
progress of education, and to every department of literary
pursuit.

Respecting the physical condition of the Icelanders in
this remarkable period of their history, we derive our information chiefly from the poems, histories, and tales of the
country, which incidentally furnish many interesting facts
connected with this subject. The ancient Icelanders possessed, as is still the case with their posterity, few of the
luxuries or more refined conveniences of life; and were
occasionally exposed to severe privations from the nature of
their soil, and the seasons under which they lived. There is
some reason, however, to believe, though the fact cannot
be regarded as positively ascertained, that the climate of
Iceland was once considerably less austere than at present.
From many sources of information it appears certain that
corn was formerly grown upon the island, though in later
periods, as a native produce, it has been utterly unknown.*
Of the fact that the trees and shrubs formerly attained a
much larger size, and were more numerous than is now the
case, there is satisfactory evidence in the discovery of trunks
of such trees among the morasses; and in the frequent mention which is made in the ancient writings of houses and
even ships, constructed of the timber which the country itself produced. It is probable also that the other internal
supplies and means of subsistence which the Icelanders of
this age possessed, were more abundant and various than at

* The evidences of the former growth of corn in Iceland are collected in a
treatise by Snorreson, an Icelander, *De Agricultura Islandorum priscis temporibus.*
Hafniæ 1757.

the present time. We find in the Laxdæla Saga, the narrative of a feast in the western part of the island, at which nine hundred persons were assembled, and which continued with much splendour and ceremony during fourteen days; and the Landnama Book affords another instance of an entertainment given by two brothers at Hialtadal, in the northern province, where there was an assemblage of more than fourteen hundred guests. Many examples occur in the histories of the country of the liberal hospitality which was exercised towards foreigners; who, coming to the island for the purposes of traffic, were received into the houses of the principal inhabitants, and frequently dwelt with them during the long winter of this northern region.

It is not easy to ascertain with exactness the population of Iceland at this period; but many circumstances render it probable that it considerably exceeded the number now on the island.* As at present, however, the people were much dispersed over the country; their habitations being seldom grouped together, but placed wherever the situation and nature of the soil were propitious. Simplicity in all their habits and modes of life was a necessary effect of situation and circumstances. The houses even of the wealthiest of the community were constructed of wood and turf; and the great annual assembly of the people was held at Thingvalla beneath the open canopy of heaven; the chiefs and civil officers dwelling at night under the covering of rude huts,

---

* Arngrim Jonas, in his Brev. Comment. de Island. sect. 4, mentions the fact of an estimate being made in the year 1090, by Gissurus, a bishop of Skalholt, of the number of those who, from the amount of their property, were enabled to pay tribute to the state. They were ascertained to be about 4,000. This estimate, however, does not afford the grounds of even a probable conjecture as to the total number of inhabitants.

erected near the place of public meeting. The occupations of the people appear to have been nearly the same as those common to the present race of Icelanders; and in these every class of the community, in greater or less degree, partook.* Fishing was necessary as a principal means of their subsistence: they clothed themselves by the manufacture of the wool of their sheep; while their cattle, which were numerous, afforded at once a regular employment, and the most valuable addition to their domestic comforts. A further occupation was furnished by their traffic with foreign countries, which even at this time seems to have been very considerable. The moral habits of the people bore a favourable proportion to their intellectual qualities, and were doubtless fostered and improved by the latter. Certain superstitious and unnatural usages, which belonged to their ancestors, and to the age, were blended with their early condition; but these speedily yielded to the influence of reason and of the Christian doctrines, and left behind few vestiges of their former existence. A very remarkable instance of this kind occurred in 1011; when by a single and unanimous act of the public assembly, the trial by duel or single combat was abolished in Iceland; though the practice was then almost universal throughout Europe, and sanctioned for more than a century afterwards by the law and usages of many of the continental nations.†

* See Vatns dæla Saga, cap. 22; Liosvetninga Saga, cap. 25, &c.

† The exposure of children was one of the barbarous customs which the Icelanders derived from their Norwegian descent. Though this practice was not formally prohibited when the Christian religion was adopted in the island, yet it did not long survive this event; and it appears to have been extinct in Iceland nearly a hundred years before it was finally abolished in Norway. Hist. Eccles. Isl. Tom. I. p. 71. Among the superstitions connected with the Scandi-

One of the most important events during this period of Icelandic history, was the establishment of Christianity in the island. This momentous change was effected in a manner strikingly accordant to the genius of the government and of the people. From the year 981, when the knowledge of the Christian doctrine was first introduced by Frederic, a bishop from Saxony, to the close of this century, the number of those embracing the new faith progressively increased; and many missionaries, both foreigners and natives of Iceland, who had been converted abroad, came over to the island to aid its propagation by their efforts. They experienced much opposition from those who still adhered to the superstition of their ancestors: the invectives of poetical satire were poured forth against them, and even personal violence occasionally attempted by their opponents. These contests, and the growth of the new religion, at length engaged the attention of the government, and at the national assembly, in the summer of the year 1000, a formal discussion took place between the contending parties. While yet the subject was in agitation before the assembled people, a messenger hurried into the place of meeting with the intelli-

navian mythology, one of the most singular was the Berserkine, as it has been called; a treatise concerning which is subjoined to the Copenhagen edition of the Kristni Saga. The Berserkir were wrestlers or warriors by profession, who were believed by magical means to have hardened their bodies, so that they could not be injured by fire or sword. These men, roused at times by their incantations into a sort of phrenzy, committed every species of brutal violence; rushed naked into battle, and overpowered and slew all who ventured to approach them; till deserted by the paroxysm, their supernatural strength left them, and they immediately sunk into a state of extreme debility and wretchedness. Many records of this strange superstition occur in the old Icelandic and Norwegian writings. It gradually disappeared together with other practices of magic and divination, frequent among the northern nations of this age.

gence, that subterranean fire had burst out in the country to the south, and was consuming every thing before it. The heathens exclaimed, that it was not wonderful the gods should burn with anger at the new and detestable heresies which were thus introduced into the country —" But wherefore," cried Snorro, a zealous advocate for the Christian cause, " wherefore was the anger of your gods kindled, when the " very rock was burning, on which we now stand." The lake of Thingvalla is in the midst of a volcanic country, and lofty cliffs of lava environed the place of public assembly. The promptitude of the reply had its full effect; the heathen party were repulsed; and though the discussion still continued, the ardour and abilities of the Christians triumphed over all obstacles, and procured a final decision in their favour, which was pronounced with much solemnity by Thorgeir, the chief magistrate of the island. Upon the promulgation of this act, all religious contests were suspended, and the whole people espoused the faith, to which the wise and the learned among them had given their assent.*

Christianity being thus introduced into Iceland, the forms of a religious establishment were soon afterwards adopted: numerous churches were erected; a provision made by tithe for the support of the clergy, and two bishops created, one for the southern, the other for the northern district of the island. Isleif, the first bishop of Skalholt, was ordained in 1057: Jonas Ogmundson was fifty years afterwards invested with the same office at Hoolum in the northern province. During the early periods of the Icelandic church, the bishops were chosen by the collected voice of the people at the great public assembly: they were men eminent for piety, talents,

* See the Kristni Saga, and Hist. Eccles. Island. vol. I. p. 60.

and learning, and their influence was successfully exerted to maintain the purity of the religion over which they presided. The superstitious practices of the ancient mythology were abolished, and the church of Rome not having yet acquired sufficient influence to substitute its errors in their stead, a simple and undisturbed exercise of religion was enjoyed by the Icelanders for nearly two centuries after the first introduction of Christianity into the country.

Another event, connected with the history of Iceland at this period was the discovery and colonization of Greenland, effected about the year 972, by a Norwegian named Eric, who had settled in Iceland a short time before. Desirous of establishing a colony there, he called the country *Groenland*, with the design, as it would seem, of alluring settlers by the idea thus given of the country. In this project he succeeded. A year or two afterwards, twenty-five vessels were fitted out on the western coast for an expedition to Greenland; of which number it appears that fourteen reached the newly discovered shores. * A colony was soon established; the population rapidly increased; and in the progress of the ensuing century, a great extent of the eastern coast, opposite to Iceland, became inhabited. Christianity was introduced there at an early period, and the bishop of Garde, the principal establishment of the country, was known even to the Roman pontiffs of the age. The colonists maintained a constant commercial intercourse with Iceland and Norway; and the records of the settlement come down uninterruptedly to the beginning of the fifteenth century, when at once every trace and vestige of it are lost. The causes of this singular fact have never yet been fully ascertained. It is the most

* Landnama Book, Part II. cap. 14.

probable supposition that an accumulation of ice took place about this time on the Greenland coast, preventing the access to it from the sea; and this idea is confirmed by the narratives of later voyagers in these seas, and by the failure of several expeditions sent out to discover the settlement, all of which have been thus intercepted. Of the fate therefore of this ancient colony, commonly called by distinction Old Greenland, nothing is yet known. The same accumulation of ice, which separated it from the rest of the world, was probably the cause of the unfavourable change which appears about this time to have occurred in the climate of Iceland; the breadth of the sea intervening between the two countries not exceeding three hundred miles.*

Another maritime adventure is due to the enterprising age before us, more remarkable in itself, though less important in its consequences, than the one just mentioned. It is a fact well ascertained, though not generally known, that the north eastern part of the American coast was discovered at this period by the voyagers of these northern countries; and that during two centuries it continued to be frequently visited by the Icelanders and Norwegians, for the purposes either of curiosity or commerce. This singular discovery was made A. D. 1001, by Biorn Heriolfson, an Icelander, who on a voyage to Greenland was driven by unfavourable winds towards the south, and reached a flat woody coast, which, from several circumstances in the original narrative, we may presume to have been that of Labrador. Attracted by the report of this voyage, Leif, the son of Eric the discoverer of Greenland, fitted out a vessel

---

* See Egede's History of Greenland; Torfæi Groenlandiæ Antiq. Descript.; and the Hist. Groenlandiæ of Arngrim Jonas.

with the design of pursuing the same adventure; though un-provided with any of those aids which science furnishes to the navigators of modern times. Passing the coasts which Biorn had before seen, and continuing his course towards the south-west, he reached a straight which separated a large island from the mainland. Near this place, finding the country fertile and pleasant, he and his companions dragged their vessel on shore, and building huts, remained there dur-ing the winter. From the observation they made that wild vines grew in the country, they gave it the name of Vinland. They remarked also that the days during the winter were much longer than in Iceland, and the weather considerably more temperate. In the spring, Leif returned to Greenland, and was succeeded in the enterprize by his brother Thorvald, who arrived in safety at Vinland, and remaining two winters there, explored a considerable extent of the country and coasts. In the course of the third summer, however, he was killed in a combat with the natives, who appear now to have been seen for the first time, and who attacked the Icelanders with arrows and darts, irritated by an act of barbarous cruelty which Thorvald had committed towards some of their number. Soon after this time it would appear that a regular colony was established in Vinland by a wealthy Ice-lander, called Thorfin; and that the colonists, increasing in numbers, carried on with the natives a regular traffic in furs, skins, &c.* Thorfin himself, having remained there three years, returned to Iceland, greatly enriched by his adventure, and making a very favourable report of the climate and pro-ductions of the new country. Few particulars, however, are

---

* The natives were called by the Icelanders *Skrælingar*, signifying feeble or diminutive men.

afforded us of the after progress of the settlement, and though we have a record of it in the early part of the twelfth century, when a bishop of Greenland went over to promulgate the Christian faith among the colonists, scarcely a vestige of its existence occurs beyond this time, and the name and situation of the ancient Vinland are now entirely unknown to the world. Whether the colonists left the country at any particular time, or whether, separated from their connexion with Europe, they were gradually blended with the savage tribes surrounding them, must for ever remain a matter of doubt. *

* The reader who wishes farther to investigate this singular subject may consult the Landnama Book, the Eyrbyggia Saga, the Annales Flateyenses, the Heimskringla of Snorro Sturleson, and, among more modern writings, Arngrim's Hist. Gronlandiæ, Torfæus's Hist. Vinland. Antiq., the Hist. Eccles. Islandiæ, &c. The well-known Venetian narrative of the voyages of Nicolo and Antonio Zeni, at the close of the fourteenth century, might be admitted as a farther evidence, did it not bear the character of one of those maritime romances, which were so common among the Venetians during the period of their commercial greatness. This narrative, however, in describing an extensive country, called Estotiland, situated to the south-west of Greenland, and which had before been visited by the Icelanders for the purposes of traffic, proves at least that the discovery of the northern navigators was not unknown by report to the people of the south of Europe. So many testimonies, indeed, direct or indirect, have come down to us on the subject of the ancient Vinland, that it is impossible not to admit the fact of their general authenticity. There is more room for doubt as to the exact situation of the place, thus named. By some it has been supposed (Hist. Eccles. Isl. Tom. I. p. 4.) that it might be as far towards the south as Virginia : others have conceived, with more reason, that it was situated on some part of the coast of Labrador, probably near to the island of Newfoundland. Mallet, who, in his Introduct. a l'Hist. de Dannemarc, has an ingenious disquisition on this subject, adopts the latter opinion ; and in the first edition of his work, cites the evidence of Father Charlevoix, a traveller into these countries, and of Dr Baumgartens, a learned German writer, to prove that the Esquimaux Indians, in this part of Labrador, differ materially in person, habits, and language, from the other North

The sketch which has now been given of the habits, institutions, and arts of the ancient Icelanders, is by no means
an exaggerated picture of this singular and interesting people.
The comparative eminence, however, to which in this age
they attained, was not destined to be permanent; and the
rapid advancement of other states towards civilization, concurred with changes in their own condition to effect an entire alteration in the balance subsisting between them. Even
the period of the commonwealth, though the most brilliant
and remarkable in the history of Iceland, presents not
throughout the pleasing features which have just been de-
lineated. In the progress of time, numerous intestine evils
sprung up to disturb the repose of the people; and the
middle of the thirteenth century is signalized in their history
by the transference of the island to the power of the Norwegian kings, three hundred and forty years after the establishment of the free constitution, under which they so
greatly flourished. Several probable causes may be assigned
for this change, some of them collateral, others perhaps
connected with the nature of the constitution itself. There
appears to have been a constant leaning of the aristocracy,
which formed the basis of the government, towards an oli-
garchy; and in the later periods of the commonwealth, disturbances were excited by ambitious individuals, who aim-
ed at the possession of more influence in the state than
the constitution allowed. Where large feudal property and

American tribes; from which the possibility is inferred that these may be the
remnants of the ancient European colonists. That wild grapes were found in
Vinland, cannot be considered an objection to this idea of the situation of the
country, since modern travellers have ascertained that a species of wild vine grows
native on the American coast, even as far to the north as the shores of Hudson's
Bay.

hereditary rights were connected with talent, ambition, and enterprize, it was natural to expect that efforts would be made to infringe the aristocratical equality, which existed in the spirit and design of the commonwealth. Accordingly we find the relation at this period of numerous contests between the more powerful chieftains of the state; and the annals of Iceland are for a time disgraced by the record of sedition, rapine, and bloodshed. *

The liberties of the Icelanders might possibly, however, have survived these intestine feuds, had not other circumstances co-operated with their effect. The Norwegian monarchs, though making no direct attempts to subjugate the island, yet appear to have contemplated at an early period its annexation to their power. This desire was doubtless confirmed by the increasing prosperity of the Icelanders during the eleventh and twelfth centuries; and the means of accomplishing their design were afforded by the disturbances which afterwards occurred. These broils appear to have been fomented by the concealed interference of the Norwegians, who were admitted to a constant intercourse with the island, and who, while aggravating the internal evils under which it suffered, held out to the people the most specious promises of assistance and protection. By such promises, the kings of Norway gained over some of the most eminent of the Icelanders; and persuaded them to urge, even in the councils of the nation, the necessity of composing their feuds by giving themselves to the dominion of a single potentate.

---

* These contests between the chieftains were not always trivial or unimportant. Instances are related where bodies of twelve or thirteen hundred men, and fleets of twenty vessels, were engaged on one side in such conflicts. Hist. Eccles. Island. Vol. I. p. 103.

The celebrated Snorro Sturleson, who resided two years at the Norwegian court, and was received there with many dignities and honours, was suspected of having aided this cause; a suspicion from which he incurred much odium among his fellow citizens.

The efforts of the Norwegians, protracted through a long period, were finally successful. The Icelanders, wearied of feuds and contests, consented at last to resign their independence; and, in 1261, an act of the national assembly, unattended with violence or the compulsion of arms, delivered up the greater part of the island to Haco, the reigning king of Norway. The eastern province which at first opposed this act, three years afterwards adopted the same course. It was not, however, a blind submission to arbitrary power which appeared in this revolution. Regular treaties were established between the Icelanders and their future sovereigns; and the acknowledgement of the kingly sway was preceded by conditions, which made it rather an alliance than a timid surrender of rights. All property was secured in the island; no tribute exacted; a liberal provision was made for the external traffic of the inhabitants, and a title given them to the acquirement of honours and civil offices in the kingdom of Norway itself. It was provided that the government of Iceland should be administered by a delegate from the king, either a Norwegian or a native of the island. Little change appears to have been made in the internal government of the country; and the celebrated code of laws, called the *Jonsbok*, which was given to the Icelanders in 1280, by Magnus the successor of Haco, was merely a revised and amended form of the more ancient body of laws, framed for the commonwealth of Iceland.*

* A detailed account of this change in the government of Iceland is given in

The short period during which the island remained subject to the native Norwegian monarchs, is dignified by no remarkable event. The laws were administered by the governors of the country in a mild and equitable manner; and it does not appear that any attempts were made to infringe the conditions upon which the liberties of Iceland were surrendered. The internal feuds which preceded and produced this event were in great measure composed, and the inhabitants at large remained in a state of perfect order and tranquillity.

The annexation of Norway to the power of Denmark in 1380, was an event of little importance to the interests of Iceland, and can scarcely be considered to form an epoch in its history. The island was transferred to the Danish monarchy without tumult or opposition. The laws were maintained in their former state; and their administration committed as before, to a governor appointed by the crown. These prefects of the island were sometimes natives, sometimes Norwegians, or Danes. Though it was intended they should reside in the country, this does not appear to have been generally done; and many of them visited their government only once in the year, to inspect and regulate its various concerns.

This is the last political change which occurs in the history of Iceland. The records we possess of the succeeding periods are less numerous and less valuable than those which

Torfæus's Hist. Norv. Tom. IV. lib. 5. Also in the Hist. Eccles. Isl. T. I. p. 373 et sqq. The code of laws presented to the Icelanders at this time was called *Jonsbok*, from the name of the governor, by whom it was introduced into the island. Some of the most eminent among the natives, particularly the poet and historian Sturla Thordson, assisted in its compilation from the ancient laws of the republic.

relate to the times already described. The historians of these
later ages are occupied chiefly in the detail of events, neither
very interesting in themselves, nor affecting beyond the moment the condition or interests of the people. Their narratives are remarkable for accuracy and minuteness ; but they
are spread over too broad and uniform a surface, and are little relieved by any of the ornaments of style or composition.
The history of Iceland, however, though now destitute of political event, is still the history of a people ; and the four last
centuries have exhibited some features which are not wholly
unworthy of attention.

The change in the constitution of the island, from its annexation to a European monarchy, produced, as might have been
expected, a corresponding change in the character and habits
of the people. Before this event, each individual, possessing
property, formed an integral part of the government of his
country. Definite objects of ambition existed to every member of the community ; and vigour, activity, and talent, gave
political importance, as well as private influence, to those in
whom these qualities appeared. This, in the same degree,
could no longer be the case, when the island was subjected
to the government of a foreign power. The great assembly
of the people was still summoned to its annual meeting at
Thingvalla ; and still, under the cognizance of the Governor,
enforced the execution of the laws ; but its national deliberations had now lost much of their spirit and importance. The
influence of property and of personal merit were diminished in
the same proportion ; and the efforts of individual ambition
tacitly and without violence repressed. Had the foreign yoke
been a tyrannical one, the primeval spirit of the Icelanders
might possibly have been maintained by the persecution
which laboured to suppress it. But the case was far other-

wise. The Norwegian, and subsequently the Danish monarchs, exercised their sway with a lenient and forbearing hand; not merely refraining from oppression, but giving much attention to the interests and welfare of this remote part of their dominions. The customs and feelings of the people were respected; the laws administered with equity, and tranquillity maintained throughout all classes of the inhabitants.

To these circumstances we may chiefly attribute the change which appears about this period to have taken place in the national character of the people, and the distinction existing between the ancient Icelanders and their posterity of the present age. Repose and security, succeeding to internal broils, produced a state of comparative apathy and indolence. The same call was not made for individual exertion, nor the same rewards proposed to its successful exercise. Rank and property became more nearly equalized among the inhabitants; and, all looking up to a superior power, the spirit of independence declined, and they expected from others the support and protection which they had once afforded to themselves. Their ardour in maritime adventure was checked at the same time by the revolution which took place in the government of the island. The trade which they had formerly carried on in the products of their country, was now gradually transferred to the natives of other kingdoms; and a copious source of activity and exertion thus in great measure extinguished. It appears, too, that about this period, the agriculture of the country declined; owing either to a change in the nature of the climate, or to diminished industry on the part of the inhabitants. These combined causes had a permanent influence upon the character of the Icelanders. The simplicity and warm social affections, which belonged to the ancient race, were still preserved unimpaired; but their independence, vigour, and activity, were now in a great degree lost, and will probably never be regained.

The period directly succeeding the union of Iceland to the Danish crown, was more especially unfortunate for the welfare of the country. In 1402, a plague, (the nature and causes of which are not distinctly explained) broke out in the island, and in the course of this and the two following years swept away, if the accounts preserved may be depended upon, nearly two-thirds of the whole population.* This tremendous affliction was succeeded by a season of such inclemency, that scarcely a tenth part of the cattle on the island escaped destruction. Another epidemic pestilence prevailed towards the close of the century, which, though less disastrous than the former, carried off a large part of the population, and produced much general distress. The calamities of the island at this period were further increased by the occasional incursions of English pirates ; who landing on different parts of the coast, plundered the property of the natives, committed frequent murders, and carried many persons into captivity.†

* " Anno 1402, atrox Islandiam pestilentia pervagari cœpit, qua multi morta-
" les ita subite extincti sunt, ut quidam dicto citius perirent: puerique, adulti, et
" senes indifferenter animam efflarent ; tantusque fuit contagionis furor ut sæpe ex
" 12 vel 15 qui unum mortuum sepultum ibant, vix duo aut tres domum incolumes
" redirent." *Hist. Eccl. Isl. Tom. II. p.* 135. Very few particulars are transmitted to us by the Islandic writers with respect to this dreadful disease, but it may be presumed with some probability to have been the same epidemic which, about the middle of the 14th century, extended its effects over a great part of the European continent. Some accounts say that it was introduced into Iceland from Britain.

† See the Annals of Biorn de Skardsaa, and various parts of the Hist. Eccles. Islandiæ. On this subject, a long and curious document is given in Vol. IV. p. 162. of the latter work, in which are preserved the names of many of the English pirates who infested the coasts of Iceland between the years 1419 and 1425 ; and a minute narrative of the various enormities they committed. From this document it appears, that most of these piratical vessels were fitted out at Hull, Lynn, and others

These events, which concurred with the causes before described in depressing the spirit of the people, and destroying the strength and prosperity of the country, are recorded in the annals of Iceland with an affecting and almost painful simplicity. No attempts are made to excite a sentiment of commiseration, beyond what humanity would of itself yield to the recital of such complicated evils. We are told that whole families were extinguished, and districts depopulated, by the virulence of disease; that the learned, the pious, the wealthy, and the powerful, all dropt into a common grave; that the labours of industry ceased; that genius and literature disappeared; and that the wretched remnant of the Icelanders, scarcely themselves saved from destruction, sunk into a state of apathy, superstition, and ignorance. In pursuing his melancholy narrative, the historian sometimes looks back for a moment to the former celebrity and splendours of his country : but he goes no further; and all beyond is left to the feelings and imagination of the reader.

Though, during this gloomy age, the talents and literature of the Icelanders were depressed almost to extinction, yet we must look to an earlier period for the commencement and primary causes of their decline. The alteration which has just been described as taking place in the character and condition of the people, after they were annexed to a foreign power, could not occur without a corresponding change in their intellectual habits. Of the various motives to literary pursuits which before existed, some in consequence of this event were

of the eastern ports of England; and that they came to Iceland with the double view of plunder and of fishing upon the coast. In 1512, the then Governor of the island was put to death by some of these marauders. The intercourse of the English with Iceland at this period was not, however, universally thus disgraced. During a considerable part of the 15th century, they appear to have carried on a fair traffic with the inhabitants in the products of the country.

entirely lost, and others so much enfeebled, as to produce few of their original and wonted effects.   Talents and know-ledge were no longer associated with that political influence, of which before they formed the fairest ornament and the most stable security.   Though proud of the eminence which their acquirements had given them among other nations, and attached to the habits and pursuits of their forefathers, these prepossessions were not sufficient to preserve unimpaired the spirit which had once animated their career.   A circumstance which assisted to produce its decline, was the change progres-sively taking place in the customs and institutions of those countries, with which the Icelanders had before been most intimately connected.   The European nations were now be-ginning to liberate themselves from that bondage of ignorance and superstition, which during the dark period of the middle ages, had suspended all but the sterner and more impetuous qualities of human nature. The restoration of civil and social order, while it gave repose to the mind, invited the exercise of those faculties, by which leisure might at once be occupied and adorned : knowledge and the arts rapidly revived, and the native literature of every country was protected and encouraged, by those who appeared before only as the op-pressors of its growth.   Under these circumstances, the poets and historians of Iceland were received with fewer ho-nours in the courts where they had once stood so proudly eminent ; and their talents were little cherished among na-tions in which science had now made equal, or even greater progress than among themselves.   They retreated gradually into their native island ; where, in the little community of their fellow citizens, they still kept alive a feeble remnant of that reputation, which had formerly extended throughout the greater part of Europe.

In addition to the causes just mentioned, there is another, which seems materially to have affected the literature, as well as the general character of the Icelanders during this age. For some time after the introduction of Christianity into the island, the state of religion among the people was distinguished for its purity and simplicity. The active and interested spirit of the Roman church did not, however, long remain dormant, even in this remote part of the Christian world ; and about the close of the 12th century, we find that its superstitious usages and ecclesiastical tyranny began to make innovations upon the religious establishments and customs of the people. Fables of miracles came into vogue ; the worship of saints was tolerated ; and the bishops of the island, formerly chosen in consequence of their learning and piety, were now recommended chiefly by their subservience to the interests and wishes of the Papal see. It appears, too, that even the poverty of the Icelanders did not afford them a security against the pecuniary exactions of the church of Rome. Besides other tributes, the celebrated one called Peter's Pence, was collected at different times among the inhabitants, and the sale of indulgences appears to have been repeatedly carried on, both by foreign missionaries, and by the native bishops of the island. The preaching of the crusades also was attempted in 1275, 1289, and some succeeding years, but with very inconsiderable success. In the first instance, many took up the cross, purchasing, however, dispensations from bearing it to the Holy Land ; but in their latter efforts, the missionaries were much less successful ; and it does not appear, that a single Icelander was at any time drawn away from his country to join in this remote and dangerous contest.*

* See Hist. Eccl. Island. T. I. p. 571.

The effect of these various circumstances upon the literature of the island was rapidly progressive. The two brothers,
Olaf and Sturla Thordson, whose reputation during the
13th century has before been noticed, may be considered the
last of the ancient Icelanders who attained any considerable
eminence in the arts and knowledge of the age. The historical work of the latter, called Sturlunga Saga, relating
the events of his own times, was characterized by a genius
worthy of the illustrious family to which he belonged. But
he was succeeded by no writer who could claim an affinity of
talent to the great names that were extinct. The department of history now degenerated into a mere collection of
ecclesiastical fables, the lives of monks and saints, and the
stories of miracles, written in a crude style, and displaying
little of the erudition or elegance which adorned the compositions of the earlier Icelanders. In the department of poetry,
a similar change occurred: the number and reputation of
those who were attached to the pursuit, gradually declined ;
and the few remaining exercised their art in the composition
of hymns to the praise of saints and martyrs, which were
distinguished only by the rudeness of their structure, and the
absence of every beauty of imagery and taste. The study of
jurisprudence, so much cherished and so successfully cultivated by the ancient Icelanders, was now exchanged for a
laborious attention to the rites and usages of the Catholic
church ; while the knowledge of the languages, of astronomy,
and of the more rational parts of theology, sunk into a state
of corresponding depression and decline. This progressive
change was completed by the events of the 15th century.
The accumulated evils which then oppressed the country,
destroyed all that was left of its former literature and greatness, and the annals of Iceland during this period, are the

records only of mental depression, and of physical calamities and suffering.*

The introduction of printing into the island, and the reformation of religion which soon after took place, give a more pleasing character to the commencement of the succeeding century. The first printing press was erected at Hoolum, in the northern province, about the year 1530, under the auspices of John Areson, who was at that time the bishop of this see. Though himself an illiterate and uncultivated man, he was extremely ambitious; and wished to avail himself of all the means which literature might afford for the promotion of his influence in the country. With this view he procured as his secretary, a Swede of the name of Mathiesson, who coming over to Iceland, brought with him a printing press, and made a small establishment at Hoolum for its use. The types were originally of wood, and very rudely formed; and the only works issuing from the press during the first forty years after its institution, were a few breviaries, church rituals, and calendars. In 1574, however, Gudbrand Thorlakson, bishop of Hoolum, made very great improvements in the printing establishment at that place, providing new presses and types, some of which were constructed by his own hand, and bestowing the utmost care upon the correction of every work, which was printed during his life time. Before the century had elapsed, a number of valuable publications made their appearance, greatly improved in their style of composition, and displaying a neatness and even elegance of execution, very

* It has been supposed by Schlozer, that the language of the Icelanders, as well as their literature, was materially affected by the events of this age. It is more probable, however, that the alterations which have taken place in the Icelandic language were made progressively, and not at any particular period.

remarkable at this early period of the use of printing in the country.

The reformation of religion in Iceland was not accomplished without some disturbance. Early in the 16th century, the Lutheran doctrines had begun to combat the superstition and tyranny of the Catholic church ; and their influence was greatly aided by the zeal of the Danish monarch, Christian III, who having abolished the usages of popery in his continental dominions, wished to extend this reformation to the religious establishment in Iceland. His intentions, and the progressive change of opinion from the growing knowledge of the people, were strenuously opposed by those of the clergy who were attached to the former state of religion ; and particularly by John Areson, the bishop of Hoolum, whose ambitious and assuming character has already been noticed. The power which this man had acquired in the country, and the haughty violence of his temper, led him into many acts of open hostility against the reformers. Attended by a body of armed men, he left his northern diocese, and proceeding into the western province, seized the person of Einarsen, the bishop of Skalholt, who had some time before espoused the Lutheran doctrines, and was at this time engaged in visiting the different churches of his district. In the course of the following year, however, he was himself arrested by order of the king of Denmark ; and being accused of various crimes, was beheaded at Skalholt, together with his two natural sons, who had participated in the violence and usurpations of their father. After his death no opposition was made to the new doctrines, and in 1551 the Reformation was legally established, and universally received throughout the country. About the same time, the public schools of the island, which, together with its other institutions, had almost been annihilated by the disastrous events of the 15th century, were again established, under the patronage of the king of Denmark; and such funds attached to them, as afforded facilities of education to those of every class among the inhabitants. At the time of their revival, it was found difficult to obtain in the country men of sufficient learning to discharge the office of teachers; so greatly had the condition of literature been depressed. Several of the Icelanders, however, having been sent to Copenhagen, to pursue their studies at that university, the schools of the island were afterwards conducted by men, whose talents and acquirements well fitted them for this important duty.

The events which have just been described, render the period of the 16th century, a new era in the history of Iceland. Though the former condition and character of the people were never entirely restored, yet their situation appears to have been considerably improved, and their more intellectual habits again excited to that exertion which once conferred so much celebrity upon the country. But the revival of literature among the Icelanders was attended by none of those remarkable circumstances, which distinguished its original propagation and growth. The relation of their little community to the neighbouring kingdoms of the north was at this time completely changed: the disparity of their physical condition exerted all its natural influence; and the flame which was again kindled among them, shone dimly beneath the splendours of that sun of science which had now risen over the nations of Europe. In later periods, the literary fame of the Icelanders has rarely been extended beyond the limits of their native island: and though the progress of their knowledge has, in a certain degree, kept pace with that of other countries, yet this must be regarded rather as an extension of the growth of the latter, than as the effect of any internal powers of acquisition or improvement.

During the century which succeeded the restoration of literature in Iceland, several individuals appeared, whose abilities and learning gave them considerable celebrity in their native island. It may be remarked, however, that almost all these eminent characters were either bishops, or masters of the public schools, the diffusion of knowledge not having yet taken place to such an extent, as to include those belonging to inferior classes of the community. The person whose name is most conspicuous among the restorers of learning, was Gudbrand Thorlakson, bishop of Hoolum. Born in 1542, he studied for some years in the school of Hoolum, and afterwards at the university of Copenhagen, where his talents and industry gained him the intimate friendship of Tycho Brahe, Resenius, Paul Matthias, and other celebrated men in the Danish court. When yet only thirty years of age, he was appointed to the see of Hoolum; an office which he sustained during the long period of fifty-six years, in a manner most honourable to himself, and advantageous to his country. His labours for the promotion of knowledge were unwearied and incessant. Having reformed the printing establishment of the island, he occupied himself in the superintendance of the press; and as the best testimony of his diligence in this office, we have a catalogue of between eighty and ninety works, which were either written by himself, or published under his immediate patronage and direction. The greater part of these publications were of a theological nature; and many of them translations of the more eminent works in divinity which at this time appeared on the continent of Europe. To the zeal and learning of Thorlakson himself, the Icelanders were indebted for the first translation of the Bible into their native language; which was published in folio in 1584, and afterwards under other forms better adapted to the common use of the people. About the same time an edition appeared of the *Log-bok*, or Icelandic code of laws; and succeeding it, several other works of much value in reference to the history, and other circumstances, physical as well as political, of the country.

Another very eminent individual of this age was Arngrim Jonas; the intimate friend and, for many years, the coadjutor of Bishop Thorlakson in the duties of the episcopal office. He was associated in all the schemes for the promotion of literature, which so much distinguished the career of the latter; and twenty-six different works in various branches of divinity, history, jurisprudence, and philology, attest equally the extent of his acquirements, and his zeal for the progress of general knowledge. The most valuable of his writings are those which relate to the history of his native island; the early condition of which, especially during the period of the commonwealth, he has illustrated with singular diligence and success. His works are for the most part composed in Latin, and are remarkable for the purity and elegance of their style, in which he appears greatly to have excelled all his contemporaries.* Of the other historical writers who distinguished themselves at this period, the most eminent was Biorn de Skardsaa, whose annals of Iceland from the year 1400 to 1645, exhibit an extreme minuteness of narrative; animated, however, by few interesting or important events, and deficient

---

* The most important of the writings of Arngrim Jonas were the Crymogæa; the Brevis Commentarius de Islandia; Anatome Blekfeniana; Historia Groenlandiæ; Specimen Islandiæ Historicum; Compendium Historiæ Norvegiæ; Tractatus de successsione ab intestato; Discursus de Literis Runicis, &c.

in all the ornaments of composition and style. His work, nevertheless, is valuable, as filling up an interval in the history of Iceland, which has been less dwelt upon than any other by the native writers of the country; and its singular simplicity of character affords an important evidence of its truth and authenticity.*

The history of the island during the 17th century is almost wholly destitute of remarkable events. The condition of the people, their laws and government, continued nearly in the same state. The commercial connections of the country underwent some change about the beginning of the century, in consequence of an edict of Christian the Fourth, which conveyed a monopoly of the traffic with the island to certain commercial towns within the dominion of Denmark. For some time prior to this regulation it had been in the hands of the merchants of Hamburgh and Bremen, who appear to have carried it on with considerable success. The piratical incursions of foreigners, which during the unfortunate period of the 15th century, had added to the other afflictions of the Icelanders, were still frequently continued; and little opposition being made to their lawless attacks by a timid and unarmed people, the banditti carried rapine and oppression along every part of the coast. Even as late as 1616, the English and French nations bore a part in these enormities; which the more engage detestation, as being exercised against those who were subject from their situation to all the evils of poverty and want. The most calamitous event of this kind occurred in 1627, when a large body of Algerine pirates landed on various parts of the southern coast of the island; and not

---

* The other writings of Biorn de Skardsaa are, the Tractatus de Groenlandia; Glossarium Juridicum; Tractatus Juridici; Illustratio odarum in Edda Sæmundina, &c.

satisfied with the booty they obtained, murdered between forty and fifty of the inhabitants, and carried off nearly four hundred prisoners of both sexes. These unfortunate captives transported to Algiers, were exposed there to so much wretchedness, that nine years afterwards, when the king of Denmark obtained their liberty by ransom, only thirty-seven out of the whole number were found to be surviving. Of these, thirteen succeeded in reaching their native island.*

Though the feelings and practices of superstition have never gained more ground in Iceland than among the greater communities of Europe, yet at some periods they appear to have existed to a considerable extent; and the 17th century is remarkable for many excesses derived from this source. Numerons individuals, both of the clergy and laity, were accused of dealing in the arts of magic; and several of these, being pronounced guilty of the offence, were sentenced to be burnt alive. For some time the belief in necromancy was so general, and its supposed practices held in so much horror, that in the course of sixty years, not fewer than twenty persons perished in the flames. The superstition afterwards gradually declined; and at present few of its vestiges are to be found in the country.

The later periods in the history of Iceland are too much distinguished by the record of physical calamities. The 18th century was ushered in by a dreadful mortality con-

* The Westmann Islands, situated on the southern coast of Iceland, suffered more particularly from the Algerines; almost their whole population being destroyed or carried into captivity. Olaus Egilson, a priest in these islands, who had been made a captive, but obtained his release from Algiers in 1629, left a manuscript relation of this event, which has since been published in Danish. Biorn de Skardsaa, and other writers, have also left narratives of these piracies ; and on the same subject a poem was composed by Gudmund Erlendson, the author of a translation of Æsop's fables into Icelandic verse.

sequent upon the small pox ; which in 1707, raged with such
epidemic virulence, as to destroy more than 16,000 of the
inhabitants.   The years intervening between 1753 and 1759
were so exceedingly inclement, that the cattle perished in
vast numbers from the scarcity of food, and a famine ensuing carried off nearly 10,000 people.   The year 1783 was
signalized by an event, more alarming in itself, and not less
disastrous in its consequences. Several volcanic eruptions had
already occurred in different parts of the island during the
preceding periods of the century ; but without producing, in
a country so thinly peopled, any very extensive devastation
or distress.   At this time, however, the great eruption took
place in the Skaptaa-Syssel, the most tremendous perhaps in
its nature and extent by which Iceland, or any other part
of the globe, has been afflicted.   The sudden extinction of
a submarine volcano near Cape Reikianes, which during
some months had continued to burn with extreme violence,
was succeeded by frequent and dreadful earthquakes, and by
the bursting out of the volcanic fire, in a tract of country
nearly two hundred miles distant.   The scene of the latter
eruption appears to have been among the lofty mountains,
called the Skaptaa Jokull, situated in the interior of the
island, and known to the natives themselves only by the remote view of their summits, clad in perpetual snows.   From
this desolate and unfrequented region, vast torrents of lava
issued forth, overwhelming all before them, and filling up the
beds of great rivers in their progress towards the sea.   For
more than a year, a dense cloud of smoke and volcanic
ashes covered the whole of Iceland, extending its effects
even to the northern parts of continental Europe ; the cattle, sheep, and horses of the country were destroyed ; a
famine, with its attendant diseases, broke out among the in-

habitants, and the small-pox invaded the island at the same time with its former virulence and fatal effects. From these combined causes more than eleven thousand people perished during the period of a few years; an extent of calamity which can only be understood, by considering that this number forms nearly a fourth part of the whole population of the country. The destruction of the fishery upon the southern coasts of the island, by the volcanic eruptions just described, was another more permanent source of distress, which even at the present time is not entirely removed.

The literature of the Icelanders in later times, though affected in some degree by the various evils of their situation, has nevertheless been preserved from decline; and may perhaps be considered to have made a certain progress, in its connection with the general advancement of knowledge among the nations of Europe. The names of numerous poets and historians still appear in the literary records of the island, and the introduction of new and important branches of science has given to the learning of the country a more extensive and diversified character than it possessed, even in the most splendid periods of its ancient history. It is probable, however, that the proportion of the inhabitants, devoted to such pursuits, has in later periods been considerably diminished; and this change may be regarded perhaps as forming the most remarkable distinction between the present and former state of literature in the country. The few are still not unworthy of the names of their ancestors; but the people at large, though possessing a mental cultivation far above their physical circumstances, have probably declined in a certain degree from that spirit of progress and improvement, which so much distinguished the early condition of their community.

The names of Thormodus Torfæus, Arnas Magnæus, and Finnur Jonson, are the most celebrated of those which have adorned the modern literature of Iceland. Torfæus, who was born in 1636, and educated first at the school of Skalholt, and afterwards in the university of Copenhagen, acquired a high reputation at the Danish court from the extent of his erudition and acquirements; and in 1682 was appointed the historiographer of Norway; a situation for which the number and value of his historical writings shew that he was peculiarly well qualified. Of these writings, the most important are the ' Series of Dynasties and Kings of Denmark,' and the ' History of Norway;' both published at Copenhagen in the beginning of the last century.* Still more conspicuous for his devotion to literary pursuits was the learned and eminent Arnas Magnæus, the son of an obscure country priest in the western part of Iceland. Raised from the original lowness of his situation by extraordinary efforts of industry and talent, he attained in 1694, when only thirty-one years of age, the honourable situation of professor of philosophy in the university of Copenhagen; and a few years afterwards was invested with the further offices of professor of northern antiquities, and secretary of the royal archives. His exertions for the progress of knowledge were laborious and incessant. Besides composing himself several important works, he collected at great labour and expence a magnificent library, illustrative of the literature and antiquities of the north; and especially valuable in reference to the literature of his native island, which he visited several times with the view of collecting all the books and manuscripts extant

---

* The other most valuable writings of Torfæus are the Historia Færoensium; Historia Orcadensium; Vinlandia Antiqua; Groenlandia Antiqua, &c.

in the Icelandic language. The greater part of this library
was unhappily consumed by the fire which happened at
Copenhagen in 1728; and the unfortunate Magnæus, present
on the spot, saw the fruits of his long continued industry
and toils, in the course of a few hours, almost entirely destroyed. He died two years afterwards, and bequeathed to the
library of the university the remant of his literary treasures,
under the superintendance of certain trustees, selected from
among the most learned men in the Danish metropolis.*
His friend, and favourite pupil, Finnur Jonson, is another of
the eminent Icelanders who, during the last century, have
contributed to preserve unimpaired the character and respectability of their country. Created bishop of Skalholt in
1754, he retained this office during the remainder of a long
life, devoted entirely to the promotion of happiness and
improvement among the community of his fellow citizens.
In his admirable work, the Ecclesiastical History of Iceland,
he has bequeathed to them a monument of extensive erudition, genuine piety, and warm patriotic feelings, which will
long continue as one of the fairest and most illustrious ornaments of their literature.†

It would be impossible to mention here even the names of
all the Icelanders who, in these later periods, have distin-

---

* The works of Arnas Magnæus are chiefly historical and critical. In his
early youth he was a pupil of the celebrated Bartholin, and assisted him in the
composition of his great work on Danish antiquities. His character is admirably
drawn by his friend Bishop Jonson (Hist Eccl. Island. V. III. p. 576), to whose
intrepidity, at the time of the great fire in Copenhagen, we owe the preservation
of some of the manuscripts from the Magnæanian library.

† The Historia Ecclesiastica Islandiæ was published at Copenhagen in four
volumes quarto. It is written in Latin of remarkable elegance; and is replete
with valuable information, not solely in relation to the ecclesiastical affairs, but
also to the political history and literature of the island.

guished themselves in the departments of history, poetry, theology, criticism, and physical science: but in a succeeding chapter, on the present state of Icelandic literature, some remarks will be found, illusrative of the actual progress of the people in these various branches of knowledge. Though among themselves a careful record is preserved of all the authors and learned men who have appeared in the country, yet the reputation of few of these individuals has been conveyed beyond the limits of their native island, and their views towards posterity have for the most part been bounded by the small and remote circle of society in which their destiny was cast.* It must be mentioned, however, to the honour of the court of Denmark, that during the last century, considerable encouragement has been given to the progress of knowledge among the Icelanders; and much pains bestowed upon the revival of the various records of their ancient history and learning. During a considerable period, several of the most eminent literary characters of the Danish metropolis were associated together, under royal patronage, for the purpose of illustrating the antiquities of the north; and to their industry and research, aided by the manuscripts which were preserved in the library of Magnæus, we owe very excellent editions of several of the most important of the early Icelandic writings. The editions of the Sagas which came out under their superintendance, are rendered particularly valuable by Latin translations of the text, and

---

* The names might be recited of between two and three hundred authors in different departments of literature, who have appeared in Iceland during the period intervening between 1650 and the present time. Besides the three distinguished individuals who are mentioned above, there are many others who appear to have merited well the reputation which they enjoyed among their countrymen. Such are, John Haltorson, Paul Vidalin, Paul Biornson, Jonas Vidalin, Eggert Olafson, Biarne Paulson, Halfdan Einarson, John Finsson, &c.

by the very copious notes and illustrations which are sub-joined to them. Among the distinguished men who were en-gaged in this office, we find the names of Luxdorph, Suhm, Langebeck, and of several Icelanders who had acquired re-putation in Copenhagen from their abilities and acquire-ments.* The most eminent among the latter was the learned Professor Thorkelin, whose exertions in behalf of the litera-ture and other interests of Iceland, are happily yet continued to his country.

The government of Iceland has undergone no material change during the last century. The country is still attached to the dominion of Denmark; and the charge of its adminis-tration is committed, as formerly, to governors appointed by the crown, who have generally resided in the island, and ad-ministered its laws in a mild and equitable manner. The change which was introduced some years ago into the judi-cial establishments of the country, when the courts of law were transferred from Thingvalla to Reikiavik, will be spoken of in the chapter on this subject; and the alterations which have taken place of late in its commercial system, will else-where be detailed at length.

It is much to be lamented that the history of Iceland may not close here. A calamity, however, remains to be record-ed, under which the people of this island are still suffering, and the termination of which is yet uncertain and obscure. Secluded from the rest of Europe, bearing no part in the contentions of more powerful states, gentle and peaceable in all their habits, the Icelanders are nevertheless exposed at the present time to the evils and privations of war. The unhappy contest which has now for some years subsisted

---

* By the lovers of northern literature, the name of Count Suhm, as one of its most active and generous patrons, will ever be held in veneration.

between England and Denmark, by intercepting the trade to the island, has abridged the few comforts the people before possessed, and deprived them of many things, which might almost be considered the indispensable necessaries of life. An attempt was made some time ago by the British government to obviate this evil ; but hitherto, from particular circumstances, without the success which such an effort deserved. It will surely be viewed as one of the most lamentable features in the history of the times, that a people on whom nature has bestowed so few of her blessings, should be despoiled even of these amid the ruthless and injurious contests of their fellow men. The mind, while it recoils from such a picture, will the more earnestly look forward to the period, when these complicated social calamities may have an end ; and when the desolate scenes of nature, as well as the fairest regions of the earth, may no longer echo to the continual tumults of war.*

* The singular incidents which occurred in Iceland in the summer of 1809, are yet too recent to be recorded in the general history of the island, even had they been less transient in their duration, and more important in their consequences. A short account of them will be found in the succeeding narrative.

TRAVELS IN ICELAND

DURING THE SUMMER OF THE YEAR

1810.

TRAVELS IN ICELAND, &c.

CHAP. I.

JOURNAL.

HAVING had the good fortune to procure accommodations on board the ship Elbe, belonging to Messrs. Phelps and Company of London, I sailed with my friends from Leith, on the 18th of April 1810, for Stromness, where we were to meet that vessel. A favourable wind enabled us to accomplish this part of our voyage in forty-four hours. The Elbe had not arrived; but in the harbour was a brig, the property of the same Company, and also bound for Iceland; which vessel had been blown off the east coast of that island in the month of November.

We had heard of a great many ships for which licences to trade with Iceland had been obtained; but had met with every kind of obstruction in our endeavours to secure a passage in any of them; on account, as we afterwards found, of the licences having been made use of to conceal a trade with Norway and Denmark.

Expecting every moment the arrival of the Elbe, we could not venture to explore any part of the Orkney Islands, except the immediate neighbourhood of Stromness. It happened, when we approached the Pentland Frith, to be the time of slack tide, so that our passage across it was very pleasant. There is nothing particularly striking in the first view of the Orkneys. On entering the narrow sounds, I was surprised by the great number of ships assembled in the harbour called the Long Hope. These, together with the rocky shores of the islands, and the hills of Hoy (the highest groupe in Orkney), formed a very pleasing scene. The security of the Long Hope, the easy access to it and the excellent harbour of Stromness, together with the light-houses on the Pentland Skerries, seem effectually to remove the terrors of the Pentland Frith; the navigation of which, from the violence and rapidity with which the tide flows through it, had long been considered as extremely hazardous. The town of Stromness is pleasantly situated along the foot of a hill, on the west side of the bay which forms the harbour. The houses, of which some are very good, are crowded together in the utmost confusion. What is called the street, is a long, narrow, dirty lane, badly paved with flag-stones. It is so narrow in some places that it seems impossible for two wheel-barrows to pass each other. In walking along, it is not unusual to be stopped by the operation of slaughtering a pig, a sheep, or a calf, in the street, which is never cleaned but by heavy rain. The inn is very comfortable, and we had no cause to complain of what was provided for us. There is often a scarcity of wheaten bread, arising from the uncertainty and irregularity of the supply of flour. The water is excellent; and there is abundance of it for the supply of ships frequenting the harbour.

We were invited by the Reverend Mr Clouston, the minister of Stromness, to pass an evening at the manse, and were hospitably and agreeably entertained. Before and after tea, brandy, and some excellent cinnamon water manufactured by Mrs Clouston, were handed about.

The state of agriculture in the neighbourhood of Stromness is most wretched; the cottages are filthy; and the inhabitants are very indolent. The black cattle, sheep, and horses, are miserable looking creatures; and the implements of husbandry are of the rudest construction, especially the plough. It seems that the people of Orkney are extremely averse to any innovation on their old practices, and exceedingly jealous of strangers. Of the latter quality we had one proof, while examining some cottages, and taking sketches of them. An old man who was busy in planting potatoes at a little distance, on seeing us thus employed, left his work, and walking up with as much fierceness as his weather-beaten countenance could express, roughly demanded what we were doing, and why we dared to go into the houses to frighten the children.

The population of the Orkneys is considerable; and merchant ships find no difficulty in getting men, who are generally landed on the return voyage, and thus escape being impressed. The Hudson's Bay Company has hitherto obtained workmen from these islands; but the increasing price of labour will probably soon close this source of supply, or diminish the profits of the Company.

The proprietors of the Orkney Islands, as they have many difficulties to encounter, will have the greater merit when they shall have improved the condition of the peasantry, as well as the productiveness of the soil.

On the 25th, the Elbe arrived at Stromness. Immediately on her arrival, we went on board, and were received with great kindness by Captain Liston, and Mr Fell, who was going out as agent for Phelps and Company To these gentlemen, as well as to their employers, I am under very great obligations. Not only did they incommode themselves that every thing might be comfortable to me and my companions, in point of accommodation, but they continued to make every exertion to render the voyage agreeable.

The island of Iceland is placed in the Atlantic ocean, in a direction nearly north-west from the continent of Europe. Its position has not been very accurately ascertained; but, from the best authorities in our possession, it appears to lie between the 12th and 25th degrees of longitude west from Greenwich; its extreme breadth from east to west being nearly 300 miles. From some notes which were obligingly sent to us by the Danish officers employed in surveying the coast, the latitude of the northern extremity of the island, Cape North, is 66° 30', which nearly coincides with the Arctic Circle. In the common maps of the country, it is laid down a degree farther north. The most southern part is in north latitude 63° 40' nearly; and the broadest part of the island from north to south, is probably not much more than 180 miles.

We sailed on the 28th for Reikiavik, the capital of Iceland. A fine breeze carried us about twenty miles to the westward of Orkney, when we were becalmed; and several ships, bound for America, which had sailed the day before, were seen in the same situation. On our return to Orkney in autumn, we were informed that, at Stromness, the same favourable wind which had carried us only twenty miles, had continued several days, and had given our friends reason to believe that we should have a very short voyage.

The weather soon became unsteady; and a heavy gale of wind overtook us on the 3d of May; but it was not against our course. Early in the morning, one of the sailors, an elderly man, fell from the main yard upon the deck. On hearing of this accident we got out of bed, and, though it was hardly possible to stand, we contrived to reach the place where he lay, and to bleed him. He died in the evening. At night on the 4th, the wind became more violent, accompanied with snow; and the rigging was stiffened with ice. Moderate weather now succeeded, and as soon as it was clear day-light, we had the pleasure of being summoned upon deck to enjoy the first view of Iceland. The land first in sight was that called by navigators Cape Hekla, though it is at a considerable distance from the celebrated volcano of that name. The range of enormous mountains, which now appeared soaring above the horizon, was entirely covered with snow; and though we felt considerable joy on finding ourselves so near the end of our voyage, we could not help being impressed with the very uninviting appearance of the country;—

> Where undissolving, from the first of time,
> Snows swell on snows amazing to the sky;
> And icy mountains, high on mountains pil'd,
> Seem to the shivering sailor from afar,
> Shapeless and white, an atmosphere of clouds
> Projected huge, and horrid o'er the surge.

In the evening we passed to the southward of the Westmann Islands, a fine groupe of rocks, extending to a distance of about twenty miles from the most southern part of Iceland. The navigation round Cape Reikianes, the south-west point of the island, not being deemed safe during the night, we lay to a little to the westward of the rocks till daylight, when a fresh breeze came off the land.

In sailing along towards the Cape, we had a fine view of the mountains of the south-west part of Iceland. They are not very high, and the snow had almost disappeared from them. Their rugged summits, and the desolate appearance of their sides, seemed to indicate a volcanic origin; and with the assistance of a telescope we thought we could distinguish the places whence lava had flowed. We also saw the vapour rising from the sulphur mountains, which are in this range. Towards Cape Reikianes, the mountains gradually decrease in height, and become more conical; and at length the country is low, and rocky, which renders an approach to the shore very dangerous. Off the Cape are some small rocks, about six miles apart from each other; and beyond them is a sunken reef extending about ten miles farther, and terminating in a rock called, in the Danish chart, *Blinde Fugle Skier*, or the blind rock. We steered for the passage between the Cape and the nearest rock; and just as we got into it, were becalmed. This circumstance excited considerable anxiety in the captain; but he was soon relieved by perceiving the tide to be in our favour, which carried the ship safely from the rocks. We now turned northwards, and observed that the flat country, between the Cape and the Skagen Point, had the appearance of having been desolated by volcanic eruptions. We sailed backwards and forwards during the night, and began to beat up the Faxè Fiord* at sun-rise, on the 7th of May. The day proved clear, and we had a distinct view of the amphitheatre of mountains which bound the Faxè Fiord. On one hand, the view terminated with the bare, rugged, and gloomy hills stretching towards the east from Cape Reikianes; on the other, with the lofty Snæfell Jokul† towering above the neighbouring snow-covered mountains,

---

* Fiord means, bay, or frith.

† Jokul (pronounced Yokul) is a name given to such mountains as are perpetually covered with snow.

which rose in a variety of shapes, forming a most magnificent scene; but such a one as seemed to forbid the approach of man. Where no snow appeared, hideous precipices overhung the sea, or the destructive effects of subterraneous fire were visible on the more level country, where alone an adventurer could hope to find access.

A great number of boats were seen coming from different parts of the coast to fish. Passing near one of them, we hailed it, and took the people and their fish on board. They had caught about thirty cod, halibut, and tusk or cat fish, for the whole of which they demanded four shillings. The people were clad in sheep skins, which they took off before coming into the ship. This covering has a very singular appearance; but it keeps the fishermen dry, and preserves their clothes from being spoiled while they are hauling in the fish. In the north of Iceland they wear seal skins, with hoods fitted to the jackets. The dress of the men in this, and in another boat which afterwards approached the ship, consisted of blue, grey, or black cloth jackets and breeches, and coarse woollen stockings. Their shoes were made of undressed seal skin. Some of them wore woollen caps, with a tassel, varying in colour, hanging at the end. The owner of the first boat had his jacket trimmed with red cloth: he officiated as our pilot. Most of them had on round slouched hats; their hair was long and lank, and several of them had long beards. On our approaching nearer to Reikiavik bay, the pilot of that place, who had been fishing, came on board, and superseded the other who belonged to a place called Kieblivik. The Reikiavik pilot was a tall, stout, good looking man, but his sheep-skin dress gave him rather a savage appearance.

Viewed from the sea, the capital of Iceland has a very mean appearance. It is situated on a narrow flat, between

two low hills, having the sea on the north-east, and a small
lake on the south-west side.   We landed for a short time in
the evening; and had I not previously seen the fishermen, I
should have been a good deal surprised at the odd figures
that flocked about us.   The Danish inhabitants, who seldom
stir without tobacco pipes in their mouths, were easily distinguished.   The beach slopes rapidly; but is extremely convenient for boats at all times of the tide.   It is composed
entirely of comminuted lava.   There were two large wooden
platforms, made to be occasionally pushed into the water, for
the purpose of loading and unloading the larger boats.   The
anchorage is good; and the bay is defended from heavy seas
by several small islands, which render it a very safe harbour.

   The houses, with the exception of one that is constructed
of brick, and the church and prison which are of stone, are
formed of wood, coated on the outside with a mixture of tar
and red clay. The storehouses, some of which are very large,
are built of the same materials, which in every case are put
together very neatly.   The longest range of houses extends along the beach; the other stretches at a right angle
from it at the west end, and is terminated by a house
which is used by the merchants as a tavern.   At the
east end of the town is the Toght-huus, or prison, which
having been white-washed by the usurper Jorgensen,* is

---

* Mr Hooker, well known as an eminent botanist, who was in Iceland during part of the summer of the year 1809, has printed an account of the revolution which happened at that time, and of the usurpation of Jorgen Jorgensen, a
Dane. This affair was slightly noticed in the public prints of the day. As Mr
Hooker has not published his work, it may be proper for me here to give a short
account of the affair. Though there was no doubt of the humane intentions
of the English government towards Iceland, no public intimation had been
given of them. In 1808, Captain Gilpin, the commander of a privateer, robbed

very conspicuous among the other dark brown buildings. Behind this end of the street, which is on the beach, stands the house of the late governor, Count Trampe; and beyond that, near the lake, is the church; a clumsy building covered with tiles. It is in a sad state of dilapidation, the winds and rain having free access to every part of it. Though sufficiently large to accommodate some hundreds of persons, it is not much frequented on ordinary Sundays. On particular occasions, such as a day of confirmation, it is much crowded. In the neighbourhood of the town there is a considerable number of cottages, all very mean, and inhabited for the most part by the people who work for the merchants. The whole population amounts to about five hundred. On

the public chest of upwards of 30,000 rix-dollars, which had been appropriated to the maintenance of the schools and the poor. From the information of a Danish prisoner of war, named Jorgensen, Mr Phelps, an eminent merchant in London, formed a project of trading with Iceland, and bringing home the produce, particularly tallow. He accordingly sent out a ship, the Clarence, with Jorgensen, and a person of the name of Savignac, in the capacity of supercargo. Jorgensen had never been in Iceland, and, from my information, it would appear that he had deceived Mr Phelps as to the amount of goods which might be in this island. It was through Mr Savignac that Jorgensen conveyed his information to Phelps and Company, the former being a clerk to one of the partners. The laws of the country prohibiting any intercourse with strangers, permission to land the cargo was refused. But, on the captain of the Clarence taking possession of a Danish brig, which had just arrived with provisions, leave was granted. It was found impossible, however, to sell any part of it, and Mr Savignac resolved to remain in Iceland, while the Clarence returned to England in ballast, the Danish brig having been restored. Count Trampe, the governor, arrived from Copenhagen in June; and the British sloop of war, the Rover, Captain Nott, arrived soon after. A convention was entered into between the Governor and Captain Nott, by which it was stipulated, that British subjects should be allowed to trade with the natives during the war, but that they should be subject at the same time to the Danish laws. Owing to some neglect, the publication of this convention was

the top of the hill, to the westward, is an observatory, in which a few instruments are usually kept. At present they are in the hands of two Danish officers, who are employed in surveying the coasts.

The drawing from which the engraving is taken was made from some rocks above a cottage a little to the south-east of the town. This is the most favourable view; and it includes part of the range of mountains in the Snæfell Syssel, which is terminated by the Snæfell Jokul.

The spring-tides often rise so high as to overflow the ground between the governor's house and the church, and the street which runs up from the beach. The water sometimes enters the little gardens which are behind most of the

delayed; and this unlucky circumstance led to disagreeable events. A proclamation still remained posted up in the town, forbidding any native to trade with the English under pain of death. While things were in this state, Mr Phelps had equipped a second expedition, and, along with Jorgensen, arrived in the ship Margaret and Anne, with a cargo selected by Mr Savignac, and provided with a letter of marque; and another vessel, the Flora. They arrived a few days after the Rover had sailed. Having been informed by Mr Savignac that obstructions were thrown in the way of the British, Mr Phelps ordered Captain Liston, who commanded his ship, to seize the Governor, and make a prize of the Orion, a vessel belonging to Count Trampe. This was done; the Count was conveyed on board the Margaret and Anne; and Mr Jorgensen was installed into the chief command of the island. He soon proclaimed the cessation of all Danish authority. A body of natives were formed into a regular armed force, and Jorgensen proceeded to seize upon all public and private property, and travelled about the country for that purpose. The arrival of Captain Jones, of the Talbot sloop of war, was unfortunate for Mr Jorgensen, who was soon stripped of his power, and sent, with Count Trampe, to England. The only result of the Count's representations to our government, that is known, is an order in council dated February 7, 1810, of which I shall probably have to take some notice in the sequel. Jorgensen was sent to the hulks at Chatham, where he remained for a year, and is now on his parole at Reading.

Sketched by Sir G. M.

Engraved by B. Scott

houses. A small stream runs from the lake, and were its channel deepened, a very useful and commodious harbour might be made.

On my shewing some letters I had received from his Excellency Count Trampe, to his agent Mr Simonson, and to Mr Frydensberg, the landfoged, or treasurer of the island, they, without the least hesitation, permitted us to take possession of the count's house, in which we were afterwards confirmed by a letter, from the count himself, to Mr Simonson, desiring him to provide accommodations for us in this habitation, which, though small, is very comfortable. It consists of three rooms below, one of which opens into the kitchen, and another is occupied as a public office. There is a pantry well fitted up with shelves, presses, and drawers. The kitchen has a fire place like a smith's forge, with a small grate in the middle, in which the fire is kindled. When a dinner is preparing, different fires are lighted to suit different utensils. The rooms are heated by close stoves connected with the kitchen vent. Above, there is a loft, the access to which is by a narrow trap stair-case. In this there are three apartments, one of which has a stove; the rest of the space is left open for lumber. Adjoining the house is an open court, beyond which are a stable, a cow house, with a hay loft above them. About a quarter of an acre of indifferent soil, behind the house, is inclosed by a paling, and used as a garden. Soon after we went to lodge here, Mr Fell and I sowed some seeds of turnip, radish, cabbage, pease, cress, and mustard. What we left unoccupied was filled by Mr Simonson with potatoes, and Swedish turnip seed.

The first visit we paid, after landing on the 8th of May, was to the bishop, Geir Vidalin, who received us with great kindness. He is a good looking man, above the ordinary

stature; corpulent, but not unwieldy; with an open countenance, which seems to declare his feelings without disguise.
He is an excellent classical scholar, and speaks Latin fluently;
and his general knowledge is equal, if not superior, to that
of any person in Iceland. Considering the high rank he holds,
I was at first sight surprised at the poverty of his dress, as
well as that of his habitation. He wore an old thread-bare
great coat, over a waistcoat of the same description; and a
pair of dark grey pantaloons that had seen better days. I
soon found, however, that he was not worse clothed than
others who could better afford to be neat, and who had as
much leisure to attend to their persons.

Having learned that we should be obliged frequently to
return to Reikiavik, after making excursions into the country,
our first care was to arrange our household affairs, and to
inquire for horses, which, at this season, were difficult to procure. The grass not having begun to grow, they were still
very lean, and unfit for hard service. In several districts a
disease had attacked the horses, and carried off great numbers of them. Every body told us that it would be in vain
to attempt travelling so early in the season; and as we saw
that delay was unavoidable, we resolved to employ ourselves
in forming acquaintance, and in observing the manners of the
people in the capital.

We thought it our duty to pay our respects to Mr Olaf
Stephenson, who has the title of Geheimè Etatsrood, and was
formerly governor of the island; and having been informed
that he would be glad to see us, we went to his house, which
is on the island of Vidöe, about three miles from the town.
It is built of stone, and bears evident marks of decay on the
outside. The situation, between two green hills, the ground
in front sloping towards the sea, is very agreeable. On the

west side is a neat chapel, where the minister of Reikiavik performs divine service once in three weeks. Before the chapel is a small garden, inclosed by a turf wall. Behind the house are cottages for the accommodation of servants; and farther off, are the cow and sheep houses.

The old gentleman, dressed in the uniform of a Danish colonel of the guards, received us at the door with great politeness, and seemed to be exceedingly gratified by our visit. He ushered us into a large room, furnished with the remains of ancient finery, some prints, portraits, and a number of profile shades, which afforded little relief to the eye while wandering over the damp, decaying walls. The house altogether appeared as if it would not long survive its venerable inhabitant. The next room we entered was our host's bed-chamber, which was very comfortable, and well warmed by a stove placed in a corner. After a little conversation on indifferent subjects, the old gentleman talked of his health, and seemed quite delighted to find that we could give him some medicines as well as advice. We had the pleasure of being told, after the lapse of a few weeks, that Mr Holland's prescriptions had been attended with the best effects.

We had no intention of remaining here to dinner; but, on proposing to take leave, we soon perceived that it would give great offence, to withdraw without partaking of his hospitality. Mr Stephenson spoke affectionately of Sir Joseph Banks, who is much and deservedly esteemed in Iceland; and he shewed us, with much apparent satisfaction, some diplomas which he had received from different societies. In due time, the repast which had been prepared, was announced by a good looking girl, dressed in the complete Icelandic costume. The dress of the women is not calculated to shew the person to advantage. The long waist, bunchy petticoats, and the fashion of flattening the bosom as much as possible, together with the extraordinary head-dress, excited rather ludicrous emotions at the first view; but there is a richness in the whole that is pleasing. A dress which I procured, consists of a blue cloth petticoat, with a waist of scarlet woollen stuff ornamented with gold lace, and silver loop-holes on black velvet, for lacing it. On the back of the waist are stripes of black velvet, which cover the seams. Over the petticoat is tied an apron of blue cloth, having a silver gilt ornament hanging from the middle of the upper part; and along the bottom, several stripes of light blue stamped velvet. Over the waist is put on a jacket of black cloth, having two stripes of black velvet in front, and next to them two of gold lace. On the back and shoulders are stripes of orange-coloured velvet. Round the waist is buckled a girdle of black velvet, covered with rich silver-gilt ornaments. A collar or ruff of black and crimson velvet ornamented with silver lace, and having attached to it a sort of tippet of black cloth, adorned in front with light brown velvet, is put round the neck. Silver-gilt chains of various forms, and medals, are worn suspended from the neck. On going out of doors, to church, or on any occasion of ceremony, the women wear a cloak or mantle of black cloth, called wadmal. That belonging to the dress above described is ornamented in front with two large hollow con-vex buttons, made of silver gilt and richly ornamented. The cloak is trimmed with two broad stripes of figured woollen stuff resembling black stamped velvet. The head-dress is formed of white linen or cotton cloth, shaped like a large flat horn bending forwards, and made stiff with a quantity of pins, fastened on the top of the head with a coloured silk or cotton handkerchief, which entirely conceals the hair. The petticoat, the apron, and the jacket, are of different colours,

Plate 2.

1   2   3   4   5

ICELANDIC COSTUME.

I. Clark direxit

though blue and black are the most common; and the ornaments of silver are variously shaped. The general fashion of the dress has long remained the same, though the head-dress has undergone some alterations in its dimensions, and in the mode of bending it. Little girls are dressed in the same manner, only they wear a cap variously ornamented. Boys have more gaudy caps. One in my possession is made of blue silk, surrounded with gold lace, and has a green silk tassel on the top.

When a lady goes abroad on horseback, the head-dress is covered by a hat of a very curious shape, which is seen in the engraving, where the different dresses are better explained than by words.

No. 1, is a lady in full dress; No. 2, the ordinary dress of the women of all ranks; No. 3, a lady in a riding dress; No. 4, an Icelander in his best clothes. The jackets of the men are frequently made of black wadmal. No. 5, is the figure of the Reikiavik pilot in his sheep-skin dress, as he came on board the Elbe.

The names of the different parts of the dress are,—*faldur*, the head-dress, of which the upper part is called, *skoit; upphlutur*, the waist of the petticoat; *fat*, the petticoat; *nær-pills*, these when joined together; *svinta*, the apron; *trojè*, the jacket; *hempa*, the cloak; *skirta*, the shift; *hals-festi*, the neck-chain; *kragè*, the collar; *herda-festi*, the shoulder-chain; *milnur*, the ornamented loop-holes of the waist; *lindè*, the girdle; *sockar*, stockings; *socka-bond*, garters; *skior*, shoes. On one side of a medal in my possession, is a representation of God, with three faces, sitting in a cloud, and holding a globe surmounted by a cross in one hand. The following inscription is above in Roman characters, TETRAGRAMMATON. JEHOVAH. ADONAY. ELOY. On the reverse is a representation of the

Trinity, with the inscription BENEDICTA. SEMPER. SANCTA. SIT. TRINITAS. The figures are in relief, and well executed.

On entering the room into which we had at first been introduced, we found a table neatly covered, and a bottle of wine set down for each person. This alarmed us a little, as we feared that the old gentleman intended, according to the ancient custom of Denmark, to ' keep wassel.' The only dish on the table was one of sago soup, to which we were helped very liberally. The appearance of a piece of roasted, or rather baked, beef, relieved us considerably; and we submitted, as well as we were able, to receive an unusual supply of a food to which we were accustomed. We had drank a few glasses of wine, when a curious silver cup, large enough to contain half a bottle, was put upon the table. Our host filled it to the brim, and put on the cover. He then held it towards the person who sat next to him, and desired him to take off the cover, and look into the cup; a ceremony intended to secure fair play in filling it; after which he drank our healths, expressing his happiness at seeing us in his house, and his hopes that we would honour him with our company as often as we could. He desired to be excused from emptying the cup, on account of the indifferent state of his health; but we were informed at the same time, that if any one of us should neglect any part of the ceremony, or fail to invert the cup, placing the edge on one of the thumbs as a proof that we had swallowed every drop, the defaulter would be obliged by the laws of drinking to fill the cup again, and drink it off a second time. He then gave the cup to his neighbour, who, having drank it off, put on the cover, and handed it to the person opposite to him. Being filled, the cup was examined by the person whose turn it was to drink next, and thus it went round. In spite of their utmost exer-

tions, the penalty of a second draught was incurred by two
of the company. While we were dreading the consequences
of having swallowed so much wine, and in terror lest the cup
should be sent round again, a dish of cold pancakes, of an
oblong form, and covered with sugar, was produced; and
after them sago puddings floating in rich cream. It was in
vain that we pleaded the incapacity of our stomachs to contain any more; we were obliged to submit to an additional
load; when a summons to coffee in an adjoining room,
brought us a most welcome relief. Our sufferings, however,
were not yet at an end. On first entering the house, I had
noticed a very large china tureen on the top of a press; and
as it had not been used at dinner, I concluded that it was a
mere ornament. We had scarcely finished our coffee, when
the young woman who had waited at table came in with this
tureen, and set it before us. It was accompanied by some
large glasses, each of the size of an ordinary tumbler. I
looked at my companions with dismay, and saw their feelings very expressively painted in their countenances. This
huge vessel was full of smoking punch; and as there was no
prospect of being able to escape, we endeavoured to look
chearful, and accomplish the task required of us. Having at
length taken leave, our hospitable friend insisted on attending us to the beach.

Our next visit was to the minister of Reikiavik, and to the
physician, both of whom live at some distance to the westward of the town. The former, Mr Sigurdson, met us at the
door of a miserable hut, and led us through a long, dark and
dirty passage, obstructed by all sorts of utensils, and by a
man beating stock fish, into an obscure room. The apartment into which we were ushered was the family bed room,
and the best in the house. The roof was so low that a person

M

could hardly stand upright, and there was scarcely room for any thing beside the furniture, which consisted of a bed, a clock, a small chest of drawers, and a glass cup-board. My surprise at finding the minister of the only town in Iceland so ill accommodated, ceased on discovering that a stipend of one hundred and twenty dollars, not very regularly paid, with pasture for a cow or two, and a few sheep, were all that he had for the support of himself and his family. We were presented with a bason of very good milk, and, after some conversation on indifferent matters, we proceeded towards the house of Dr Klog. * Nothing can be more dreary than the face of the country hereabouts; and how an elevated and exposed situation, at a distance from the town, came to be chosen for the residence of the physician, seems difficult to explain. The doctor's house, and that of the apothecary-general, are under the same roof; and with respect to size, furniture, and cleanliness, were the best we had yet seen. The house is built of stone, and white-washed.

Dr Klog having been informed that we had brought some vaccine virus with us, said that there had been none in Iceland for two years, and rejoiced that such a blessing was renewed to the country. He was very impatient to have some children inoculated. Having taught him to use the crust, we had soon the satisfaction of seeing a supply of virus sent off to different parts of the island; and before we left it, we learned that it had reached the most remote corners. The people have implicit faith in the virtues of this mild substitute for small pox.

Madame Klog soon made her appearance, and brought some chocolate, which, we were told, had been made by the

* Pronounced Klo.

apothecary, and had some of the Lichen Islandicus mixed with it. We found it to be remarkably good, but could not distinguish the addition of the Lichen. On examining the laboratory, it was observed to be well stored with old fashioned drugs of all sorts, most of them quite useless. The compounds were prepared according to the directions of the Danish pharmacopœia. From the large supply of medicines, one might suppose that they were much called for; but there are few physicians in the world less troubled with practice than Dr Klog. There are five other medical practitioners in the island, who have salaries from the Danish government. The little practice they have is very laborious, on account of the very scattered state of the population; and their fees are extremely small.

We had an opportunity of seeing the funeral service of the Icelandic church performed, at the burial of the sailor who lost his life on the voyage. The minister, dressed in a gown of plain black cloth, and with a band, met the corpse on the beach. and walked before it. On entering the churchyard, he began to chaunt, and was joined by many of the people. This continued till the coffin was laid in the grave, when the priest took a small wooden spade, and pronouncing some words of the same import as ' dust to dust,' &c. in the English service, threw in a little earth. The chaunting then recommenced, and continued till the grave was filled up; after which all present put their hats before their faces, and seemed to pray. A general obeisance followed, which closed the ceremony. The whole was conducted with solemnity, and the people seemed very serious and earnest in their devotions. I did not admire the music of the Icelanders, who seem to have no idea of harmony.

On the 13th of May, we were honoured with a visit from Mr Magnus Stephenson, son of our friend at Vidöe. He presides in the highest court of justice, and styles himself Lord Chief Justice of Iceland. He is also a counsellor of state, and was invested with the government at the time Count Trampe went to England in 1809. He was accompanied by one of his sons, and two of his brothers, one of whom is Amt-mand, or governor, of the southern district, and the other secretary to the court of justice. They were all extremely polite, and, through the Chief Justice, who spoke English tolerably well, expressed their desire to be of use to us in very handsome terms. We found every one emulous in offering his services; and I shall ever remember with gratitude the kind attention and hospitality I experienced during my stay in Iceland, both from the natives and from the Danes. The latter had really little cause to make Englishmen welcome. But on all occasions they spoke of what had happened at Copenhagen in the most liberal manner; and one gentleman observed, that, though our government had used them ill, we were not to blame; and had as much right as strangers from any other country, to expect and receive such attention as they had it in their power to bestow.

On the 14th a court was held, in order that the deputy-governor might pass sentence in some civil causes. He was dressed in a red coat adorned with gold lace and embroidery. The meeting of the court was announced by the beating of an old drum, the prelude to a short proclamation. In the room was a table covered with a tattered green cloth, and a few chairs. Beside the deputy-governor, the secretary and half a dozen other persons were present. There was no sort of ceremony, or appearance of dignity. Mr Stephenson

took up a book, read the sentences as fast as he could, signed them, and then the whole ended.

We went to Mr Frydensberg's to breakfast, between eleven and twelve o'clock, where we found a table covered as if
for dinner, having bottles, glasses, &c. upon it. Madame
Frydensberg brought in a dish of mutton dressed somewhat
in the manner of currie. It is customary in Iceland, either
for the lady of the house, or one of her daughters, to place
the dishes on the table, and to remove them, the plates,
knives and forks, &c.; though sometimes the housekeepers,
who are on a very familiar footing with their employers, perform these offices. There are no men-servants. After the
currie, or ragout, came roasted mutton, cheese, and bread
and butter. After a few glasses of wine, coffee was served,
and concluded the entertainment; when all rose up, and
bowed to the lady.

Next day, we went to see some hot springs, about two
miles to the eastward of the town, which the people frequent
for the purpose of washing clothes. A rivulet runs past them;
and, by a little management, a person may plunge his hand
into the water, so that one part of it may be subjected to any
degree of heat up to 188°, while another is chilled. The water of the springs mixing with that of the rivulet, every desirable temperature is to be found within a short space. The
women who go out to wash clothes, boil fish, or meat of any
sort, in the places where the water is hottest. The plants
which we took out of the water had a disagreeable, somewhat
sulphureous smell, as well as the water itself; but it was not
strong. There are some large and deep tepid pools, in which,
on a certain day in June, all the girls in the neighbourhood
bathe.

On the 15th we gave a ball to the ladies of Reikiavik, and

the neighbourhood. The company began to assemble about
9 o'clock. We were shewn into a small low roofed room, in
which were a number of men ; but to my surprise I saw no
females. We soon found them, however, in one adjoining,
where it is the custom for them to wait till their partners
go to hand them out. On entering this apartment, I felt considerable disappointment at not observing a single woman
dressed in the Icelandic costume. The dresses had some resemblance to those of English chambermaids, but were not
so smart. An old lady, the wife of the man who kept the
tavern, was habited like the pictures of our great grand-
mothers. Sometime after the dancing commenced, the bishop's
lady, and two others, appeared in the proper dress of the
country.

We found ourselves extremely aukward in dancing what the
ladies were pleased to call English country dances. The music, which came from a solitary ill scraped fiddle, accompanied by the rumbling of the same half rotten drum that had
summoned the high court of justice, and by the jingling of a
rusty triangle, was to me utterly unintelligible. The extreme
rapidity with which it was necessary to go through a multiplied series of complicated evolutions in proper time, completely bewildered us ; and our mistakes, and frequent colli-
sions with our neighbours, afforded much amusement to our
fair partners, who found it, for a long time, impracticable to
keep us in the right track. When allowed to breathe a little,
we had an opportunity of remarking some singularities in the
state of society and manners, among the Danes of Reikiavik.
While unengaged in the dance, the men drink punch, and
walk about with tobacco pipes in their mouths, spitting plen-
tifully on the floor. The unrestrained evacuation of saliva
seems to be a fashion all over Iceland ; but whether the na-

tives learned it from the Danes, or the Danes from the natives, we did not ascertain. Several ladies, whose virtue
could not bear a very strict scrutiny, were pointed out to us.
One was present, who, since her husband had gone to Copenhagen on business, had lived with another merchant by whom
she had had two children. Another, thinking her husband
too old, had placed herself under the protection of a more
youthful admirer, and left the good easy man to brood over
his misfortune, or to find a partner more suited to his age.
These ladies, and others who paid as little regard to character, were received into company, and treated with as much
complaisance and familiarity as the most virtuous. This total
disregard to moral character, and the rules of decorum, may,
without breach of candour, be regarded as impeaching the
virtue even of those who maintain the appearance of greater
strictness in their behaviour. It is no overstrained inference,
that their associating with such ladies as those whose conduct
has been described, is owing to some fellow feeling, some necessity for keeping secrets which it might be dangerous to
divulge. Where no guardian of morals is present ; or where
there is one, if he winks at such indecorum ; if he converses
with those who have broken the dearest ties of affection ;
there may, indeed, be some excuse. Here we saw the bishop
himself countenancing vice in its worst shape, and appearing
perfectly familiar with persons who, he must have known,
bore the worst characters. I was informed, that when a
couple are dissatisfied with each other, or when a lady chuses
to change her helpmate, the separation is sanctioned without
any inquiry into the cause, and new bands solemnly unite those
who have most openly slighted their former engagements.
Such are the morals of the people of Reikiavik.

During the dances, tea and coffee were handed about ;

and negus and punch were ready for those who chose to partake of them.   A cold supper was provided, consisting of hams, beef, cheese, &c. and wine.   While at table, several of the ladies sung, and acquitted themselves tolerably well. But I could not enjoy the performance, on account of the incessant talking, which was often loud enough to overpower the harmony.   This was not considered as in the least unpolite.   One of the songs was in praise of the donors of the entertainment ; and, during the chorus, the ceremony of touching each other's glasses was performed.   After supper, waltzes were danced, in a style that reminded me of soldiers marching in cadence to the dead march in Saul.   Though there was no need of artificial light, a number of candles were placed in the rooms.   When the company broke up, about three o'clock, the sun was high above the horizon.

During our stay in the town, my friends had an opportunity of seeing the marriage-ceremony of the Icelanders, which was performed in the church.   The bride, in full dress, was seated on one side of the church, accompanied by an elderly woman, probably her mother.   Opposite to her, on the other side, was the bridegroom. His seal-skin shoes were fastened by cross bands of white tape ; and his striped garters were crossed about his legs.   He was attended by several of his friends, who, during the whole of the ceremony, indulged themselves with a profusion of snuff.   The priest standing at the altar opposite to the party, began the ceremony by chaunting, in which he was joined by all present. This was followed by a prayer, and a long exhortation to the bride and bridegroom, who were now brought forward.   Three questions, similar to those used in the English service, were then put to them ; first to the man.   The priest afterwards joined their hands, laid his hands upon their shoulders, and gave them his blessing. They were then conducted to their respective seats, and the service concluded by chaunting. In going from the church, the bride preceded the bridegroom, both being attended by their friends of the same sex. They usually go, on such occasions, to the house of some relation. When the bride retires after supper, she is accompanied by her female friends. When the husband arrives, he finds them all seated by his wife's bedside, and is refused admittance. On his persisting, he is told he must pay; and he offers a snuff-box, or any trifle he may have in his pocket, which is refused. At last he promises some present of value, from twenty to a hundred dollars, according to his circumstances; and the women tell him that he must give it to his bride. This altercation sometimes continues for an hour, in perfect good humour. In the morning, the husband makes a present to his wife of some articles of dress, money, or silver spoons. They now go to their own house. We did not see the procession to the church; but were informed, that from the house of the minister, or some cottage near the church, girls go first two and two, then the bride attended by a female relation, or the most respectable woman in the company. She is followed by the women, after whom goes the bridegroom with a friend. Next in order is the priest, and the men close the procession.

A day was spent in examining the rocks of the island of Vidöe; and we again dined with the old governor. We now found that the young woman who had attended at table on our former visit, was his niece; and that an elderly female who had appeared at the same time, was his sister-in-law. We had sent some trifling present to these ladies; and, on this account, as soon as we entered the house, it became necessary to submit to the customary salute denoting the gratitude of those who receive presents. On many occasions, we could well have dispensed with the ceremony ; and our talents were often exercised in contriving means of evasion or escape. On meeting after a short absence, and on taking leave, the Icelanders take each other by the right hand, remove their hats with the left, and stroking back their long hair, kiss each other with much apparent satisfaction.

Our reception at Vidöe was as cordial as on our former visit, and the entertainment more agreeable, as the ceremony of the cup was not repeated, and we were not obliged to over-eat ourselves. The old gentleman told us of his having been robbed by the people of a ship that had come to Iceland two or three years before ; and I was afterwards informed that this outrage had been committed by direction of Baron Hompesch, who after his wanton attack on the Faro islands, sent a vessel to Iceland to plunder. It must be mentioned, however, that Mr Stephenson obtained the restitution of part of his property by an application to the commander of the ship.

We now became very anxious to commence our travels, though all our Icelandic friends endeavoured to dissuade us from undertaking any expedition so early in the season. The horses were still very lean, and the grass had hardly begun to grow. Observing, however, that there was no part of the Guldbringè Syssel* covered with snow, we resolved to travel through it, and to visit the sulphur mountains without delay. Having mentioned the Guldbringe Syssel, it may be proper to take notice here, that Iceland is divided into four large departments called Amts; the northern, southern, eastern, and

---

* *Gold-bringing* district, a name probably derived from the principal fishing stations being situated on its coasts.

western Amts. These are subdivided into districts called Syssels, and the syssels into parishes. The number and names of the districts will be found in the Statistical Tables, in another part of the work. Not being able to procure riding horses, we determined to walk; and this resolution seemed to astonish the people not a little, as the meanest person in Iceland never travels on foot. A young man who had been educated as a priest, offered his services; he spoke Latin tolerably well; and as he seemed active, and disposed to be useful to us, a bargain was soon made with him. Early in the morning of the 20th, the preparations for our departure commenced; but the motions of the Icelanders were so extremely slow, and they had so many discussions about distributing the loads on the horses which we had procured for carrying our baggage, that it was past two o'clock in the afternoon before all was ready. The packsaddles consist of square pieces of light spungy turf, cut from the bogs. These are tied on with a rope; a piece of wood made to fit the horse's back, with a peg projecting from each side, is fastened over the turf, and on these pegs the baggage is hung by means of cords. The Icelanders pretend to be very nice in balancing the loads; but I do not recollect ever having travelled two miles, without stopping ten times to rectify the baggage. When all the horses are loaded, they are fastened to each other, head to tail. A cord is tied round the under jaw of the second horse, and the other end of it is joined to the tail of the first; and thus I have seen thirty or forty marching through the country. The Iceland horses, though very hardy, and patient of fatigue, are easily startled. When any one horse in a string is alarmed, it often happens that the cords break, and the whole cavalcade is put into confusion. The poor animals, however, never fail to stop where they can get any thing to eat; and at all times they are easily caught.   A well broke riding horse will wait on the spot where his master leaves him, for any length of time.   If any grass is near him, he may feed; but if there is none, he will stand perfectly still for hours. Many horses will not even touch grass when under their feet. Every Icelander, of whatever rank, can shoe a horse.   The shoes are plain; and the nails, which are very large, are driven firmly through the hoof, and carefully doubled over; and in this simple state the shoes remain firm till completely worn, or accidentally broken.   Travellers always carry a supply of shoes and nails, when going long journeys.   For a short journey it is customary to put shoes only on the fore feet of the horses.   When iron is scarce, the horns of sheep are made use of for horse shoes.

It often happens, when horses are heavily laden, especially when they are in low condition, that their backs are galled. By way of preventing this accident, or curing any tumour or ruffling of the skin, the Icelanders insert one or more setons of horse hair into the breasts of the animals. This cruel practice, instead of alleviating the pain which the horses suffer from their burdens, only serves to add to their torments; and the artificial sores thus produced, soon become very disgusting.

We left Reikiavik with the intention of stopping at Havnefiord, about six miles distant.   The day was fine, but we saw snow showers falling on the mountains towards which we were travelling, and expected to meet them.   We passed through a bare dismal looking country, over some low hills; till, not far from Havnefiord, we entered a rough path, and got upon a tract which bore dreadful marks of fire.   As we approached this scene of desolation, snow began to fall so thickly, that we could not see more than fifty yards distance; and this ad-

Plate 3

Sketched by Sir Geo. Mackenzie.

HAVNEFIORD.

I. Clark direxit

ded not a little to the awful impressions which the first sight of a stream of lava made upon us. The melted masses had been heaved up in every direction, and had assumed all sorts of fantastic forms; on every side chasms and caverns presented themselves. As we advanced the scene assumed a more terrific aspect; and when we least expected it, we descried Havnefiord situate directly under one of the most rugged parts of the lava; and so placed, that the houses obtained the most complete shelter from masses of matter, that had formerly carried destruction in their course.

There are only two merchants' houses here, and a few store houses, all constructed of wood, and placed close to the sea, which here forms a small but very snug bay, in which there is good anchorage. There is a dry harbour, into which a sloop of forty or fifty tons may be brought for repair. The wooden buildings, the cottages scattered among the lava, the sea, and the distant country, form altogether a singular scene. The view of this place was taken at a little distance from the path. It is extremely difficult to express the appearance of a rough stream of lava by the pencil; and the representation given conveys but a faint idea of its terrific appearance.

We stopped at the house of Mr Sivertson, who had gone to England with Count Trampe, and had not yet returned. His family received us in the kindest manner, and every civility that could be shewn, was bestowed upon us by his son and daughter. The house is one of the cleanest and most comfortable that we saw in Iceland. We were regaled with excellent fish, ragout of mutton, and pan-cakes. Here we endeavoured to sleep under Eider down for the first time. To a stranger, crawling under a huge feather bed seems rather alarming. But though very bulky, the down of the Eider duck is very light; and a bed which swells to the thickness of two or three feet, weighs no more than four or five pounds. At first, the sensations produced by this light covering were very agreeable ; but the down being one of the very worst conductors of heat, the accumulation soon became oppressive ; and at length I was under the necessity of getting rid of the upper bed, to escape the proofs of the good qualities of Eider down, which I now experienced to an intolerable degree.

On the 21st we went to visit the school at Bessestad, the only one now in Iceland. We had been invited by the bishop to be present at the examination, which was to take place previous to the boys dispersing to their respective homes for the summer. On our way we visited Mr Magnuson, the minister of a place called Gardè. This gentleman is styled Provost of the Guldbringè Syssel, and superintends all the ecclesiastical concerns of that district. It is to be regretted, that the poverty of the clergy in Iceland has never been taken into consideration by the Danish government. Their influence over the people, by whom they are highly respected, would, we might suppose, be a sufficient reason for not leaving them to subsist on miserable pittances, hardly sufficient to keep them from starving. Knowing his poverty, we were not surprised that this dignitary of the church exhibited in his person and habiliments, a figure, the description of which I shall spare my readers, that they may not partake the pain and disgust inspired by the most squalid indigence in a clerical garb. This gentleman, however, has a considerable collection of books, among which we observed German translations of some of the works of Pope, and Young's Night Thoughts. There were a number of volumes written in Latin, chiefly on theological subjects. We were obliged to leave Mr Magnuson sooner than we wished, because the hour of commencing the examination at Bessestad was already past; but we promised to call for him on our return.

The following is the invitation we received, and the order in which the examination was to take place. The first will serve as a specimen of the familiar style in writing Latin; and the other will shew what are the principal branches of education at the school.

' Hic tibi, vir nobilissime! exhibeo exscriptum invitatio-
' nis publicæ lectoris nostri ad examen, quod in scholâ nos-
' trâ (quæ nunc unica in Islandia est) sequenti hebdomade
' habendum est. Si tibi, tuisque, vel unica hora a propriis
' negotiis vacaverit, summo nobis honori ducemus, si nos tua
' et illorum præsentia dignari velis.

' Tuæ singularis humanitatis,

' Observantissimus cultor,

' GEIRUS VIDALINUS.

' 17. *Maii*, 1810.'

Ad

Examen Publicum

Alumnorum Scholæ Bessestadensis,

Audiendum d. 21, et seq. Maii 1810,

ita ordinatum:

Die Lunæ, hora antemer... 8—12. 11 Class. in auth. Lat. interpretandis.

    h. pomerid..... 2—6. 11 et 1 Cl. in Stylo Latino.

Die Martis, h. antemer.... 8—10. 11 Class. in Theologicis, secundum Niemeirum.

       10—12. 1 Cl. in auth. Latin. interpretandis.

> h. pomerid.... 2—4. 11 et 1 Cl. in Lingua Danica.
>
> 4—6. 11 et 1 Cl. in Stylo Danico.

Die Mercurii, h. antemer.. 8—10. 11 Cl. in auth. Græcis interpretandis.

> 17—12. 1 Cl. in auth. Græcis interpretandis.
>
> h. pomerid.. 2—3. 11 et 1 Cl. in Declamatione.
>
> 3—6. 11 et 1 Cl. in Stylo Islandico.

Die Jovis, h. antemer..... 8—10. 11 Cl. in Novi Fæd. Exegesi.

> 10—12. 11 Cl. in Analysi Hebraica.
>
> h. pomerid.... 2—4. 1 Cl. in reliq. dogmatica juxta Niem.
>
> 4—6. 11 et 1 Cl. in Arithmetica.

Die Veneris, h. antemer... 8—12. 11 et 1 Cl. in Historia, et Geographia.

Omnes rei scholasticæ patronos, fautores, et amicos, qua par est observantia, invitamus.

STEINGRIMUS JONÆUS, *Lector Theologiæ.*

Before going to the schoolhouse we called at the house of the head master, Mr Steingrim Jonson, who received us with great kindness. His countenance displayed a degree of intelligence not common in the physiognomies of Icelanders; and it was not long before the first impressions we received were amply confirmed. We have often regretted that we had but little of this gentleman's company. His learning is untinctured by pedantry; and his communicative disposition renders him a most agreeable companion. His room was well furnished with books; and the library he possesses as Lector Theologiæ of the school, is the best collection of theological works in the island. Mr Jonson married the widow of the late learned and worthy Bishop Finsson. The lady politely brought us coffee and milk, and after this refreshment we went to examine the church. It is a building of some size, roofed with tiles, and is in much better repair than the church at Reikiavik. At the door is a grave-stone, with the effigy of Paulus Stigotus, a governor of the island, who died in the year 1566, carved upon it. He is represented in armour, leaning on a two-handed sword. In a vault opposite to the door we saw two large coffins, containing the bodies of two ladies, mother and daughter. The date of the inscription was 1788. They were wrapped in white sarsnet, and enveloped in a quantity of vegetable matter, resembling hop flowers. The bodies had the appearance of mummies, the flesh being completely dry and black. Above the altar is an indifferent painting of the last supper, and representations of some saints. On each side is an inscription to the memory of a young man who died in the year 1707. One of them is an acrostic. On the altar were a gilt chalice, wafer box, and two small brass candlesticks; and in a press behind, two very large ones. The space where the altar is placed, is separated from that part of the church, in which are the pews. At the entrance to these, there is a stone inscribed to the father and mother-in-law of our friend at Vidöe. The entrance is formed by two painted pillars, surmounted by two large gilded balls, and between them is an arch supporting the crown and cypher of Christian the Seventh. The pulpit is gaudily ornamented, and has figures of the evangelists

o

painted on the pannels, and inscriptions above and below
them; a dove is represented on the sound-board. Opposite to
the pulpit is an elevated seat, closed all round with glazed
windows, in which the governor formerly sat.

Having satisfied our curiosity in the church, we proceeded
to the school-house, and were received by the bishop in his
full dress.   His under robe was of black velvet; over which
was a silk one bordered with velvet; the cuffs were of white
cambric. Round his neck he wore a white ruff. In this dress
our good friend had a most venerable and dignified appearance.

The school rooms are lofty, but rather confined and dirty.
Above these are barrack-rooms for the scholars and attendants, all in a very uncleanly condition. The lector lamented
this, and the bishop pronounced with emphasis, ' Bella, hor-
' rida bella!' which at once explained the evident state of
decay into which this only remaining seminary of learning
seemed to be falling.   There is a considerable collection of
books, a few of them curious and rare; but they are piled
confusedly together in a miserable garret, and covered with
dust and cobwebs. An account of the institution and management of this school will be given in the chapter on the
present state of Icelandic literature.

Bessestad was formerly the abode of the governors of Iceland, and was defended by several small batteries, the remains
of which we saw. During the usurpation of Jorgensen in 1809,
some guns were removed from the sand, near one of the batteries, to Reikiavik, where they were planted so as to command
the harbour. The fort has been demolished, and the guns are
now lying on the beach.

From Bessestad we went to visit Mr Einerson, a member
of the court of justice, with the title of assessor, who received

us with great cordiality. While drinking coffee, we received much useful information from this gentleman. He seems to have set about improving his farm in earnest, and has inclosed a small space for a garden ; but he complained much of the want of seeds. He is levelling the ground, and inclosing it with the stones he takes up in that operation. If Mr Einerson perseveres, there is no doubt of his rendering his farm much more productive in grass; and he may even attempt to cultivate rye, barley, and potatoes, with a reasonable hope of success. Had I met with Mr Einerson any where else, I should have concluded from his complexion, that he had spent the former period of his life in some of the tropical regions. As well as I can recollect, we did not meet with any other Icelander, who had black hair ; certainly none that had hair curled somewhat like that of a negro. Mr Einerson may possibly be descended from a Lapland family. There is nothing in the general appearance of the natives of Iceland, that can discriminate them in the natural history of Man. If there is any circumstance which can be deemed peculiar, it seems to be that the spine is generally long in proportion to the limbs. Were those to be proportioned to the length of the body, numbers of the Icelanders would become giants. At first this peculiarity appeared to be a deception occasioned by dress ; but though it does not hold universally, it occurred so often, that I am inclined to state it as characteristic.

Having returned to Mr Magnuson's, we were treated with a dram and with coffee. He told us, that by going a few miles out of the direct road to Krisuvik, we might see a curious cave ; and his description of it determined us to visit it.

On the 22d we left the hospitable house of Mr Sivertson. Having passed a low ridge of hills opposite to Havnefiord, we descended into a valley filled with lava, which is connected

with that about Havnefiord, and has evidently proceeded from
the same source.   Along the edge of this we travelled for
about two miles, and then began to ascend a ridge covered
with light slags.   We observed that the lava had run down
on the east side of the valley, and in some places it appeared
as if it had ascended.   The ascending of lava is a well known
fact, though in examining a cold mass, this circumstance
strikes an inexperienced observer as something wonderful;
more so than the ' random ruin' it so awfully displays.   It is
caused by the formation of a crust on the cooling of the
surface, and a case or tube being thus produced, the lava
rises in the same manner as water in a pipe.   Beyond this spot
we saw the most dreadful effects of subterraneous heat all
around us; and as far as the eye could reach over a wide extended plain, nothing appeared to relieve it from the black
rugged lava, which had destroyed the whole of the district.
The surface was swelled into knobs from a few feet in diameter to forty or fifty, many of which had burst, and disclosed
caverns lined with melted matter in the form of stalactites.
The day being very warm we relieved our thirst, which began
to be troublesome, by means of snow and icicles which we
found in these cavities.   We met with some deep clefts ; and
soon afterwards reached a green spot on the bank of the river
Kald-aa,* or the cold river.   Here we resolved to pitch our
tent, being informed that there was no other place on the road
to Krisuvik where the horses could pick up any food.   Having
unloaded the horses, we proceeded towards the cave described
by Mr Magnuson, which lay about two miles to the eastward.   It was nothing more than an extensive hollow formed
by one of those blisters or bubbles, hundreds of which we had

---

* The double a, or ä, is pronounced like *ow* in *how.*

walked over.   Many of these are of considerable depth, and great length.   The bottom of this was covered with ice, and numerous icicles hung from the roof.   Having lighted our lamps, we went to the end of the cave, the distance of which from the entrance we found to be 55 yards, the height not being in general more than seven or eight feet.   The inside was lined with melted matter, disposed in various singular forms.

On leaving the cave, we took a circuit in order to examine some of the hills in the neighbourhood.   In our progress we saw the source of the river Kald-aa, which is a large bason at the bottom of a hollow, into which numerous springs empty themselves, and at once form a considerable stream.   After running about two miles the water entirely disappears among the lava, and is lost.   We now proceeded towards the hill called Helgafell, and passed over a lava, which from the comparative smoothness of its surface, and the evident marks of fusion which it bore, struck us as being something uncommon.   After traversing the plain formed of this lava, we met with a number of little craters, which, by giving vent to the vapours produced by the heat, may have prevented the usual heavings and burstings of the surface. One of these craters presented a very singular spectacle.   The melted matter had formed a sort of dome, about twenty-five feet in diameter, and open at one side.   Within, it was lined with an assemblage of stalactitical forms, hanging in a very curious and fantastic manner.   After visiting several of these craters, we returned to the place where we had left the baggage, and pitched our tent.   By means of the apparatus called a conjurer, we cooked a tolerable mess of portable soup and fish ; which with biscuit, cheese, and pure water, composed a very comfortable dinner.   The night became so cold that we enjoyed but little sleep.   Towards morning it began to rain, the wind having gone round to the east; but before we had accomplished three miles of our day's journey, the clouds broke, and the rest of the day was fine.

We crept along the foot of the hills by the edge of the lava for some miles, and then began to ascend near several craters, larger than any we had yet seen.   While examining some fissures, we found the remains of a woman who had been lost about a year before, and of whom there had hitherto been no tidings.   Her clothes and bones were lying scattered about; the bones of one leg remained in the stocking.   It is probable that she had missed the path during a thick shower of snow, and had fallen over the precipice, where her body was torn to pieces by eagles and foxes.   It is astonishing how the Icelanders find their way during winter across these trackless deserts.   Even with the assistance of marks, which are set up here and there, unless the snow be perfectly firm, there is danger at every step of being swallowed up in a fissure.   Accidents of this kind, however, very rarely happen.

We all now became exceedingly thirsty; and our guide having informed us that we should not get any water for some hours, we were very uneasy.   Scrambling among lava is very hard exercise; and this, with the uniformity of the prospect, presenting nothing but an extensive country literally burnt up, occasioned our feeling thirst more painfully than usual.   We kept up our spirits as well as we could; and while meditating how to get the horses over a great heap of slags at the foot of a crater, we descried a small wreath of snow on the side of a hill about half a mile distant.   We instantly left our guides and the horses to manage matters as they could; and rushing over slags, lava, and mud, fell upon the snow like wild beasts upon their prey.

My enjoyment was excessive; and the very recollection of it
is so gratifying, that I must be excused for recording a circumstance of so little importance.

We now proceeded through this dismal country, without
any gratification but that occasionally afforded by views of
the romantic summits of the mountains. The relief experienced from a further supply of snow, raised our spirits and
strength so much, that we had got a considerable way before
our guide, when he came hallooing after us, and making signs
for us to return. Having kept close to the foot of the mountain, and being persuaded that nothing worthy of observation
had escaped our notice, we were at a loss to conceive the occasion of this alarm, and feared some disaster among the
horses. We were somewhat surprised on being told that we
had passed the road. Already accustomed to the risk of missing
our way, we had flattered ourselves with being sufficiently
cautious; and when we came to the place where the horses
had halted, we could not see any other track than the one
we had followed. The guide pointed to the side of the mountain; but no outlet was visible. The steepness of the mountain
we might overcome; but how the horses loaded with baggage
could gain the summit, was to us utterly inconceivable. After
ascending a little way by a winding path, we saw a large hollow, the sides of which were very steep, and composed of loose
sand and large stones which required very little force to be
rolled from their places. Along one side of this hollow we
ascended; and the poor horses, with a degree of steadiness
and perseverance truly astonishing, accomplished their arduous task. The scene now before us was exceedingly dismal.
The surface was covered with black cinders; and the various
hollows enclosed by high cliffs and rugged peaks, destitute of
every sign of vegetation, and rendered more gloomy by float-

ing mist, and a perfect stillness, contributed to excite strong feelings of horror. After a pause, we proceeded, and our eyes were soon gratified by a small lake coming into view. From the general appearance of the whole of this mountain, and the situation of this lake, it is more than probable that the hollow now filled with water had been formerly a crater from which the profusion of burnt matter which we had seen had been thrown. The horses made several attempts to reach the water, but in vain: the clay banks sunk under them, and the poor animals were obliged to proceed on their journey, after being thus cruelly tantalised.

From this place we saw vapour arising behind a hill at a short distance. We approached, and beheld it ascending with impetuosity from a circular bason, in a hollow near the summit of the mountain. Having advanced to the spot, we were surprised to see no water running from what was supposed to be a boiling spring. On advancing nearer we heard loud splashing, and 'going close to the bason, which was twelve feet in diameter, we perceived it to be full of thick black mud, in a state of very violent ebullition. This singular phenomenon seemed to be occasioned by steam escaping from some deep seated reservoir of boiling water, and suspending the mud, which was probably produced by the action of the steam, in softening the matter through which it forced its way. We discovered a number of little fissures in the sand round about, from which steam rushed with a hissing noise. Though the splashing of the mud was incessant and violent, we did not observe any of it to escape from the boundary it had formed for itself.

Having satisfied our curiosity here, we descended towards the valley of Krisuvik, and soon saw so large a quantity of vapour below, as effaced the wonder excited by the extraordi-

Plate 4

KRISUVIK & the SULPHUR MOUNTAINS.

nary appearance we had left. This proceeded from various cavities, from banks of clay and sulphur, and chinks in the rock. Knowing, however, that we had more curious things of the same kind to examine the next day, we did not spend much time here, but proceeded through a long valley, forming one continued swamp, which forced us frequently to take very circuitous tracks to advance. In going along we had a striking view of the Sulphur Mountains, to visit which was the object of our coming to this part of the country. There is a large lake in this district called the Kleisar Vatn, a part of which we saw when descending from the mountains. There are two small ones in the valley, one of which derives its name from the green colour of the water, which resembles that of the sea. The colour is owing to that of the bottom, which seems to be covered with clay, such as is met with in abundance on the mountains. The water has no peculiar taste or smell. I may here remark, that the want of curiosity, and the want of observation, so conspicuous in the people of Iceland, often occasioned us much loss of time. We had to explore a country already, one would think, well enough known to the inhabitants to enable them to give some general directions concerning our journey through it. We now reached Krisuvik after two days of fatigue, which might have been saved by our being told, that by taking the direct road, and marching eight hours, we should see just as much, as by making a journey that lasted two days.

At Krisuvik there is a farm-house with a few cottages. It was proposed at first that the party should occupy the church, but, on examining it, we preferred our tent. The length of the church, which is constructed of wood, is 18 feet, the breadth 8 feet, and the height, from the floor to the joist, 5 feet 8 inches. Near the door, in the inside, is suspended a bell, large enough to make an intolerable noise in so small an apartment. About ten feet from the door is the division between the rest of the church and the altar. The space between this and the door was occupied by large chests, filled with the goods and chattels of the farmer, many of which were also piled under the roof. The pulpit, raised about two feet, stood in a corner on one side of the division ; and it was evident that, if the priest were a man of ordinary size, his audience would be totally eclipsed from his view, and that he would have to address himself to lumber and stock fish, in the loft. There were seats before the pulpit that, with a little squeezing, might be capable of accommodating half a dozen persons. Beyond the pulpit was a space of about eight feet square, in which the altar was placed, between two small windows. The altar was merely a wooden press or cupboard, seemingly destined to serve many unhallowed purposes. It contained various household utensils. The farmer and his wife cleared away several articles on the top, and, placing some milk on the altar, invited us to eat. There was nothing so sacred in the appearance of this church, as to make us hesitate to use the altar as our dining table. The open space round the altar being rather small, the floor perfectly rotten, and very damp, we could not think of sleeping in the church; and as to the house, exposure to frost and snow would have been preferable to spending a night in such a place. Varying a little in size, all the houses of the Icelanders are constructed on nearly the same plan. An outer wall of turf, about four feet and a half high, often six feet thick, encloses all the apartments. On one side, generally that facing the south, are doors, for the most part painted red, surmounted with vanes. These are the entrances to the dwelling-house, the smithy, dairy, cow-house, &c. From the door of the house is a long, narrow, dark, and damp passage, into which, on each side, the different apartments open. Between each of these is a thick partition of turf, and every one has a separate roof, through which light is admitted by bits of glass or skin, four or five inches in diameter. The principal rooms of the better sort of houses have windows in front, consisting of from one to four panes of glass. The thick turf walls, the earthen floors kept continually damp and filthy, the personal uncleanliness of the inhabitants, all unite in causing a smell insupportable to a stranger. No article of furniture seems to have been cleaned since the day it was first used; and all is in disorder. The beds look like receptacles for dirty rags, and when wooden dishes, spinning wheels, and other articles are not seen upon them, these are confusedly piled up at one end of the room. There is no mode of ventilating any part of the house; and as twenty people sometimes eat and sleep in the same apartment, very pungent vapours are added in no small quantity, to the plentiful effluvia proceeding from fish, bags of oil, skins, &c. A farm house looks more like a village than a single habitation. Sometimes several families live enclosed within the same mass of turf. The cottages of the lowest order of people are wretched hovels; so very wretched, that it is wonderful how any thing in the human form can breathe in them.

The weather being warm and calm, we slept very comfortably in our tent, which was pitched near the banks of a small stream, at a short distance from the church. The 25th was a delightful day, and having taken an early breakfast of biscuit, cheese, and milk, we set out towards the Sulphur Mountain, which is about three miles distant from Krisuvik. At the foot of the mountain was a small bank composed chiefly of white clay and some sulphur, from all parts of which steam

issued.   Ascending it, we got upon a ridge immediately above
a deep hollow, from which a profusion of vapour arose, and
heard a confused noise of boiling and splashing, joined to the
roaring of steam escaping from narrow crevices in the rock.
This hollow, together with the whole side of the mountain
opposite, as far up as we could see, was covered with sulphur
and clay, chiefly of a white or yellowish colour. Walking over
this soft and steaming surface we found to be very hazardous; and I was frequently very uneasy when the vapour
concealed my friends from me.   The day, however, being
dry and warm, the surface was not so slippery as to occasion
much risk of our falling.   The chance of the crust of sulphur
breaking, or the clay sinking with us was great, and we were
several times in danger of being much scalded.   Mr Bright
ran at one time a great hazard, and suffered considerable
pain from accidentally plunging one of his legs into the hot
clay.   From whatever spot the sulphur is removed, steam
instantly escapes; and in many places the sulphur was so
hot that we could scarcely handle it.   From the smell I perceived that the steam was mixed with a small quantity of sulphurated hydrogen gas.   When the thermometer was sunk a
few inches into the clay, it rose generally to within a few degrees of the boiling point.   By stepping cautiously, and avoid-
ing every little hole from which steam issued, we soon discovered how far we might venture.   Our good fortune, however,
ought not to tempt any person to examine this wonderful
place, without being provided with two boards, with which
any one may cross every part of the banks in perfect safety.
At the bottom of this hollow we found a cauldron of boiling mud, about fifteen feet in diameter, similar to that on
the top of the mountain, which we had seen the evening
before; but this boiled with much more vehemence.   We

Plate 5.

CAULDRON of BOILING MUD on the SULPHUR MOUNTAINS.

went within a few yards of it, the wind happening to be remarkably favourable for viewing every part of this singular scene. The mud was in constant agitation, and often thrown up to the height of six or eight feet. Near this spot was an irregular space filled with water boiling briskly. At the foot of the hill, in a hollow formed by a bank of clay and sulphur, steam rushed with great force and noise from among the loose fragments of rock.

Further up the mountain, we met with a spring of cold water, a circumstance little expected in a place like this. Ascending still higher, we came to a ridge composed entirely of sulphur and clay, joining two summits of the mountain. Here we found a much greater quantity of sulphur than on any other part of the surface we had gone over. It formed a smooth crust from a quarter of an inch to several inches in thickness. The crust was beautifully crystallized. Immediately beneath it we found a quantity of loose granular sulphur, which appeared to be collecting and crystallizing as it was sublimed along with the steam. Sometimes we met with clay of different colours, white, red, and blue, under the crust; but we could not examine this place to any depth, as the moment the crust was removed, steam came forth, and proved extremely annoying. We found several pieces of wood, which were probably the remains of planks that had been formerly used in collecting the sulphur, small crystals of which partially covered them. There appears to be a constant sublimation of this substance ; and were artificial chambers constructed for the reception and condensation of the vapours, much of it might probably be collected. As it is, there is a large quantity on the surface, and by searching, there is little doubt that great stores may be found. The inconvenience proceeding from the steam issuing on every side,

Plate 6.

Sketched by Sir Geo. Mackenzie.

I. Clark direxit

GREAT JET of STEAM on the SULPHUR MOUNTAINS.

wonders, or its terrors. The sensations of a person, even of firm nerves, standing on a support which feebly sustains him, over an abyss where, literally, fire and brimstone are in dreadful and incessant action; having before his eyes tremendous proofs of what is going on beneath him; enveloped in thick vapours; his ears stunned with thundering noises; these can hardly be expressed in words, and can only be well conceived by those who have experienced them.

Earthquakes are said to occur frequently at Krisuvik, li-mited, however, to a small district in their extent and effects. It was remarked to us, also, that they happen generally after a continuance of wet weather; but whether these statements are accurate or not, we had no means of ascertaining.

On returning to our tent, we were agreeably saved the trouble of cooking, by the farmer and his wife, who brought in a large pewter dish full of boiled rice and milk, and some slices of smoked mutton. The 26th was spent in drawing, and examining the cliffs on the coast about four miles distant, where we found some curious mineralogical appearances, which will be afterwards described. While we were sitting at dinner in our tent, a woman came from an adjoining cottage, having a wooden vessel full of milk in one hand, and a snuff box in the other. These she alternately held out to us. We did not at first understand her gestures; but afterwards found that she wished to barter milk for snuff. Before I left the place, I gave her some tobacco and snuff, with which she ap-peared to be highly gratified. She took each of us by the hand, but we dexterously evaded the usual proffered salute.

I here met a man who was travelling eastward. He had an old gun in his hand, and a white feather was stuck in his hat. I made earnest enquiries about this man's occupation, and how he came to be travelling with a gun; but I could get no satisfactory answer. He may have been going in search of swans or foxes.

On the 27th, we set out for Grundevik, and walked about fifteen miles through a perfect desert of lava, slags, and sand. We saw hardly the slightest appearance of a vegetable, dead or alive. The mountains were of the most dreary aspect: nothing appeared to relieve the eye, or cheer the spirits, till we descried some cottages on the coast, to which we made all possible haste. Our eagerness to get out of this dismal country, made us walk so fast that our guides and horses were left far behind. On coming towards the houses, the people, men, women, and children, came crowding out like ants from a disturbed hillock, to gaze at us. We were the first strangers that had ever been in that part of the island. On enquiry we found that the place we had to go to, was yet some miles distant. We struck into a path, and at length came to a bay which from a chart we had with us, we knew to be that of Grundevik. The cottages here were of mean appearance, and I could not persuade myself that this was the place where the person, to whom we had an introduction, dwelt. There was one of the same name, however, and I produced our passport. This was given to an old man who had thick grey hair, and a bushy grey beard. He sat down, and putting a pair of spectacles on his nose, proceeded to read the paper with all the gravity of a patriarch. This done, plenty of milk was brought to us; and we were informed that the Jon Jonson we wanted lived a mile or two farther on. This was by no means very agreeable news. It had become cloudy, and the wind was blowing strongly from the east. At last we reached a part of the coast well studded with cottages. Heaps of fish were every where piled up; and though it was Sunday, this seemed to be a place of considerable business, and to contain a large population. Our surprise, on finding so dreary a region so well in-habited, ceased when we were informed that, at the commence-ment of every fishing season, numbers of people came from different parts of the country to the fishing stations, to pro-vide fish for winter use. Cottages are set apart for their ac-commodation, and we here saw a large building destined for this purpose, called the *bud*, or booth.

One of the chief cares of an Icelander is the laying in a stock of provision for the winter season. Towards this object almost his whole time and exertions are directed, and the sea is his great and sure resource. About the beginning of February, the people of the interior, and even those inhabiting the north-ern parts of the island, begin to move, and a great part of the male population emigrates towards the western and south west-ern coasts. The farmers send their servants, and frequently accompany them. They take a small stock of butter and smoked mutton, and sheep skin dresses. The farmers are sometimes distinguished from their servants by having a small quantity of rye bread and a little brandy. Before commencing his journey, an Icelander takes off his hat, places it before his face, and repeats a prayer prescribed for such occasions. He is welcomed at every cottage he stops at, and it is seldom that any remuneration is required for his entertainment. Many travel two hundred miles amidst snow and darkness, to the place they chuse for their fishing station; and if they have not previously made some agreement with the proprie-tor of a boat, a bargain is soon entered into, the terms of which have long been established by custom. By these, a person coming from the interior, engages himself from the 12th February to the 12th May, (the period varies a little) to be ready to obey the call for fishing, and to assist in the management of the boat. In return for his labour he receives forty pounds of meal, (if any is to be got) and five gallons of sour whey, besides a share of the fish that are taken. The fish are divided into two shares more than the number of men employed, and these belong to the owner of the boat, who provides lines and hooks. When he furnishes nets, which are generally used during the early part of the season, he receives one half of the fish caught. All the people engaged for one boat generally live together in the same hut. The previous arrangements being made, a long period of hardship and privation begins. In darkness, and subjected to intense cold, these poor people seek from the ocean the means for subsisting their families during the following winter. As soon as the boat is pushed from the shore, the man at the helm takes off his hat and repeats a prayer for success, in which he is joined by the crew. They generally remain at sea from eight to twelve hours, during which time they taste nothing but a little sour whey, which is the only provision ever taken out with them. The women assist in cleaning and splitting the fish, after they have been brought to land. When the weather is so stormy that the fishermen cannot venture to sea, they frequently amuse themselves by wrestling, or playing at leap frog, and other diversions. Their mode of wrestling is somewhat peculiar. The two men who are to try their strength and skill, lay hold of each other in the way they think best. As soon as they have secured their grasp, each endeavours by sudden jerks, or by lifting him from the ground, or by quick turning, to throw his antagonist; and the dexterity they display is often extraordinary. About the beginning of May, the fishermen return home, leaving their fish, which are not by this time perfectly dried, and which may amount to five or six hundred for the share of each individual, to the care of some person who resides on the spot. About the middle of June,

when the horses have got plump and strong, the farmers set
out on their second annual journey, carrying with them all
their marketable commodities, which they dispose of, and return home with their fish and such things as they may have
purchased.   When the stock of fish is thought too small, they
barter wool, tallow, or butter, for any additional quantity
they may require, and pay a small sum for the lodging and
food of their servants.

Jon Jonson was not at home when we arrived; but he soon
made his appearance, and pressed us to go into his house. It
is somewhat remarkable that the Icelanders should display
considerable industry, and even ingenuity, in making the outside of their houses neat, while they keep the inside in such a
state of dirtiness as to be truly disgusting. We complied
with Jonson's request, but were glad to seize the first opportunity of escaping into the open air.

Our baggage arrived just as it began to rain. Upwards of
thirty men, women, and children, gathered round, and disturbed us greatly while pitching our tent, which, had it not
rained all night, would have exhibited lasting tokens of the
pawing of the crowd that assembled about it. Even after
we had gone to bed, they still continued at the door of the
tent, and we were not suffered to sleep quietly, till our tor-
mentors thought of rest themselves. With every desire to indulge the curiosity of the natives, and a readiness to submit
to various privations and hardships, we could not endure to
be touched ourselves, or to have any thing belonging to us
handled by them. For this nicety we had another reason
besides dislike of filth : the cutaneous eruptions from which
very few Icelanders are free, more powerfully deterred us
from any near approach. Poor Jonson did all he could to
serve us, and gave us what milk he could spare, and some

fish. Milk is extremely scarce in this district, and it was lucky that we had filled some bottles with it at Krisuvik.

We rose early in the morning, and found the rain falling heavily, accompanied by a gale of wind from the eastward. It was our original design to go to Cape Reikianes, where there are some hot springs; but having been informed that no food could be got for our horses, we resolved to proceed to Kieblivik. The country towards the Cape becomes low, and is entirely of volcanic formation; and this part of Iceland has been rendered famous on account of a remarkable event in the history of volcanoes, which happened in the year 1783, a description of which will be found in the chapter on Mineralogy; for which it is reserved on account of the phenomena being important in relation to the origin of pumice.

We now crossed the peninsula to the opposite shore, a distance of about fifteen miles, through a wild and dreary tract of lava; meeting with nothing remarkable except a considerable extent of gravel, which had every appearance of having once formed the sea beach. We reached the northern side of the Guldbringè Syssel at Niardivik, a large fishing village, much frequented by the inhabitants of the interior, who come down to procure fish. There are not fewer than three hundred boats, of different sizes, belonging to this place; and it is said that the population, during the fishing season, sometimes amounts to two thousand, while at other times it does not exceed two hundred. The fishing banks are at a very short distance from this place, and the fish are esteemed finer than on any other part of the coast.

A mile or two westward is Kieblivik, situate at the head of a small bay, in which there is very good anchorage. Two or three merchants are established here, who reside in houses constructed of wood, resembling those at Reikiavik. We were very cordially received by Mr Jacobæus, a Dane, one of the most respectable as well as wealthiest merchants in Iceland; who, with his lady, entertained us in the most hospitable manner for three days, during which there was a violent gale of wind and heavy rain. Soon after our arrival at his house, we got excellent coffee, rye bread, and butter. Dinner was not served up till nine o'clock; but our patience was well rewarded with soup, mutton, and pan-cakes; and we were treated in the same luxurious manner during the whole time of our stay with the family. It was a matter of much regret to us that, from our ignorance of the Danish language, we had no direct means of conversation with Mr Jacobæus, who appeared to be a particularly intelligent man, had resided a long time in Iceland, and was well informed in many circumstances relating to the statistics and commerce of the country. It may be mentioned as a singular fact, that Madame Jacobæus, though she has now lived sixteen years in Iceland, has actually never been out of Kieblivik; not even to visit the metropolis of the island.

On the first of June we took leave of our hospitable friends. As we had about thirty miles to walk, and could not foretel what obstacles we were to meet with, we made a very hearty breakfast. The day was cloudy, but favourable for walking. After having advanced a mile or two, Mr Holland returned for some things he had forgot. Mr Bright and I walked slowly on, and not seeing our friend coming up, we concluded that he had resolved to remain with the baggage. After passing Niardivik, we were bewildered by the number of tracks, and at length lost them altogether. As we knew the general direction in which we were to go, we proceeded, till after four hours walking, we began to feel the painful sensation of thirst. By good luck we found a puddle of rain water, near which we sat down and refreshed ourselves. Soon afterwards, we found a path, but lost it at the edge of a stream of lava. After a fruitless search to recover it, or find the proper place for entering the lava, we left all to chance, and ventured forward. We kept our course as well as we could, among heaps of loose slags, rugged lava, and deep fissures. The moss which grew in some places on the lava, often gave way, and we slid down among the slags. However trying this was to the limbs, we felt no inconvenience till thirst once more distressed us. The soles of Mr Bright's shoes having been torn by the lava, he sat down to cut away the loose pieces which were troublesome. On rising he neglected to take up his great coat, which he had laid down beside him. Only a few minutes had elapsed before he discovered that he had left it behind, and we had not yet proceeded two hundred yards from the place. I ascended a peak of lava in order to direct his course, but very soon lost sight of him. He was away half an hour, and I remained all the while at my station; and was beginning to feel some uneasiness, when I heard him hallooing very near me. On my answering, he scrambled to the top of a mass of lava, and was surprised to see me so near him. So rugged was this lava, and so circuitous the way to get through it, that he could not retrace a single step, and failed in his attempts to recover his coat.

It had never occurred to us to enquire whether it would be necessary to carry a little water with us; and the people being accustomed to perform their journeys on horseback with great speed, had no idea that we had any thing else to do than to make the best of our way through the country. Our examination of the rocks, and picking up plants, seemed to afford much amusement to the natives. Delays occasioned by such occupations often produced some inconveniences, and thirst was by far the most considerable. At this time it was almost intolerable ; but after some hours, the path appeared, and in the hollows there was a sufficient quantity of rain water to afford relief.

After a very fatiguing walk of twelve hours, we arrived at Mr Sivertson's, and found Mr Holland already there. On comparing notes, it appeared that he had got into the same track of lava in which Mr Bright had lost his coat, and that he must often have been not more than fifty yards from us. This was discovered by means of a few green knolls that had attracted our attention. He fared worse upon the whole than we did, for he had been alone during the whole day, his shoes were completely destroyed, and his feet very much cut. In one respect, however, he was more fortunate ; as he had found a peasant's cottage on the edge of the lava, where by the aid of signs and a few Icelandic words, he had contrived to obtain a draught of milk, and a direction for the remainder of the way.

Every thing that could administer to our comfort was provided at Havnefiord. Our horses arrived four hours after us. Next morning we felt completely refreshed, and able to undergo a renewal of our fatigues. Though the appearance of the lava about Havnefiord is sufficiently terrific to a person who has never before seen any, yet on my return to it, I wondered how it had made so strong an impression upon me. It is not to be compared in any respect with the horrible scenes we had passed through, nor were these equal to some lavas we afterwards encountered. No portion of the inhabited part of Iceland is so dreary and barren as that we had now travelled over. The only inducement, (but in this country it is a powerful one) for people to settle here, is the vast abundance of fish obtained on every part of the coast.

On the third of June we returned to Reikiavik, after having been absent nearly a fortnight upon our journey. We were much gratified to observe the progress that vegetation had made during our absence. We now saw some patches of green scattered here and there; and the vegetables in the little gardens about the houses were beginning to appear. On the 4th, we went on board the Elbe, from which a salute was fired in honour of our Sovereign's birth-day, and the day was spent in showing every mark of loyalty and affection for our gracious king, that our means allowed.

On the 8th we went to Vidöe to see the Eider-ducks, which had now assembled in great numbers to nestle: at all other times of the year these birds are perfectly wild. They are protected by the laws, a severe penalty being inflicted on any person who kills one. During the breeding season, the fine is thirty dollars for each bird. As our boat approached the shore, we passed through multitudes of these beautiful fowls, which scarcely gave themselves the trouble to go out of the way. Between the landing place and the old governor's house, the ground was strewed with them, and it required some caution to avoid treading on the nests. The drakes were walking about, uttering a sound very like the cooing of doves, and were even more familiar than the common domestic ducks. All round the house, on the garden wall, on the roofs, and even in the inside of the houses, and in the chapel, were numbers of ducks sitting on their nests. Such as had not been long on the nest, generally left it on being approached; but those that had more than one or two eggs sat perfectly quiet, suffering us to touch them, and sometimes making a gentle use of their bills to remove our hands. When a drake happens to be near his mate, he is extremely agitated when any one approaches her. He passes and repasses between her and the object of his suspicion, raising his head, and cooing. The nests were lined with down, which the duck takes from her own breast; and there is a sufficient quantity laid round the nest, for covering up the eggs when the duck goes to feed, which is generally during the time of low water. The down, which is a valuable article of commerce, is removed at two different times from the nest. Sometimes the poor duck is compelled to provide a fourth lining, and when her down is exhausted, the drake supplies the deficiency. A certain number of eggs is also removed, as they are esteemed a great delicacy. Our good friend at Vidöe used to send us two hundred at a time. When boiled hard they are tolerably good, but much inferior to the eggs of common poultry. Swans eggs, of which we got a few, are superior, and really excellent when boiled hard.

When taken from the nest, the Eider down is mixed with feathers and straws. To separate them, and make the down fit for market, is part of the employment of the women during winter. As soon as the young birds leave the eggs, the duck takes them on her back, and swims to a considerable distance from the shore. She then dives, and leaves the little ones to exercise themselves in swimming about. As soon as they have got the use of their feet in this way, the duck returns, and becomes their guide. Several broods, often great numbers, join company, and are seen quite wild for a few weeks; after which, they totally disappear. Long before we left Iceland, there was not a single Eider-duck to be seen. Whither they retire is not known. These birds are found in the Flannel Isles, to the west of the Island of Lewis. They are sometimes seen in Shetland and Orkney, but seldom farther to the south.

A few days after our return to Reikiavik, a Danish galliot arrived in the harbour, which had sailed from Liverpool about three weeks before, and which brought over as passengers from England, Mr Sivertson of Havnefiord, and Mr Floed, a young Norwegian, who had been private secretary to Count Trampe. The latter was the bearer to me of a packet which the Count had been polite enough to send, containing letters of introduction to several of the principal people in Iceland. Besides those directed to individuals, there was one addressed generally to all persons holding offices or authority in the island, desiring them to facilitate our travels through the country, and to assist us in the prosecution of the objects we had in view. The arrival of this vessel effected a sort of miniature revolution in the government of Iceland; which was transferred from the hands of the Chief Justice Stephenson to those of three deputy governors; Mr Thoranson, the Amtmand of the northern province; Assessor Einarson; and our friend Mr Frydensberg of Reikiavik. At a ball which we attended a few days after this political change had taken place, the health of the new Governors was drunk with many demonstrations of satisfaction. The name of Count Trampe was also given as a toast, with much applause; and a poem in his praise, composed for the occasion by Mr Magnuson, one of the most celebrated of the Icelandic poets of the day, was sung in chorus by the whole party.

During our present stay in Reikiavik, we frequently visited the Bishop, and continued to be greatly pleased with him. We saw once or twice at his house a Mr Paulson, one of the medical practitioners of Iceland, who lives on the southern coast, at the distance of about a hundred miles from Reikiavik. He is a man of much information, and particularly conversant in the natural history of his own country, which he has studied with great attention. A few days after our return from the Guldbringè Syssel, Mr Holland received from him a note, written in Latin, * requesting his medical attendance upon one of the daughters of Mr Sigurdson, the minister of Reikiavik. Mr Holland visited Mr Sigurdson's family frequently in this capacity, not only at the present time, but also upon our return from the journies we subsequently made ; and his offices were rewarded by that expression of gratitude, of which its simplicity was the strongest recommendation.

The 11th of June was a holiday ; and the ceremony of confirmation was performed in the church. The ordinary service began with prayer and singing. Lessons were then read from the Bible, and the conclusion was a sermon, the delivery of which occupied more than half an hour. During the first part of the service, the minister was dressed in a sort of surplice, ornamented with broad blue bands, and with gold lace. The singing, or rather roaring, was performed by ten or twelve men, standing round the space enclosed about the altar. The sermon was read from the pulpit with much emphasis and gesture. The Bishop entered the church just before it began, and took a seat near to the altar. The minister, having resumed his surplice and his station at the altar, read a long exhortation to the children who attended for confirmation, and were ranged round him, the boys apart from the girls, all of them dressed in their best clothes. After an examination from a printed form of catechism, the children received confirmation from the minister, who laid his hands upon them. The whole was concluded with ano-

---

* Dr Holland, ut filiolam pastoris templi Reykiavianam, domini Sigurdson, ægrotantem, data occasione, concomitante interprete, visitare velit, humillime rogatur   Paulsonio.

ther exhortation, and a prayer by the minister, kneeling in the midst of the children. During this service, the Bishop was a mere spectator : his only occupation consisted in taking snuff, and chewing tobacco. A great number of Icelanders had come to Reikiavik from the adjoining country, to attend this ceremony ; and the church was crowded with people, all in their finest suits. The women, who were habited in the proper costume of the country, sat together on the left side, and formed a singular and interesting assemblage. They were, for the most part, rather tall than otherwise ; their features in general well formed ; and their complexions fair and florid. The men were seated on the opposite side of the church. None of the Danish inhabitants appeared at this ceremony ; nor is it customary with them to attend any of the religious services of the Icelandic church.

CHAP. II.

JOURNAL OF SECOND EXCURSION.

THE object of our second journey in Iceland, was to explore the peninsula on the western side of the island, which is terminated by the remarkable mountain called Snæfell Jokul. We were assisted in fixing our route, by the Chief Justice Stephenson, who was well acquainted with the district called the Borgarfiord Syssel, and by Mr Clausen, a Danish merchant, settled at Olafsvik, not far from the extremity of the peninsula. This gentleman, who came over to Reikiavik soon after our return from the Guldbringè Syssel, we found to be remarkably intelligent. We received from him every necessary direction as to our route, and also some account of the natural curiosities we should meet with in the tract of country through which we were to pass.

Having purchased five more horses, and hired two men, Gwylfr and Gudmundr by name, to attend our cavalcade, we made preparations proportioned to the length of the journey we were about to undertake, which we calculated to be between three and four hundred miles. Our baggage-horses

cost from eight to ten rixdollars each ; and those we intended for our own use, about twelve.   They were by no means
of the best description of riding-horses ; but sufficiently good
for the rough work they had to encounter.   An exceedingly
good horse may be procured for twenty or thirty dollars ; a
sum, according to the rate of exchange at this time, equivalent to two or three guineas.   All our cattle were rather lean ;
and they had not yet lost their rough winter coats.   Our
servants professed to be well acquainted with the country we
wished to examine ; and, being young and stout, we flattered ourselves that we should have little occasion to reproach
them with laziness ; but we soon found that, like all their
countrymen, they were systematically slow in their movements ; and that every attempt, either in the way of entreaty
or of threat, to make them alert, was quite fruitless.   One of
them, however, who had been a servant to the Danish officers surveying the coasts of Iceland, was somewhat more disposed to activity than the other; and we were gratified to
find, on more than one occasion, that this superiority was
the cause of high words between them, and of a little saving
of time.   Every one who undertakes to travel in Iceland,
must resolve to submit with patience to the tardiness of his
attendants.   The young man who had accompanied us during our first journey, had left our service on being refused an
increase of wages ; and he went to cut hay in the northern
part of the island, where labourers are very scarce.   He addressed a long letter in Latin to Mr Holland, which was so
well written, that we suspected it was not his own composition ; and our suspicions were confirmed, on learning that
he had gone to the school at Bessestad a few days before.

Early in the morning of the 15th June, we sent off the
horses with our baggage, as they had to go round the bay; and

we crossed it ourselves in the afternoon, in a boat belonging
to the captain of the Danish galliot, then in the harbour, expecting to meet our horses at the opposite side.   The wind
was strong from the eastward, and it began to rain violently,
when we were about half way across the bay, which is here
six or seven miles in width.   The rain and the water of the
sea, which frequently broke over the boat, in a short time
completely wetted us through.   Our voyage, however, was not
long, and we landed in safety near the foot of the mountain
called Esian.   We now found ourselves in a country very different from that we had before traversed.   Here we saw none
of the desolated appearances which had marked the progress
of volcanic fire in the Guldbringè Syssel.   From the shore to
the base of the mountain, was a flat green country, about a
mile in breadth.   It was, however, almost one continued bog,
in many places nearly impassable.   The mountain rose precipitous from the flat, the lower part being covered by the
debris of the beds, which we saw ranged horizontally above;
and its bold and lofty front was broken into gulleys of various dimensions.   The height of Esian is about 1500 feet,
which it preserves for an extent of several miles, without varying.   It forms a very fine object viewed from Reikiavik,
and is seen from the sea southward of the Guldbringè Syssel,
over the mountains of that district.

Our horses not being in sight, we proceeded along the
shore, examining the rocks that appeared.   We then left the
shore, and crossed the swamp towards the foot of the mountain.   The rain still continued heavy, but we went on with
the intention of stopping at the first convenient spot where
we might pitch our tent.   After walking to a considerable
distance, and descrying no cottages, we waited for more than
an hour in momentary expectation that our horses should

make their appearance. Being disappointed, however, in this, and dreading lest we should by any accident miss them, we returned towards the shore, and, in crossing the bog, were unfortunate enough to break one of our barometers. We soon reached some cottages near the sea, and having waked the inhabitants, who had already retired to rest, a man very civilly undertook to guide us to Brautarholt, a place where we had been recommended by some of our Reikiavik friends to pass the night. Here we found a farm-house, and a church of similar construction, but considerably larger than that at Krisuvik. Our horses did not arrive till eleven o'clock, owing to the extreme difficulty of guiding them across the bogs; and this lateness of the hour, as well as the difficulty of finding a convenient place in which to pitch the tent, led us to determine upon passing the night in the church. The people who inhabited the adjoining house readily acceded to this plan, brought us the keys of the church, and prepared for our supper a large dish of eider-duck eggs, which was placed before us on the altar. We found the portable bedsteads, which we had with us, extremely well adapted to the form and dimensions of the church; and placing them in the space before the altar, which was just large enough to receive them, we slept most comfortably in this new situation, undisturbed either by mortal or spiritual visitants.

Next morning, we were supplied with boiled fish, rice, and milk from the farm-house, which, with the addition of biscuit we had brought with us, made a very excellent breakfast. The morning being stormy, we delayed our departure till noon, when the weather became more favourable. We proceeded along the shore of the Hval Fiord (Whale Frith), which runs up the country about twenty miles in a north-easterly direction, preserving nearly an uniform breadth of about three miles. The scenery was similar to that of many of those arms of the sea, called Lochs, in the highlands of Scotland; the grandeur and variety of which it rivalled in every respect, except that its shores were wholly destitute of wood. The precipices are magnificent; and the eye is carried to the extreme distance, by mountains assuming every variety of romantic form, many of them capped with snow.

Having walked about six miles, we mounted our horses, and left the Fiord, passing into a valley to the right. This valley is high, and little more than a mile in breadth. The mountains on each side are lofty, bold, and rugged; and the patches of snow which yet remained upon their sides added much to the wildness of the scene. Though the ground is swampy, it affords much excellent pasture, and we observed several farm-houses in different parts of the valley.

Turning round the mountains on our left, we came to an open country of considerable extent, but so boggy, that it was with great difficulty we got through it. We now found how necessary the provision of riding horses was to our present journey, and had at the same time an opportunity of ascertaining the confidence which might be placed in these animals. In going through a bog, an Iceland horse seems to know precisely where he may place his foot in safety, and where he cannot venture to pass. If in doubt, he will feel the ground with his foot before he attempts to place his whole weight upon it. If convinced that there is danger, neither coaxing nor whipping will induce him to go forward. When left to himself he will find his way, and carry over his burden in safety. It sometimes happens, though very seldom, that in traversing an extensive bog a horse will sink to his belly, but he soon extricates himself with apparent ease.

So very indifferent do the horses appear to such an accident, that we have seen them begin to eat the grass within their reach, till reminded of making their way forward by the application of a whip.

Having left our baggage and our guides behind, we should have been greatly perplexed in crossing the bogs, had not a peasant, whom we accidentally met, very kindly assisted us by leading the way.   He appeared to be active and intelligent, though we inferred all this from his motions and signs alone, as we had no other means of conversation with him.   Though we had a perfect confidence in his guidance, we were not a little astonished when he conducted us through places where we were in continual expectation of being swallowed up.   Our astonishment was not lessened, when we observed him plunge into a small inlet of the sea connected with the Hval Fiord, and proceed directly across it.   A river of considerable size empties itself into this bay ; and being much swollen by the rain which had fallen during some days before, we were fortunate in arriving at a time when the tide was out ; the river itself being impassable.   The breadth of the water was little less than half a mile ; but by scrupulously following the steps of our guide, we passed over in safety, and soon afterwards arrived at the farm-house called Houls.   This we found to be much superior in appearance to such houses of the same description as we had already seen.   The general construction of the habitation was the same ; and though it was very far indeed from being neat, yet it was some degrees nearer to that desirable state than most others. Our baggage horses did not arrive till a very late hour ; and the good people of the house having given us their best apartment, we removed some of the furniture, which appeared not to have been displaced since the house was furnished, and put up our own beds, for which there was just sufficient room. The floor was earthen, and extremely damp, and exhaled no very agreeable odour. This we endeavoured to correct by smoking segars.

The proprietor of this house, by name Gudmundson, was the Hreppstiore, or constable of the parish; an office next in rank to that of Sysselman, or Sheriff. The general duties of the Hreppstiore, relate to the preservation of the peace, and the superintendance of the poor. The farm attached to the house supports ten cows, several horses, and above a hundred sheep. Beside a certain proportion of the produce of the cattle, twenty-seven rixdollars per annum are paid as land rent. For this sum, however, independently of his farm, the tenant is entitled to a part of the profits of a salmon fishery in the adjoining river; it being the custom, in the case of most of the salmon rivers in Iceland, to divide the profits of the fishery among the different farms, which are situate on their banks. We had some salmon for breakfast at Houls, caught in the river the preceding evening, and found it to be excellent. No other mode of taking the fish is here practised, than that of constructing dams under the falls. The number of salmon caught does not much more than suffice for the use of the different farm-houses in the neighbourhood.

At Houls we saw a small patch of ground laid out as a garden, with regular beds. A young woman was planting small cabbages in it: by means of a plank she preserved the beds from being spoiled by her feet; and making holes in the earth along the edge of it, she placed the plants in regular and equidistant rows.

Next morning, we gave to the farmer's wife, who was a neat, good looking woman, a trifling present of needles, pins, scissars, thread, &c. with which she appeared highly pleased; and having breakfasted, we took our leave. Mr Gudmundson showed his kindness in accompanying us as a guide. We returned to the shores of the Hval Fiord, and proceeded to a place called Huamr, where we procured a small fishing boat, and crossed over to Saurbar, leaving our baggage horses to go round by the head of the Fiord, and taking only our beds with us. Two men rowed us across, who appeared perfectly indifferent to the swell of the water, which sometimes came over the gunnel of the boat, and which now and then occasioned us considerable alarm.

Saurbar is situate on a rising ground, at a short distance from the shore of the Hval Fiord, and is the residence of the parish priest, Mr Hialtalin; from whose house there is a striking, and somewhat picturesque view of the upper part of the Fiord, which has here the appearance of an extensive lake. Mr Hialtalin had just alighted from his horse as we arrived, and received us in the kindest manner. He had more the appearance of a gentleman, both in dress and manner, than any of his order whom we had yet met with ; and we found him possessed of considerable information. He had been settled at this place twenty-four years, with a stipend of thirty dollars, and as much land as maintains a small stock of cows and sheep. Upon this slender provision he has contrived to support a very numerous family ; having had altogether, from two matrimonial engagements, not fewer than twenty-three children ; thirteen of whom are still living. One of his daughters is married to Mr Gudmundson, our host of the preceding night. Mr Hialtalin's habitation entirely resembles the common farm-houses of Iceland, except that it is somewhat cleaner and more comfortable in the interior. The sitting room, which is very small and ill lighted, is furnished with a stove, an article not common in the houses of Icelanders, and possesses a considerable collection of books ; among which we met with a sort of *catalogue raisonnée* of all the Icelandic authors, which we wished very much to obtain, but found the owner unwilling to part with it.

Having perceived the church to be tolerably clean, we asked permission to sleep in it, which was cheerfully granted. The altar was covered with crimson silk, ornamented with gold lace; and above it, was a very indifferent painting of the Last Supper, surmounted by a crucifix. The dimensions of the church we ascertained to be thirty-five by sixteen feet, and about seven feet in height to the joists. I have chosen a representation of this building, in order that the reader may form some idea of the general exterior appearance of the country-churches in Iceland.

For supper, we had coffee, mashed fish prepared with butter, milk, and spices, and rice-milk. This last dish was given us for breakfast the next morning, with cakes made of the same materials.

Sketched by R. Bright   E. Mitchell Sculp.

CHURCH OF SAURBAR.

In the course of the evening, we had much conversation with our worthy host, who spoke Latin exceedingly well. We obtained from him some interesting particulars relative to his parish, and had much reason to admire his paternal care of the flock committed to his charge. In a population varying, in different years, from two hundred to two hundred and ten, there are fifteen married couples. The average annual number of births is seven; and of deaths, six or seven; of marriages, below one. The extent of the parish is sixteen English miles in length, and ten in breadth; so that the population does not exceed $1\frac{1}{4}$ to a square mile.

We were gratified with a sight of Mr Hialtalin's parish register; a very interesting book, in which, for his own satisfaction, he makes an annual record of the state of each family within the district of which he has the pastoral charge. He permitted us to copy part of this book; and the following is a translation, made by his assistance, of the first page of the register for the year 1805. This example of the attention and pious care with which the duties of a country priest are performed, in so remote a corner of the Christian world, may excite a blush in many of his brethren in more fortunate countries, and amid more opulent establishments.

Names of Habitations	Names of the People in the Family.	Situation, Occupation, &c.	Age.	Confirmed.	Communicants.	Whether able to read.	Conduct.	General abilities, &c.
Storibotn	Gudrun Sigurdardottir,	Widow, and owner of the house,	57	Yes	Yes	Yes	Clean and industrious,	Well informed.
	Oddur Jonsson,	Widow's son,	19	Do.	Do.	Do.	A good boy,	Well educated.
	Hans Jonsson,	Do.	19	Do.	Do.	Do.	Clever at work,	Not so good an understanding as his brother.
	Ingiborg Jonsdottir,	Widow's daughter,	18	Do.	Do.	Do.	A hopeful girl,	Well informed.
	Gudrun Jonsdottir,	Do.	17	Do.	Do.	Do.	Equally good,	Above mediocrity in her abilities.
	Wigfus Gudmundson,	An orphan kept by the widow,	15	No.	No.	Do.	A tractable boy	Good understanding.

The books in this house are, the New Psalm-Book; Vidalin's Sermons; Thoughts on the Nativity of Christ; Psalms relating to the Passion of Christ; the Conversation of the Soul with itself; Thoughts on the Passion; Diarium; Thordir's Prayers; the New Testament, and a Psalm-book.

Thyrill.	Jorundr Gislasson,	Hreppstiore, Elder or Constable,	41	Yes	Yes	Yes	Well disposed and clean,	Moderate abilities.
	Margret Thorstensdottir,	His wife,	53	Do.	Do.	Do.	Good character,	Piously disposed.
	Gudrun Eireksdottir,	Her daughter by a former husband,	19	Do.	Do.	Do.	A hopeful girl,	Well informed.
	Gudrun Grimson,	Servant man,	25	Do.	Do.	Do.	A faithful labourer,	He has neglected his improvement, and is therefore admonished.
	Thorsdys Sæmnsdottir,	Maidservant,	42	Do.	Do.	Do.	Neat & faithful,	Well informed.
	Jarfrudr Stephansdottir,	Her child,	3	—	—	—	- - -	- - - -
	Hristin Jonsdottir,	A female orphan,	8	—	—	—	A tractable child,	Has finished her catechism;—to be confirmed.
	Waldi Sterinderson,	A male orphan,	6	—	—	—	Tractable and obedient,	Is learning the catechism.

The books in the house are, the Old Psalm-book and the New one; Vidalin's Sermons; Vidalin's Doctrines of Religion; Fast Sermons; Seven Sermons; Psalm-books; Sturm's Meditations (translated into Icelandic); Bible Extracts; Bastholm's Religious Doctrine; a Prayer-book; and a New Testament belonging to the church.

This table is extremely interesting in many points of view. Besides showing the great attention of Mr Hialtalin to the duties of his office, it exhibited in some degree the character of the people, the importance they attach to religious and moral dispositions, and the attention which is paid to education even among the lower classes. By attending to the list of names, the manner of forming the sirname among the Icelanders may be observed. The son takes the Christian name of his father, and adds *son* to it for his sirname; and the daughter annexes *dottir* in the same manner. A similar custom is well known to have given rise to many English sirnames. In Scotland, the word *Mac*, signifying son, was prefixed. The inconvenience of this mode is not felt in Iceland, where the population is so much scattered. There are, however, instances here also, in which a sirname has been perpetuated, as in the family of the Stephensons.

We slept very comfortably in the church at Saurbar. Before leaving the place the following morning, I inoculated, with the vaccine virus, the minister's eldest daughter, a fine healthy-looking girl, about twenty years of age. As soon as I had performed the operation, she kissed me. I also inoculated a younger child; and a third was brought to me, but finding his arm covered with itch, I declined wasting matter on him, having formerly had experience of the inefficacy of inoculating, when any considerable cutaneous eruption existed. The itch is a very common disease in this country, and it seems to be thought conducive to general health. Having informed Mr Hialtalin how to treat the cowpox in its progress, and to save the crusts, and showed him how he might inoculate from them with a penknife or a needle; we took leave of his hospitable family. Our host himself, putting on a cocked hat, and taking his staff in his hand, accompanied us about a mile, in order to show us the road.

We travelled along the shore of the Hval Fiord, through a flat swampy country, lying between the mountains above Saurbar, and Akkrefell, a mountain which forms the headland between the Hval Fiord and Borgarfiord. On the low ridges above Saurbar, and stretching westward, we observed some patches of stunted birch-wood, the first thing of the kind we had seen in Iceland. We passed some small lakes, one of which formed a sort of bason, close to the Hval Fiord, so that a very little labour would be sufficient to drain off the whole of the water. This day was one of the warmest we experienced in Iceland. At two o'clock P. M. the thermometer stood at 65°, while a pretty strong breeze prevailed. In the sun, at the same hour, the degree was 86°.

Passing round the mountain of Akkrefell, we came in sight of Indreholm, the house of the Chief Justice Stephenson, from whom we had received an invitation when we saw him at Reikiavik, soon after our arrival in Iceland. It is situated in a large extent of flat, boggy ground, stretching from the base of the mountain to the sea, in the same manner as the tract under Esian; but containing much more verdure, and better grass. Behind is the lofty and precipitous mountain, which, in passing along, had attracted our particular notice, on account of the disclosure of its internal structure, almost from top to bottom, in a precipice not much short of two thousand feet in height. We arrived at the house about five o'clock. In appearance, it is rather a groupe of buildings than a single habitation; and, together with the outhouses and church, looks like a little village.

We were received very cordially, but with a considerable degree of form; and were ushered into the best room by Mr Stephenson, who met us at the door. Almost immediately after we had seated ourselves, the ladies of the family

T

made their appearance; and we had coffee, wine, biscuit, and English cheese, set before us. This was merely a prelude to a more substantial dinner, or rather supper, which was brought in at 8 o'clock. It consisted of boiled salmon, baked mutton, potatoes (from England), sago and cream, London porter, and excellent port wine. We had no doubt that the ladies, who had prepared and brought in the dishes, would partake of it; and, on our declining to take our seats before they had placed themselves at table, we were surprised when told that they had already dined. The females, of the highest, as well as of the lowest rank, as in former times in our own country, seem to be regarded as mere servants. During our repast, our hostess stood at the door with her arms a-kimbo, looking at us; while her daughter, and another young woman, were actively employed in changing the plates, and running backwards and forwards for whatever was wanted. Occasionally her ladyship assisted in the rites of hospitality; and next day, when restraint was somewhat worn off, she and the young ladies chatted and joked with us, laughing heartily at our broken Icelandic, which was mixed with English and broken Danish, neither of which they understood.

While busily engaged with our viands, our ears were all at once struck with musical sounds. Knives and forks were instantaneously laid down; and we gazed at each other in delight. Having heard nothing of the kind before in Iceland, except the miserable scraping of the fiddle in the Reikiavik ball-room, the pleasure we now derived from agreeable sounds and harmonious music, was very great. When our first surprise was over, and we could recollect ourselves, we thought that the music, which proceeded from an apartment above, was from a piano-forte; but we were told that

it was an Icelandic instrument, called the *Lang-spiel;* and that
the performers were the son and daughter of Mr Stephenson,
whose proficiency upon this instrument was considered to be
very great.   The *Lang-spiel*, which was now brought down
for our inspection, consists of a narrow wooden box, about
three feet long, bulging at one end, where there is a sound-
hole, and terminating at the other like a violin.   It has three
brass wires stretched along it, two of which are tuned to the
same note, and one an octave lower.   One of the two passes
over little projections, with bits of wire on the upper part.
These are so placed, that when the wire above them is pressed down by the thumb-nail, the different notes are produced
on drawing a bow across; and the other wires perform the
same office as the drones of a bagpipe.   In short, it is simply
a monochord, with two additional strings, to form a sort of
bass.   When the instrument is near, it sounds rather harsh;
but, from an adjoining room, especially when two are played together, as was the case when we first heard the music,
the effect is very pleasing.   The tunes we heard played were
chiefly Danish and Norwegian.   Mr Stephenson's daughter
made me a present of her Lang-spiel, from which this description and the drawing were taken.

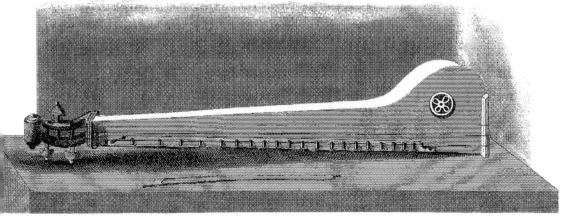

Engr.ᵈ by E. Mitchell

THE LANG SPIEL

The young ladies did their best to entertain us with singing; but the mode they had of screwing, not raising, their
voices to a pitch never before attempted, reminded me of
an error not unfrequent in my own country, where musical
proficiency is too often only a display of feats of art, which
have no reference whatever to the emotions which natural
melody, is calculated to excite.

Mr Stephenson's family is the only one in Iceland that can
be said to cultivate music at all.   He himself plays upon a
chamber-organ, which he brought from Copenhagen a few
years ago.

This gentleman, who has been already mentioned as at the
head of the Icelandic courts of justice, and a privy counsellor of Denmark, with the title of Etatsraad, certainly possesses talents; and has been very assiduous in his endeavours
to distinguish himself in the walks of literature.   He has had
great merit in recommending the pursuit of knowledge to his
countrymen; and has himself written various works on po-
litics, history, and morals.   All these amount to about twenty different books; and he had in the press, at the time we
were in his house, an additional work on the Polity of Iceland.   He is the owner of a very good library, consisting
probably of seven or eight hundred volumes, among which
are a number of English works, history, novels, and poetry;
and a valuable collection of Icelandic books and manuscripts.
In his house is also the library belonging to a society, which
will be particularly mentioned in the chapter on the present
state of literature in Iceland.

The family of Mr Stephenson consists of his lady, the
Fru Stephenson, as her title stands; his daughter, intitled
the Frukin, or young ladyship, a tall, lively girl, apparently about twenty years of age, whose stiff and formal dress,

of coarse blue cloth, but ill accorded with the laughter ever present in her countenance; another young lady who is at present under the guardianship of Mr Stephenson; two sons, both of whom appear to be clever, intelligent youths; and an elderly gentleman, the father of Mr Stephenson's lady. At the time of our visit to Indreholm, two nephews of the Chief Justice likewise formed a part of the family establishment, to whose education Mr Stephenson appears to have paid a good deal of attention.

During the three days we remained at Indreholm, we experienced the utmost hospitality and attention. Our residence here was interesting, as giving us some view of the habits and modes of life among the Icelanders of the highest class. We made a very minute examination of every part of the house; penetrating, under the guidance and authority of Mr Stephenson, even into the bed-chambers of the females of the family. While viewing these apartments, the ladies brought us various little articles of their own manufacture, in which considerable ingenuity was displayed, though not much elegance. They consisted principally of rude flower-work in coloured worsted. The extent of the house, as was before noticed, is very considerable. At a little distance is a water-mill, which is turned by the water of a small stream striking against a horizontal wheel. The dairy and other offices are detached from the house. Behind these buildings is a small smithy, where, at the time we visited it, we found the smiths busied in preparing scythes. The fuel is charcoal, made of birch-wood; great quantities of which, though the shrubs are very small, grow on the western bank of the Huitaà, and in some parts of the Borgarfiord Syssel. The making of charcoal is not the least important employment of the Icelanders during the summer. Every farmer has a smithy; and almost every man in the country knows how to shoe a horse; even the son and heir of the Chief Justice of Iceland having been seen thus occupied.

On going to bed, each night, during our stay at Indreholm, a cup or bason, full of milk and water, was set down at the bedside; a custom we had before observed at Kiebli-vik, but here for the first time in the house of an Icelander.

Indreholm is, on the whole, a very pleasant place; and is so situated as to command a fine view of the Faxè Fiord, and of the mountains of the Guldbringè and Kiosar Syssels. Did the climate permit the cultivation of corn, or the growth of trees, it might become a very beautiful residence. The pas-tures immediately round the place are very good, and adorn-ed by a profusion of the Statice armeria. Adjoining the house are two small gardens, well inclosed with walls of turf, in which cabbages, turnips, and sometimes potatoes, are cul-tivated with success, for the use of the family. At a short distance from the shore is a small island, crossing over to which, at low water, we saw vast multitudes of Eider-ducks, for whose convenience, rows of little apartments are con-structed of stones, in different directions across the island. About forty pounds weight of Eider-down are annually ob-tained from this spot.

Mr Stephenson has considerable property in this part of the country, as well as in other more remote districts of Ice-land. In his own hands he holds land sufficient for support-ing twenty-five cows and three hundred sheep. He has late-ly brought over from Norway some fine-woolled sheep of the Spanish breed, which seem likely to thrive well in the island. Connected with his property at Indreholm, there is a large fishing establishment, comprehending about twenty boats of different sizes, the use of which is given to the people coming from the interior of the country, on the terms formerly described.

The last day of our stay at Indreholm was occupied in the ascent of the mountain of Akkrefell; a labour of no small difficulty and hazard, from the excessive steepness of the face of rock which we had to climb; but one for which we were fully recompensed by the important mineralogical facts occurring to our observation, which will elsewhere be spoken of at length. The view from the pinnacle of the mountain, which seems almost to hang over the plain below, was extensive and interesting, comprehending a very considerable part of the south-western district of Iceland. We descended at another part of the mountain, but with even more risk than had attended our ascent. Availing ourselves of the conveyance by sea from Indreholm to Reikiavik, we left the specimens we had collected here, and in the preceding part of our journey, under the care of Mr Stephenson, being desirous of reserving our baggage-horses for further duties of the same kind.

On the 21st of June we left Indreholm, accompanied by Mr Stephenson and one of his nephews. The Chief Justice was, on this occasion, dressed in blue trowsers and a short jacket of the same colour, the stuff coarse and warm. On the top of his saddle was buckled a pillow of blue plush, stuffed with Eider-down. His tall figure, thus raised upon a small Iceland poney, formed rather a grotesque exhibition. We retraced our steps for about four miles, and then turning to the left, crossed the bogs towards Leira, the abode of Mr Scheving, Sysselman of Borgarfiord, who is married to a sister of Mr Stephenson. This place was formerly the residence of the latter; and the house was built by him. Externally, it is somewhat out of repair, but the

interior, and especially the sitting-room, displayed more decoration than is usual in the houses of Iceland. The cor-
nices were formed of wood, coloured red, and carved with
some degree of neatness ; and the chairs covered with a sort
of tapestry-work. Near the house there is a church, superior in accommodation to most of the edifices of the kind we
met with ; and having a gallery, which is by no means common in the country churches of Iceland. The lady of Sysselman Scheving is a tall, and rather handsome woman. She
was habited in the common dress of Icelandic females of the
higher class, except that the head-dress was merely a cap of
blue cloth, with a tassel hanging from the top. At supper,
we had a dish set before us, made of the Lichen Islandicus,
which we had hitherto scarcely ever seen employed by the
inhabitants of the country. The Lichen, chopped small, is
boiled in three or four successive portions of water, to take
off its natural bitterness, and then for an hour or two in milk.
When cold, this preparation has the form of a jelly, which is
eaten with milk or cream, and makes a very palatable dish.

On our arrival at Leira, I had observed, in a causeway
leading to the house, a fragment of stone, appearing to be
an incrustation, or deposit from water, and containing numerous vegetable petrefactions. Upon inquiry, we found
that there was a hot spring at the distance of about a mile
from the Sysselman's house. After supper, we set off, by
the light of an Icelandic midnight, to visit this spring, being
guided to it by Mr Stephenson and his nephew. We found
the water, which had a temperature of 138°, issuing from
two or three small holes in the rock, and running into a
stream which flows near the spot. A small cavity which has
been formed so as to receive the hot water, is occasionally
employed as a bath. Near the springs, we observed a con-

siderable extent of surface covered with curious petrefactions, evidently formed by deposition from some more ancient hot springs, which have now disappeared.   Our walk was finished a little before 12 o'clock at night.   Though the sky was cloudy and lowering, and a high range of mountains limited the horizon towards the north, yet the light was such as, even within the house, to be sufficient for the perusal of the smallest type, without difficulty or inconvenience.

We left Leira next morning ; and the Sysselman attended us for some miles.   We visited in our way the only printing-office now in Iceland, which is close to Leira, in a small and miserable wooden building, situated in the midst of a bog. This establishment is at present kept up by the literary society, of which Mr Stephenson is at the head.   He has the sole management of the press ; and is so fond of his own compositions, that few other people now give it employment; none liking to submit their works to so severe a censor.   This state of the press is extremely injurious to the literature of Iceland.   Two men are engaged in the printing-office : they have a press of the common construction, and make their own ink of oil and lamp-black.   There are eight founts of types ; six Gothic, and two Roman ; with a few Greek characters.   We found a small collection of books, which had been printed within the last few years, and remained here for sale.   We purchased several of these, among which was Pope's Essay on Man, translated into Icelandic verse.   During the last winter, the printing-office, with all its contents, was very nearly swept away by a flood ; and, at the present time, the building is in a state of wretched repair.

Leaving the plain in which Leira is situated, we began the ascent of the Eastern Skards-heidè, or mountain-road.   Part of this range of mountains, which divides the Borgarfiord Sys-

sel into two portions, is extremely lofty and precipitous ; and
the pass through it is very grand.   The ascent of the road is
long, and it certainly attains a height of not less than a
thousand feet : the mountains on each side of the pass,
however, have a much greater elevation, and some of them
were still almost entirely covered with snow.   In making this
ascent, we overtook several cavalcades of horses, returning
from various parts of the coast, loaded with dried fish, the
winter's stock of the farmers to whom they belonged.   The
Icelandic peasants who were guiding these cavalcades, addressed us in the language of salutation as we passed them ;
this being the invariable custom in every part of the island.
In some of the parties, we observed as many as ten or twelve
horses, each bearing its respective burthen of fish.

Just as we had gained the highest part of the mountain-
pass, we saw rain approaching, and had only time to cast
our eyes over the wide and extensive valley of the Huitaa,
or White River, which now lay beneath us, bounded on each
side by a magnificent range of hills ; and the view terminated towards the north by mountains entirely covered with snow.
The stream of the Huitaa, and of other smaller rivers winding over the broad expanse of the valley, and meeting the
eye at intervals, added much to the pleasing features of the
landscape.

Not far from the place where we began to descend, we
observed a lofty and very remarkable mountain, called Honn,
on our right hand.   It is a complete four-sided pyramid,
composed of regular beds of rock, piled one above another,
and diminishing to a point ; and forming the steps, as it
were, of a huge staircase.   When near the bottom of the hill,
we went a little out of the road to examine some hot springs.
We found the water gushing from several holes in different

parts of the side of the hill. Its temperature was from 100°
to 132°. It is pure; no incrustation being formed by it, nor
has it any peculiar taste. A little farther down, we had to
cross a deep and rapid river, which comes tumbling from the
rocks above, forming some very fine cascades.

After several tedious turnings and windings through bogs,
which we crossed not without considerable difficulty and danger, we arrived at Huaneyrè, the abode of Amtmand Stephenson, brother of the Chief Justice, and governor of the
western province of Iceland. The house of this gentleman
is situated upon an eminence in the great plain or valley of
the Huitaa; commanding, in front, a fine view of the arm of
the sea called the Borgarfiord ; behind, a still more striking
view of the Skards-heidè mountains, which we had lately
passed. There is nothing particular in the appearance or
construction of the habitation, or of the farm buildings attached to it. The only novelty to us was a small and rudely
constructed windmill, used for grinding rye, which we were
informed was the only edifice of the kind in Iceland.

The rain was now heavy ; and we were glad to enter the
house, where we were kindly received by the Amtmand's lady,
the daughter of a country priest; the Amtmand, with his
eldest son, being absent on an official tour through part of
his district.

While supper was preparing, we amused ourselves with
some English books which we found in the library in the sitting room. Our evening repast consisted of veal and salmon. The salmon of the Borgarfiord rivers are particularly
good : indeed, were we implicitly to credit the Chief Justice
Stephenson, there is nothing in this district but what is peculiarly excellent, and far surpassing any thing that is to be
met with elsewhere in Iceland. The best fish, the best cows,

the best sheep, the best horses, the best pasture, the best everything, was to be found in the favoured region of the Borgarfiord Syssel, and more especially at Indreholm. Nor were these praises bestowed without some appearance of reason. Though the extensive pastures of Borgarfiord Syssel, and particularly those in the valley of the Huitaa, are mere morasses, yet they yield a large quantity of grass, and support vast numbers of cows, horses, and sheep. The farm which Amtmand Stephenson holds in his own hands, is reckoned the best in the island. He keeps upon it between thirty and forty horses, fifty cows, and two or three hundred sheep; and gets as much hay as suffices for the maintenance of this large stock during the long winters of Iceland. The district of Borgarfiord is likewise remarkable for the vast number of swans frequenting it, which are particularly numerous in the extensive marshes below Huaneyrè. On the morning after our arrival, we counted forty of them within a short distance of the house.

Our breakfast, at Huaneyrè, consisted of salmon, boiled sorrel, sweet cakes, excellent coffee, sago jelly, a large tureen full of rich cream, rye-bread, and biscuit. We had reason to expect to find here the perfection of Icelandic cookery; for the Amtmand's first wife was the authoress of a work on that art, which is held in great esteem, and of which we each treated ourselves with a copy from the Leira printing-office. This family, with respect to manners and domestic economy, was much the same as that at Indreholm; the ladies here also performing the meanest offices of waiters and chambermaids.

After breakfast, we took our leave, still accompanied by the Chief Justice, who was resolved to attend us to the verge of the Borgarfiord Syssel. He conducted us for several miles through the bogs and swamps by which Huaneyrè is completely surrounded, till at last we got to the place called Huitar-vellir, where there is a ferry over the Huitaa.   This river, which is one of the largest in Iceland, is here contracted in its channel by the rocks, which rise abruptly on each side ; but the stream is very deep and rapid.   It is very properly named the White River, as it has very nearly the colour of milk and water, owing to the suspension of the finer particles of clay, washed down from the Jokuls among which it rises. A clergyman, Arnar Jonson, who is Provost of the Borgar-fiord Syssel, lives at this place ; and, while the people were employed in carrying our baggage over in a boat, and causing the horses to swim across the river, we went to his house and were treated with coffee.   We here parted with our friend Mr Stephenson, who had shewn us at his house, and during our short journey with him, a degree of attention and kindness, for which he will always be entitled to our gratitude.

Having crossed the river, we were told that it would be ne-cessary to send the boat up the stream, to carry us over ano-ther branch of it, which we had still to pass.   We walked to-wards this second crossing-place, where, the baggage being taken off, our horses were driven over ; the holes in the chan-nel rendering it unsafe to have them tied together.   The breadth of this part of the river was about two hundred yards ; and it was not without considerable trouble that all got safely across.   It now began to rain, which occasioned the rest of our day's journey to be very uncomfortable, as we had to pass through many swamps.   The whole of the valley of the Huitaa may not improperly be called a vast morass.   The western side of it, which we had now reached, belongs to the district called the Myrè Syssel ; a name literally signifying the Syssel of Bogs.   The southern part of this district, more especially, is so swampy that, during the summer, it is in general wholly impassable; though, in winter, it may occasionally be traversed upon the ice. We stopped for the night at a place called Svigna-skard, where we found a farm-house inhabited by the widow of a Sysselman of Myrè, who died about a year before. The poor woman was blind; but her son and daughter, who managed the affairs of the house, paid us every attention. We here found a small room, which, though by no means elegant, nor remarkably clean, we requested and obtained leave to occupy. In this apartment we discovered several old books belonging to the late Sysselman, a few of which we purchased from the family.

We remained at Svigna-skard during the whole of the next day; the rain still continuing very heavy. It being Sunday, we saw some of the people setting out on horseback for their parish church, which was at the distance of a few miles. On leaving the house, they took off their hats, and, putting them before their faces, continued for some time in the act of prayer, while the horses went on.

We had for dinner here, a dish called *skier*, which is similar to one well known in Scotland by the name of Corstorphine cream, or *Hattit kit*. In Iceland, it is made by means of sour whey; in Scotland, by butter-milk, over which cream or milk is poured, and allowed to remain till it has become sour; when the whey is suffered to run off by removing a plug in the bottom of the vessel into which the materials were put.

In proceeding towards the interior of the country, more cleanliness appears in the domestic habits of the people. Fresh fish is here an article of greater scarcity; and the offences to the sight and smell, which are always found in habitations where this is the principal food, decrease in proportion as it is less used.

On Monday morning, the 25th, we rose at two o'clock,

and commenced our journey a little after four. Two or three hours were always occupied by our Icelandic guides in loading the horses, and making the other preparations for departure; though it is probable that an active Englishman would have accomplished the same business in a third part of the time. We were attended by a peasant from the neighbourhood, as a guide, on account of a thick fog, which obscured every thing around us. Our route lay along the course of a river, which came tumbling over a rocky channel from the mountains forming the western boundary of the valley of the Huitaa. We could not see further than fifty or sixty yards before us; but the rocky sides of the river being studded with birch shrubs, three or four feet in height, the scene was, upon the whole, a pleasing one. We had to cross the river occasionally; and found the current so strong, that it was with difficulty the horses could make their way across. The pass over these mountains, which are called the Western Skards-heidè, must be very grand in fair weather. Several small lakes appeared in the hollows between the hills, from which there issued large and rapid streams. The fog did not begin to clear away till we found ourselves amongst lava, and were beginning to descend. We had before observed slags scattered about the sides of the mountains; and were now in a hollow on the summit, bearing all the characters of the volcanic country we had seen in the Guldbringè Syssel. After winding amongst some steep and broken hills, we descended into a valley completely filled with lava, which, from its being in many places covered with soil, appeared to be very old. The rude and irregular rocky masses of which it is composed, are broken into every possible variety of form; and, in some places, being partially covered with moss, strikingly resemble the ruins of old castles or fortifications. We were informed that there is a

great deal of lava amongst the swamps towards the south, which
has probably proceeded from the same source as the stream we
now followed.   At the opening from the mountains of the valley through which the lava has flowed, we saw a small groupe
of cottages, called Hraundalur,* and a wide extended plain,
stretching, on our left hand, towards the sea.   From the rugged appearance of many parts of it, we easily credited the account we had received of much lava existing in this quarter.

We now skirted along the base of the mountains; and,
after travelling a few miles, came to the opening of an extensive valley, the lower part of which is occupied by another
great stream of lava, apparently about two miles in breadth.
Having passed along the edge of this, by a very rough and
dangerous path, we at length reached a place called Stadarhraun, where we found a small church, and the priest's house,
situated on a small grassy spot, almost entirely environed by
rugged masses of lava.   The priest had just mounted his
horse for the purpose of accompanying the Sysselman of
Myre, who was at this time travelling through his district.   On
observing us, he dismounted; and, after a little conversation
with the Sysselman, who invited us to his house, the latter
proceeded on his journey alone.   The priest readily allowed
us to take up our night's abode in the church; and provided
us with plenty of boiled milk; with which, and a lamb we had
purchased and roasted at Svigna-skard, we made up a tolerable dinner.   The night became extremely cold, the wind being
from the north.   The sky was, however, remarkably clear;
and, from the door of the church, we had a fine view of the
Snæfell Jokul, which, from our nearer approach to it, now
made a very magnificent appearance.

* Hraundalur, literally translated, signifies the *Lava Valley*; lava being called
Hraun in the Icelandic language.

We were informed of the existence of a mineral water near this place, which the minister discovered, and which will be more particularly noticed in the chapter on mineralogy.

Not being able to procure a guide, the priest offered to accompany us as far as Kolbeinstadr; and we resumed our journey at seven in the morning. Our companion, who was a tall, aukward man, dressed in a very uncouth manner, exhibited a singular figure when mounted on his poney; and the effect was rather heightened by the ornamental trappings with which the animal itself was decorated. We passed through the lava by a winding and rugged path; and, as we went along, observed many conical hills, which were evidently of volcanic formation. One of these, on the west side of the valley, is remarkable. It stands alone; is about 300 feet high; and is composed of slags and sand, having a rocky, scorified looking mass at the top. On some parts of the lava, where sand and a little soil had accumulated, we observed birch shrubs growing more luxuriantly than any we had before seen. At the place where we descended and quitted the lava, it appeared to be the most considerable mass we had met with. Having crossed a deep and rapid stream which skirts it, we arrived at the foot of a lofty range of precipitous mountains; among the debris of which we found abundance of zeolite. On turning into the valley in which Kolbeinstadr is situated, which here is several miles in width, we observed it to be filled, like many other valleys in this district, partly with lava, and partly with bogs. In the flat towards the sea, we remarked a circular crater about sixty yards in diameter, surrounded by lava. Other appearances of the same kind occurred higher up the valley; and we saw likewise several conical hills resembling that just described, many of them of considerable height, and exhibiting a surface of a reddish brown colour, derived from the loose volcanic scoriæ and sand of which they seem to be composed. Before us, in the striking precipices of the lofty mountain called Kolbeinstadr-Fialla, (or Fell), we saw some horizontal beds of rock beautifully arranged in columns. On the side of one of the mountains I was fortunate enough to find a specimen of the Ranunculus Glacialis; for which plant, Mr Hooker, when in Iceland, had looked in vain. We never afterwards met with it.

We arrived at Kolbeinstadr about one o'clock. There is a very decent church here, in which we dined; the good people of an adjoining farm-house bringing us abundance of milk. They remained in the church while we were engaged with our repast; and examined our appearance and dresses with the most minute attention. The priest of Stadarhraun, who had partaken in our meal, here took leave of us. We proceeded, under the direction of a peasant, towards Roudemelr, on the opposite side of the valley; crossing, in our way, a broad and rugged stream of lava, and traversing some very dangerous and unpleasant bogs. When very near the end of our journey, our guide and his horse suddenly sunk into the swamp, but soon scrambled out. After this warning, we deemed it prudent to dismount, and proceed on foot, leaving our horses to find their own way. The house and church at Roudemelr, are placed under the termination of a vast stream of lava, which here exposes a precipitous front of considerable height. The farmer, an old man, undertook to guide us to a spring of mineral water, which has been long known, and much celebrated, in Iceland; and which Mr Chausen had recommended us to visit. Winding round the base of one of the conical volcanic hills which are so numerous in this valley, we came to the spring, which is situated about two miles to the north of Roudemelr. It is called

Öl-kilda, or the *ale-well*. We saw two small cavities full of water, which was kept in constant and violent agitation by the escape of carbonic acid gas. On tasting it, we found it to possess a strong, but grateful degree of pungency, very much like that of soda water, after it has been exposed to the air for a few seconds. As water highly impregnated, as this is, with carbonic acid gas, has been known to produce some degree of intoxication, the name may have been derived from this circumstance. No water ran from the cavities, nor was there any other remarkable appearance in the vicinity of the spring. The temperature was 45°; precisely the same with that of an adjacent stream.

Near Roudemelr there is a very fine range of columns in a bed of rock, of which from fifteen to twenty feet are visible. In general, along the whole extent, which is about half a mile, the columns are very thick ; many fragments which had fallen down, measuring five, and some six feet in diameter. How far they extended under the surface, could not be ascertained ; but, from their thickness, it is probable that their length is very considerable. Some of them were detached several feet from the rock, and stood singly. The regularity of this rock formed a striking contrast to the ruggedness of the lava which has flowed near it.

We had already found so much advantage in the plan of sleeping in the churches as we proceeded on our journey, that we did not hesitate to avail ourselves of the same accommodation at Roudemelr, though we found the church here not in the very best condition. In almost all these edifices, there is a painting of the Lord's Supper above the altar, for the most part very wretched both in design and execution. There was a painting of this kind in the church at Roudemelr ; and we saw chalked upon it the numbers and

order of the psalms to be sung during the service of the sab-
bath-day. These numbers are usually inscribed with chalk
upon the walls or beams of the building. While we made
the churches our place of nightly abode, our guides, though
they had an Iceland tent with them, always slept in the adjoining farm-house or cottage; and appeared to be everywhere received by the inhabitants with great cordiality. In
one of the apartments of the farm-house, we found two large
barrels filled with the Lichen Islandicus, which was laid up
for the use of the family during winter.

On the morning of the 27th, a very thick fog came on.
Having breakfasted on curds brought from the farm-house,
we were fortunate in being enabled to purchase three additional horses, which made the whole number in our cavalcade amount to fourteen. These were found necessary to
the convenience of our journey; the mineralogical specimens, which formed a considerable proportion of the burthen, receiving every day some increase. One of our new
horses was obtained for seven rixdollars; the other two for
eight rixdollars each. We likewise exchanged a horse here
with a country priest, who was travelling to the coast for fish,
and had pitched his tent near the church during the night.
This person was more miserable in his appearance than any
one of his profession whom we had seen in Iceland; his habiliments being such as would scarcely have distinguished him
from an English beggar of the lowest description.

During the whole of our journey to Miklaholt, which was
our next resting-place, the fog was so thick as to prevent our
seeing the country through which we passed. This perhaps
was not to be regretted, as we traversed, during the whole
way, either lava, loose stones, or almost impassable bogs,
Of all we had hitherto seen, the bogs here were the most for-

midable.   The farmer of Roudemelr was our guide; but he
was glad to procure the assistance of a peasant whom we
took up at a house in the way.   The horses were often in great
danger from sinking into the swamp; and every step was
made with apprehension.   Some idea of the difficulties and
danger of this stage may be formed, when it is stated, that
we spent five hours in going eight miles on horseback.   We
saw, in the course of our day's journey, several Icelandic
tents, pitched in green spots, where the horses employed in
carrying goods could feed.   The tents used by the natives of
the country are made of woollen stuff, and formed like a mar-
quee, but very small.   They are supported by two poles and
a rail, and stretched by means of small cords fastened by
stones, or sometimes by wooden pegs.   The baggage was
heaped up round the outside of the tents which we passed in
our way to Miklaholt.   Some of the people were going for
fish, and others returning with it from the coast.   They had
stopped on account of the weather, about which an Icelander
is much more apprehensive than his general habits would
seem to indicate.

Miklaholt is situate on a small eminence, completely surrounded by morasses; and consists of a church, the priest's
house, and a few cottages.   The priest was not at home when
we arrived, but soon made his appearance.   His wife received us with every demonstration of kindness; and, shewing
us into a small room, the furniture of which consisted of a
bed, two tables, a bench, and a few boxes, prepared some
coffee for us.   When the husband, a tall, good looking man,
returned, he displayed an equal degree of kindness and attention.   He was extremely fond of snuff; and we returned his hospitality by a present of some of English manufacture.   The Icelandic snuff-boxes are commonly made

of wood; but the better sort of people have them made of the tooth of the sea-horse, or walrus, and ornamented with silver. Their common shape is seen in the vignette, which is taken from one that was presented to me by Mr Simonson of Reikiavik. The snuff is put in at the bottom, and taken out at the smaller end, and laid on the back of the hand, from whence it is drawn into the nose. In windy weather, this end of the box is put into the nostril, and the snuff is poured out. The quantity of snuff taken by the Icelanders is very great, and it constitutes one of the most important of the few luxuries which they can be said to enjoy.

Early in the morning we began to prepare for our departure, and were as usual a little disturbed by the curiosity of the people, who crowded round us in the churchyard, and watched all our movements with the most minute attention. The minister, habited very much like an English sailor, undertook to guide us through the bogs. We were told that we should not find the way through them, on this side of Miklaholt, so bad as on the other, as there was a bridge constructed for the accommodation of passengers. This bridge we found to be no other than a narrow and deep ditch, with loose, sharp stones at the

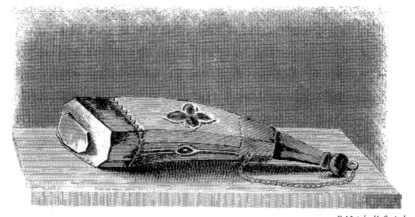

E Mitchell fculp.!

S N U F F B O X

bottom, along which we passed in a string; for if any of the
horses had stepped but a few inches to the right or left, they
would infallibly have sunk into the swamp. Having passed
the bogs, we came to a stream, which we had to cross a dozen
times at least, before we reached the sea-shore, which we
gained by passing over a broad and deep ford. Being now upon
tolerably firm sand, we put our poneys to the gallop, and soon
got over two or three miles very pleasantly. The priest of
Miklaholt, having seen us safe across some deep holes filled
by the flowing tide, took his leave.

For the last three days, we had been travelling in the district called Hnappadals Syssel. About six miles to the west
of Miklaholt, we entered Snæfell's Syssel, which forms the
extremity of the peninsula, terminated by the Snæfell Jokul.
The central part of this peninsula, for its whole length, is
occupied by a lofty range of mountains, between which and
the sea on the southern side, an extensive tract of flat land
intervenes. Beyond Miklaholt we found this plain to narrow
considerably, becoming not more than three or four miles in
width. Some part of this extent is sandy, and appears to
have been left by the sea. The greatest proportion, however,
is boggy; and the whole is covered with an abundance of fine
grass, on which we saw numbers of cattle and sheep grazing.
Here and there are small elevations, on one of which we
found Stadarstad, a group of buildings, comprising a church;
the habitation of the minister, Mr Jonson, who is Provost of
Snæfell's Syssel; and a few detached cottages. The general
appearance of this little hamlet had more the air of neatness
and respectability, than any place we had seen since we left
Indreholm; and, on inquiring for the Provost, in order to
deliver a letter of introduction from the Bishop, we were
pleased to find that his appearance was in conformity with

the impressions which his dwelling had made upon us. He
was a good looking man, apparently upwards of sixty years
of age, dressed in a gown of coarse black cloth, and a cap of
the same stuff. He received us with an air of politeness,
which corresponded, as much as any thing we had hitherto
seen in this country, with what we are accustomed to regard
as the manner of a gentleman. We delivered the Bishop's
letter, which came to us while at Indreholm, accompanied
with the following epistle.

> ' Dominum perillustrem nobilissimum Mackenzie
>   ' Salvere jubet Geirus Vidalinus.
>
> ' Exigua hæc epistolia, ut datâ occasione, Præpositis To-
> ' parchiæ Borgarfiordensis, et Snæfellnæssensis tradantur
> ' enixè rogo, certe persuasus ut, me vel non rogante, quid-
> ' quid in eorum potestate situm est, lubentissimè servient, ad
> ' iter tuum facilitandum. Arnorus Jonæ tibi quæ in Borgar-
> ' fiordo visu digna sunt indicabit. Gudmundus verò Jonæ
> ' facilem tibi præbebit antiquum montis hujus incolam domi-
> ' num Bardum Snæfellsas,* cujus sine auspiciis mons Snæfell
> ' Jokul vix ac ne vix quidem superari potest.'

The church at Stadarstad was the best we had seen, except
that of Bessestad. It is constructed entirely of wood, and has
a pretty large gallery. The weather had now become very
serene, and in the evening we had a striking view of Snæfell
Jokul, and the mountains stretching from it towards the east,
the summits of which were still capped with snow. As the
summer snow-line is in general about 2,000 feet above the

---

* A sort of tutelar saint of the Snæfell Jokul.

level of the sea, we had thus a good *datum* for judging of the height of these mountains. Those rising immediately behind Stadarstad are very lofty, and present a bold and precipitous front towards the sea. Our breakfast consisted of trout, tern eggs, and milk; and after finishing it, we spent some time in looking over the books in the Provost's house, a few of which we purchased. We found him to be an extremely intelligent man, and particularly well informed on every subject connected with the history and condition of his own country. He was formerly secretary to the late Bishop Finnsson at Skalholt, and recollected having seen Sir John Stanley there, when on his way to Mount Hekla.

Having arranged all our little affairs, the Provost left us for a while, and soon returned, so altered by his dress, that we hardly recognised him. He had on a decent black suit, with boots, and had decorated his head with a very respectable brown wig, and a hat. Altogether he looked much like a country clergyman of the Scotch church.

We took leave of his wife, who had been very attentive to us, and proceeded towards Buderstad, the next stage of our journey. The road for the greatest part of the way lay along the shore. About six miles from Stadarstad, we left the regular track, and proceeded towards the mountains (which now begin to approach nearer the sea), in order to visit a hot spring. We found it near a place called Lysiehouls. The water issues from the top of a mount about ten feet high, and fifty yards in diameter, entirely calcareous. The temperature of the spring was 96°, and the water had an acidulous taste. Not far from the mount, were great quantities of incrustations not calcareous, which had evidently been formed by some ancient springs. They were like those we had found at the hot springs of Leira. About half a mile from the spring, we came to a stream of lava that had flowed down the precipices above, and spread over the flat plain intervening between the mountains and the sea. It did not differ in any respect from the many streams of lava we had seen before. Near Buderstad, we found another mineral water, the taste of which was similar to that of Lysiehouls : the temperature was 46°.

In approaching Buderstad, it was necessary for us to cross some small inlets of the sea ; and as the tide was flowing, and the creeks were of considerable depth, we did not accomplish this without much difficulty. Our friend, the Provost, however, brought us safely to the end of our day's journey at Buderstad, which place we found situate on the edge of lava, in the same manner as Havnefiord. This is one of the trading stations of Iceland ; and consists of a merchant's house, a large wooden storehouse, a church, and a considerable number of cottages. The house is constructed of bricks, which have been brought hither from Denmark. We were received by Mr Gudmundson, its inhabitant and the principal person of the place, with every demonstration of civility. He is a merchant, and has connections at Reikiavik, and at Copenhagen.

The war between England and Denmark has been severely felt by Mr Gudmundson. No vessel has come to Buderstad for three years past, previously to which time, one or two every year used regularly to visit this station. In consequence of this interruption of the intercourse, the inhabitants are in great want of corn, timber, and iron ; and the storehouses are every where full of the produce of the country, for which no proper market can be found.

The curiosity of the people manifested itself no where in such a degree as at this place. We could not move without being closely observed; and when we applied our hammers to the lava, with the view of collecting specimens, it seemed to excite no small surprise among the groupe of people who were watching our motions. But this changed to astonishment, when following us into the house, they saw us carefully wrapping our specimens in paper. Whether they thought us very wise or very foolish, we could not ascertain.

Provost Jonson took his leave in the evening, and returned to Stadarstad, having previously made us promise that we would write to him on our return to Britain. In Mr Gudmundson's house we passed the night tolerably well, being disturbed only by the crowing of a cock, which took up its lodging in the room where we had fixed our beds. We had a plentiful breakfast the next morning, of mutton, cheese, rye-bread, and coffee, and departed highly pleased with the attention we had received.

Our next stage was to Stappen, situated farther along the coast towards the west. We found the lava of Buderstad, which it was necessary to cross in our way, far more rugged than any we had met with. Numerous rents and chasms of great depth presented themselves on every side; and it was with much difficulty and a considerable degree of danger that our horses got across by a winding path, in many places exceedingly steep and rough. On examining one of the caverns which occur in this lava, it appeared to have been formed in a manner similar to those we had seen in the Guldbringè Syssel. We penetrated into it about 40 yards upon the surface of the congealed snow, which forms its flooring or pavement.*

---

* Eghert Olafson particularly describes the Buda-hraun, or lava of Buderstad; and endeavours to account for the vast caves and fissures which appear in it, by supposing that the water of the sea obtained access to the lava, while yet in a heated state.

After much time spent in crossing this lava, which is here two or three miles in breadth, we at last reached the bay of Stappen. Here we found a large extent of flat sand forming the beach, upon which we halted, and measured a base, with the view of calculating the height of the Snæfell Jokul, from the foot of which mountain we were now not very far distant. The atmosphere being perfectly clear, we succeeded in taking angles, by a calculation from which we ascertained the height of the mountain to be 4,558 feet. We afterwards found that this estimate did not differ more than a few feet from the measurement of the Danish officers, who are now employed in surveying the coasts of Iceland. On the other hand, Eghert Olafson, one of the most eminent naturalists of Iceland in modern times, asserts that, by barometrical measurement,

Sketched by Mr. Holland.   E. Mitchell fc.

SNÆFELL JOKUL.

he found the height of the mountain to be not less than 7,000 feet; a calculation certainly differing widely from the truth. The vignette of Snæfell Jokul was taken from the sands upon which we measured the base.

On leaving the beach, the road became more and more romantic. We ascended and descended by winding paths, and crept along the edge of high cliffs overhanging the sea, over which numerous streams were dashed into spray. At one place we crossed a rapid stream within a few feet of the precipice over which the water fell. Numerous flights of seabirds rendered the scene still more lively. We found Stappen on the brow of a range of curiously columnar rocks, large insulated masses of which stood in the sea, in various singular forms. We had not expected to find the Jokul a volcanic mountain, but the observation we now made of streams of lava descending from it in various directions, left no doubt of this being the case.

Stappen, like Buderstad, is a trading station, and consists of a merchant's house, two or three storehouses, and a few cottages inhabited by fishermen. We were met at the door of the house by Madam Hialtalin, a Danish lady, whose husband, brother to our friend the priest at Saurbar, had been absent for some years. He had been taken prisoner on a voyage to Denmark, and had afterwards contrived to reach Norway; but since his arrival in that country he had not been heard of. The situation of his wife, and her family consisting of six children, was highly deserving of pity; and we had but a melancholy satisfaction in receiving the numerous marks of hospitality which they lavishly bestowed upon us. The manners of Madam Hialtalin were those of a lady, and appeared to us, who had seen no one in Iceland entitled to this appellation, to the greatest advantage. The house was perfectly clean, and the rooms neatly furnished. The principal bed-room was really a most refreshing sight to us, after the places of nightly abode to which we had for some time been accustomed. From the roof was suspended a small glass chandelier. There were three windows with festooned curtains of white muslin; a handsome canopy bed, with very neat cotton furniture, sheets white as snow, and as usual a heap of Eider-down upon it. From the window there was a fine view of the mountains; and the dashing of a little stream over its rocky bed beneath, produced a very pleasing sound. Nothing could exceed the gratification we derived from the good breeding and attention of our hostess and her family.

The coast in this neighbourhood of Stappen is very remarkable; presenting, for the extent of about two miles, striking and beautiful columnar appearances, both in the cliffs which form the shore, and in the numerous insulated rocks, which appear at different distances from the land. The ranges of columns, which in general are about fifty feet high, and perfectly regular in their forms, are variously broken, in consequence of their exposure to the action of the sea. In some places large caves have been formed, and in two of these the light is admitted by fissures in the roof, producing a very singular and striking effect.

In general the ranges of columns have a vertical position; but in different places they are disposed in bundles upon one another in all directions. In several instances they appear diverging from a centre; and they assume, in short, every form which such rocks can be imagined to take. About a mile and a half to the west of Stappen, there is a curious perforated rock, forming a detached arch of considerable magnitude, the view through which is singularly picturesque, comprehending in the foreground many of the insulated masses of colum

nar rock, and in the distance, the fine range of mountains, which stretches along the peninsula towards the east.   On the whole, it is probable that a more curious range of cliffs is no-where to be seen, both with respect to the picturesque appearances they present, and also from the interesting facts they offer to the attention of the geologist.

We proceeded along the coast, till interrupted by a broad and rugged stream of lava, the characters of which were different from any we had before surveyed.   It originated in some part of the Snæfell Jokul, and had flowed into the sea ; but we could not discover any way by which we might safely reach the place where it had met the water.   The lavas of this district are very ancient, no eruptions having taken place, either from the Snæfell Jokul, or from any part of the adjacent range of mountains, since the island first became inhabited.

Sketched by R. Bright

E. Mitchell sculp.t

CAVE AT STAPPEN.

We remained at Stappen nearly two days, occupied in the survey of the various interesting objects in its vicinity. On the second day of our stay here, we took a boat for the purpose of examining from the sea the columnar cliffs and caves already described, which in many points are more advantageously thus seen than they can be from the shore. We saw on the beach at Stappen several sharks that had been taken for the sake of the oil of the liver, and the skin: two or three of these fish were of very large size. In several parts of Iceland, particularly on the northern and north western coasts, the shark fishery is a regular occupation. Strong hooks fastened to chains, and baited with muscles, &c. are anchored a little way out at sea, and the fish when caught are towed on shore. Of the skin, shoes are made; and some parts of the flesh are occasionally smoked, and used as food by the natives. It was long before we could prevail on Madame Hialtalin to procure us a little of this delicacy; but when it did make its appearance, our noses were assailed by so horrible an odour, that we were glad to have it removed as soon as possible.

It was our original design to have attempted the ascent of the Snæfell Jokul from the side of Stappen; but having been disappointed in this by the foggy state of the weather, we took leave of our kind hostess on the 2d July, and set out for Olafsvik, situated on the northern coast of Snæfell's Syssel, to visit Mr Clausen. On our leaving the house at Stappen, we were honoured by the display of the Danish flag, which was hoisted on the roof. We retraced our steps for a few miles, and then, by a pass called the Kamskard, began to ascend the mountains. There are different routes by which one can reach Olafsvik from Stappen. That round the coast, by the west side of the Jokul, is very dangerous, and we preferred the one we took, as being the best and most secure. The ascent was long and fatiguing, but we were amply repaid for our labour, by a very fine mountain scene at the summit, and an extensive view towards the north of the great bay called Breidè-Fiord, with the mountains of the Dalè and Barderstrand Syssels in the remote distance. While descending, we came to a stream which fell over a precipice, forming an extremely fine cascade. The rocks above the channel of the stream were composed of very perfect columns, about sixty feet in height.

On reaching the shores of the Breidè-Fiord, we travelled two or three miles to the westward to Olafsvik, which place is situate on the side of a fine bay. It consists of the dwelling houses of Mr Clausen and his factor, and about a dozen cottages scattered on the rising ground behind, which is bounded by grand precipices. The beach was covered with numerous piles of dried fish; and we found the warehouses quite full of the same article, for which lately no sufficient market has been found. We afterwards learned from Mr Clausen, that his stock at Olafsvik consisted of several hundred thousand fish, salted or dried; besides very large quantities of woollen goods, stockings, gloves, &c. manufactured in the country, and ready for exportation.

Mr Clausen, who had returned from Reikiavik by sea a short time before, received us with much kindness, and introduced us to his wife, a lady whose appearance and manners we found extremely pleasing. She is a native of Denmark, and came over to Iceland only two or three years ago.

The weather having now become more favourable, the ascent of the Snæfell Jokul was accomplished by my friends on the 3d of July; and I give the following narrative of the expedition in the words of Mr Bright:

After a hesitation of an hour or two, on account of the

' doubtful appearance of the day, Mr Holland and myself, with
' our interpreter, and one of our guides, who was very desirous
' of accompanying us, put ourselves under the direction of a
' stout Icelander, who undertook to be our leader in the ascent
' of the Jokul. He, however, honestly confessed, that he had
' never been higher up the mountain than the verge of the
' perpetual snow, as the sheep never wandered beyond that
' limit ; but this was also the case with the other inhabitants
' of the district. Every one of us provided himself with an
' Iceland walking staff, furnished with a long spike at the end;
' and in case of need, we carried some pairs of large coarse
' worsted stockings of the country manufacture. We likewise
' had our hammers and bags for specimens, a compass and
' thermometer, a bottle of brandy, with some rye bread and
' cheese.

' Thus equipped, we set forward on our march ; and hav-
' ing passed two or three cottages, whose inhabitants gazed
' with wonder at our expedition, we directed our course in
' nearly a straight line towards the margin of the snow. The
' nearer we approached it, vegetation became more and more
' scanty, and at length almost entirely disappeared. After
' walking at a steady pace for two hours, in which time we
' had gone about six miles, we came to the first snow, and
' prepared ourselves for the more arduous part of our enter-
' prize. The road being now alike new to all, we were as
' competent as our guides to the direction of our further
' course. The summits of all the surrounding mountains
' were covered with mist ; but the Jokul was perfectly clear;
' and as the sun did not shine so bright as to dazzle our eyes
' with the reflection from the snow, we entertained good hopes
' of accomplishing our purpose. During the first hour the
' ascent was not very difficult, and the snow was sufficiently

' soft to yield to the pressure of our feet.   After that time
' the acclivity was steeper, the snow became harder, and
' deep fissures appeared in it, which we were obliged to cross,
' or to avoid by going a considerable way round.   These fis-
' sures presented a very beautiful spectacle : they were at
' least thirty or forty feet in depth, and though not in gene-
' ral above two or three feet wide, they admitted light enough
' to display the brilliancy of their white and rugged sides.
' As we ascended, the inferior mountains gradually diminish-
' ed to the sight, and we beheld a complete zone of clouds
' encircling us, while the Jokul still remained clear and dis-
' tinct.   From time to time the clouds, partially separating,
' formed most picturesque arches, through which we descried
' the distant sea, and still farther off, the mountains on the
' opposite side of the Breidè-Fiord, stretching northwards
' towards the most remote extremity of the island.

' In the progress of our ascent, we were obliged frequently
' to allow ourselves a temporary respite, by sitting down for
' a few minutes on the snow.   About three o'clock, we ar-
' rived at a chasm, which threatened to put a complete stop
' to our progress.   It was at least forty feet in depth, and
' nearly six feet wide; and the opposite side presented a face
' like a wall, being elevated several feet above the level of the
' surface on which we stood ; besides which, from the falling
' in of the snow in the interior of the chasm, all the part on
' which we were standing was undermined, so that we were
' afraid to approach too near the brink lest it should give
' way.   Determined, however, not to renounce the hope of
' passing this barrier, we followed its course till we found a
' place that encouraged the attempt.   The opposite bank was
' here not above four feet high, and a mass of snow formed
' a bridge, a very insecure one indeed, across the chasm,

'Standing upon the brink, we cut with our poles three or four
'steps in the bank on the other side, and then, stepping as
'lightly as possible over the bridge, we passed one by one to
'the steps, which we ascended by the help of our poles. The
'snow on the opposite side became immediately so excessive-
'ly steep, that it required our utmost efforts to prevent our
'sliding back to the edge of the precipice, in which case we
'should inevitably have been plunged into the chasm. This
'dangerous part of our ascent did not continue long ; and we
'soon found ourselves on a tolerably level bank of snow, with
'a precipice on our right about 60 feet perpendicular, pre-
'senting an appearance as if the snow on the side of the
'mountain had slipped away, leaving behind it the part on
'which we stood. We were now on the summit of one of the
'three peaks of the mountain ; that which is situated farthest
'to the east. We beheld immediately before us a fissure
'greatly more formidable in width and depth than any we
'had passed, and which, indeed, offered an insuperable ob-
'stacle to our further progress. The highest peak of the
'Jokul was still a hundred feet above us ; and after looking
'at it sometime with the mortification of disappointment, and
'making some fruitless attempts to reach, at least, a bare ex-
'posed rock which stood in the middle of the fissure, we were
'obliged to give up all hope of advancing further.

'The peak of the Jokul we had now attained, is about
'4,460 feet above the level of the sea. The extensive view
'which we might have obtained from this elevated point, was
'almost entirely intercepted by the great masses of cloud,
'which hung upon the sides of the mountain, and admitted
'only partial and indistinct views of the landscape beneath.
'It has been said by Eghert Olafson, and others, that from
'one part of the channel which lies between Iceland and

' Greenland, the mountain of Snæfell Jokul may be seen on
' one side, and a lofty mountain in Greenland on the other.
' It is difficult to ascertain how far this is an accurate state-
' ment.   The distance between the two countries at this place
' cannot be less than eighty or ninety leagues.

   ' The clouds now began rapidly to accumulate, and were
' visibly rolling up the side of the mountain; we were there-
' fore anxious to quit our present situation as speedily as
' possible, that we might repass the chasm before we were in-
' volved in mist.   Our first object, however, was to examine
' the state of the magnetic needle, which Olafson in his travels
' asserts to be put into great agitation at the summit of this
' mountain, and no longer to retain its polarity.   What may
' be the case a hundred feet higher, we cannot affirm; but at
' the point we reached, the needle was quite stationary, and,
' as far as we could judge, perfectly true.   We then noted an
' observation of the thermometer, which we were surprised to
' find scarcely so low as the freezing point; and after an ap-
' plication to the brandy bottle, began with great care to re-
' trace the footsteps of our ascent.   We found re-crossing the
' chasm a work of no small danger; for whenever we stuck
' our poles into the snow bridge, they went directly through.
' The first person, therefore, who crossed, thrust his pole deep
' into the lower part of the wall, thus affording a point of sup-
' port for the feet of those who followed; Mr Holland, how-
' ever, who was the second in passing over, had, notwithstand-
' ing, a narrow escape, for his foot actually broke through the
' bridge of snow, and it was with difficulty he rescued himself
' from falling into the chasm beneath.   We were scarcely all
' safe on the lower side of the chasm, when the mist surround-
' ing us, made it extremely difficult to keep the track by which
' we had ascended the mountain.   When we came opposite

' to a small bank which we had remarked in our ascent as
' being free from snow, we desired our guide to remain where
' he was, that we might not lose the path, while we went to
' examine that spot. We found the bank to be almost entirely
' composed of fragments of pumice and volcanic scoriæ. After
' our return to the former track, we made the best of our way
' back to Olafsvik, which we reached at about a quarter past
' six, to the great surprise of every one; for we were scarcely
' expected till the following morning; such is the reverential
' awe inspired by the Jokul. None of our party seemed more
' gratified with the exploit than our guide, who having always
' been accustomed to look upon the Jokul as some invincible
' giant, greatly exulted in this victory over him; but we after-
' wards learned, that he found considerable difficulty in mak-
' ing his friends credit his narrative of the ascent.

Thermometer at different stages.

At 11 o'clock on the shore . . . . . 58° Fahren.
— 12   on the mountain . . . 56°
— 1   verge of the snow . . . 43°
— 2 . . . . 42°
— 3 . . . . 39°
— 3 17 min. at the highest point . . 34°
On the snow at the same time . . 32°."

We remained three days with Mr Clausen at Olafsvik.
During this time several persons came to us for medical advice; and we inoculated a great number with the vaccine
virus. The nearest medical practitioner resides at Stikkesholm, about forty miles distant.

During our stay at Olafsvik the weather was remarkably
clear and serene, and even oppressively warm. Having a good

opportunity from the situation of the place, of observing the
setting and rising of the sun, it was found by Mr Holland on
the night of the 5th July, that it remained under the horizon
exactly two hours and thirty-five minutes. Previously to setting, it hung for a long time on the verge of the horizon, and
even at midnight it had sunk so little below, that the bright
glow of the luminary was completely visible, and the light
sufficient for the pursuance of the most minute occupation.
We were at this time in latitude 64° 58′ N.

While in Mr Clausen's house, we felt quite at ease. No
obtrusive curiosity, no restraint, incommoded us in our pursuits ; and our host, having taught himself English, was able,
without the uncertain assistance of an interpreter, to give us
much information relative both to the topography and commercial concerns of Iceland. He had only a collection of
voyages, and a volume of Roderick Random, in English; most
of his books in that language being at Copenhagen. He had

Sketched by Sir G. M.   OLAFSVIK   Engrᵈ by E. Mitchell

travelled through several countries of Europe, and at this time talked of going to England, which purpose he has since accomplished.

In a walk which we took along the coast towards the west, on the last evening of our stay at Olafsvik, we saw a great number of seals, at several of which we fired, but without success. These animals are particularly numerous on the shores of the Breidè-Fiord.

We left the hospitable house of Mr Clausen on the 6th of July, and pursued our journey to Grunnefiord, having the intention of varying our returning route along the peninsula, by following the line of its northern coast. We deviated from the road about four miles from Olafsvik, to examine some rocks which formed a promontory overhanging the sea. These rocks were columnar, and covered with vast numbers of kittiwakes, which took wing on our approach, and almost darkened the air. We here saw two large sea eagles, which prey upon the water fowl, and are very destructive to the Eider-ducks. Whenever the eagles passed over the rock, the noise made by the kittiwakes stunned our ears. The columns forming this promontory are for the most part vertical, and about sixty feet in height. Some of them, however, are contorted in a very striking manner. They presented some curious geological appearances, which will hereafter be noticed.

We now approached a place called Bulands-höfde, where the only means of advancing is by a path on the face of a precipice not less than a thousand feet high. To a considerable height, it is quite perpendicular above the sea, and the path was in many places obliterated by the falling of the rocks and gravel above it. A more difficult and dangerous track cannot well be conceived. We were greatly alarmed lest some of our horses should fall down; but they passed safely, having given remarkable proof of their steadiness and caution. This pass is totally stopped during winter by ice and snow; and there are several instances of people who have perished in the attempt to proceed along it.

All the way from Olafsvik to Grunnefiord, a distance of nearly twenty miles, the mountain scenery is very fine. The lofty precipices are varied in every form, and the summits of the mountains are broken into a thousand abrupt and singular shapes. Cataracts are seen foaming and dashing from the rocks, and nothing but wood is wanting to make this one of the most picturesque countries in the world.

There is one singular mountain not far from Grunnefiord, called Sukker-Toppen (Sugar loaf) in the charts, from its resemblance to a sugar loaf. Viewed from the east or west, the top appears tabular; but when surveyed from the north or south, it is seen as a cone ending in a sharp point. Thus it appears that the summit of the mountain is an exceedingly narrow ridge.

We found the merchant's house at Grunnefiord (Greenfrith), situate at the head of an arm of the sea of the same name, which is derived from the extensive green flat stretching from the sea towards the mountains. The merchant here, Mr Müller, was on the eve of departing from the place in order to settle at Reikiavik, and had got most of his goods and furniture on board a sloop which was lying at anchor in the bay. We were well received, and as well entertained as Mr Müller's present circumstances would admit. We slept very comfortably in a room in his house; and after the usual salutations had been evaded by some of the party, and submitted to by others, we proceeded the next morning towards Stikkesholm, which is about twenty-five miles distant. A day's journey of such a length was a serious undertaking, impeded as we were by a long cavalcade of baggage horses; and had any bogs been in the way, it would have been impossible to have travelled so far in one day.

Our route lay among mountains of a character similar to those we had passed. In a sort of cleft of prodigious magnitude, we saw a fall of water about 150 feet high, dashing over a rock, of a curious reticulated appearance from the veins which intersect it, and afterwards rushing violently under several arches of snow. The rock, and the whole surrounding scenery, were very magnificent.

We had now to cross a steep mountain, on the other side of which we met with a stream of lava, and the country beyond it appeared to be entirely volcanic. The path through this stream was much better than any we had seen in similar situations, and we found very little difficulty in crossing the lava, though it was fully as rugged as that at Buderstad. We observed several cones composed of slags; and the face of the mountain which we passed, after crossing the lava, was entirely covered with cinders.

Having at length come to a green spot, near a cottage, we stopped to refresh ourselves and our horses. The poor people brought us some milk, which we were glad to drink, although we had recently witnessed a very disgusting instance of Icelandic uncleanliness on a similar occasion. We made our rustic meal on the turf, our horses quietly grazing around us.

Crossing another stream of lava, near its termination on the shore, we entered a low country, leaving the mountains on our right hand. The indentations made by the sea obliged us to proceed in a very zigzag direction towards Stikkesholm, which is placed at the extremity of a small peninsula. Near the isthmus over which we passed in entering this peninsula,

is a hamlet called Helgafell, or the Holy Hill, from its situation upon an eminence, with which certain superstitious ideas
and usages were in ancient times connected.   On this spot
was established one of the earliest of those settlements which
the Norwegian emigrants made upon the coasts of Iceland.
While approaching Stikkesholm, we had several fine views of
the Breidè-Fiord, which is here completely studded with small
rocky islands.   Their number is stated to be 150, and this
does not seem to be an exaggeration of the fact.   Many of
these islands contain vast numbers of Eider-ducks.   Stikkesholm is singularly situated close to the sea amidst abruptly
precipitous rocks, some of them columnar in their form.

The houses are large, and, as well as the storehouses and
cottages, belong to Mr Thorlacius, a native of the country,
and reputed the richest man in Iceland.   He lives at a place
called Bildal, in the district of Bardestrand.   His factor Mr
Benedictson, another merchant, and Mr Hialtalin a surgeon,
occupy the houses at Stikkesholm.   The latter gentleman is
a son of the minister of Saurbar. He studied at Copenhagen,
and was about to settle in some town in Jutland about two
years ago, when he was ordered to occupy a vacant medical
situation in Iceland.   The district allotted to him is very extensive, and his salary is only about L.12 per annum.   The
profits arising from his practice during the first year of his
residence at Stikkesholm have not exceeded L.6, and on this
pittance he must support a wife and family.

Before the war between England and Denmark, Stikkesholm was a place of considerable traffic.   The fishery begins
earlier than in the Faxè-Fiord, and is very productive.   Only
one vessel came hither during the last season, from Norway;
and our entertainers were of opinion, that the connection
intended to be established with England by the late proclam-

ation, would not remove their distresses. Mr Hialtalin was in Copenhagen during the last attack upon that city, and his house, and the greatest part of his property had been destroyed by the bombardment. He shewed us an umbrella which had been broken by a shot while he was sleeping under it in a tent. It was not very agreeable to listen to these narratives, as we had nothing to say in vindication of the attack on Copenhagen. Mr Hialtalin also spoke of some bad usage he had met with from the captains of two English ships of war, while he was on his passage to Iceland. He and Mr Benedictson made many inquiries respecting the present state of Europe, and were greatly astonished when told of the marriage of Bonaparte.

ᵥ We spent the following day, which was Sunday, in the house of Mr Benedictson, at Stikkesholm. During the early part of the day, all the occupations of the people were suspended, and many of them went to the neighbouring church at Helgafell; but, at six o'clock in the evening, the storehouses were again opened, and the inhabitants of the place, resuming their common dresses, went to work as usual. This is the case in every part of the country. The sabbath of the Icelander, according to the ecclesiastical law of the island, begins at six o'clock on Saturday evening, and terminates at the same hour on Sunday; after which time any occupation or amusement may proceed as on the ordinary week days. The females of the family at Stikkesholm, as in the other houses of the higher class of people which we had visited, did not sit at the table when we were eating our meals. We observed here, however, that the master of the house always saluted his lady, when himself rising from the table; a practice which had not occurred to our notice before.

On our arrival at, and departure from Stikkesholm, we

were as usual honoured by the display of the Danish flag.
On the 9th, we left the place, accompanied by Mr Hialtalin,
who rode with us to the mountain of Drapuhlid, situated
about six miles to the south of Stikkesholm. This mountain,
from the previous accounts we had received of it, we had been
led to consider as something very remarkable in a mineralogical point of view, and in these expectations we were by no
means disappointed. Mr Hialtalin remained at the foot of
the mountain while we were examining it; and, on our return, we found the Sysselman of the district with him. This
officer had heard of our arrival at Stikkesholm, and came to
deliver a message from the Amtmand Stephenson, requesting that we would take his house in our way in returning to
Reikiavik.

We observed a stream of lava that had descended from
the mountains behind Drapuhlid, had divided, and run down
each side of a ridge which connects this mountain with those
to the south.

On a small grassy spot at the foot of the mountain, we
packed up the specimens we had procured; and enjoyed a
draught of excellent milk, brought to us in a wooden vessel
from an adjoining cottage. In this repast the Sysselman and
Mr Hialtalin partook with us, the whole party sitting on the
grass. Having taken leave of these gentlemen, we pursued our
journey to Narfeyrè, situate on the east side of the Alpta-
Fiord; an arm of the sea which runs up several miles into
the country, and is bounded on each side by lofty and magnificent mountains. We had to make a large circuit towards
the head of the Fiord, and should have had a longer journey,
had not the tide luckily been out; so that we were enabled
to cross the mud, by the help of a guide whom we took with
us from the cottage near Drapuhlid. We found a very good

church at Narfeyrè, and took up our quarters for the night
in it.   On examining the loft of the building, we saw lying
on a chest a mass of human fat that had been taken out of a
grave.   It appeared, however, to be the muscular substance
converted into the matter so much resembling spermaceti, a
change effected by water.   This is reckoned a very precious
article as a medicine, and is frequently used by the Icelanders
in pulmonary complaints.

Before our departure from Narfeyrè, we went into the
farm house adjoining the church, to see the mode of weaving
commonly employed in Iceland.   The whole process is awkward and laborious.   The threads for the woof hang perpendicularly, being stretched by stones tied to them on a wooden
frame.   No shuttle is used; but a thread is passed across the
woof by the hand, and is stretched by rubbing a little piece
of wood upon the threads.   A portion of the rib of a whale,
nicely polished, and shaped somewhat like a broad sword, is
then introduced between the threads, and with this the warp
is struck forcibly. In this way, a woman can weave a yard of
stuff in a day.   The stuff we saw in this loom, was composed
of red and yellow threads, and was intended for a bed-cover.
These articles are exchanged for fish; and the value of three
yards is reckoned to be somewhat less than two dollars.

Having procured a guide, we set out for Snoksdalr, the
next stage of our journey.   The wind blew sharply from the
north-east, and the day was the coldest we had yet experienced.   The country through which our route lay, was for
the most part low, and totally uninteresting.   Not far from
Narfeyrè, there was a considerable track covered with small
birch shrubs.   Excepting the hamlet called Breidabolstadr,
where there is a church and a solitary cottage, no human
habitation appeared to enliven the dreary scene.   We kept

along the shore of the Fiord, and after a tedious and unpleasant ride of about twenty-four miles, reached Snoksdalr at
eight o'clock in the evening, where we found a good farm-
house and a small church. The latter did not differ materially
from those we had already seen, except in having a different
subject for the altar-piece, which was the stoning of St Stephen. On each side of the altar was a large chair, in the
bottom of which were kept the habiliments of the priest.
There were two dresses; one of them of a red stuff, having a
large cross embroidered in white, on each side of the robe.
The other was made of a sort of crimson velvet or plush, the
crosses being formed of silver lace. Over the door-way to
the altar was a curious groupe of male figures carved in wood,
which appeared as if intended to represent Christ disputing
with the doctors.

Snoksdalr formed the limit of our northern journey. It is
situate at a short distance from the extremity of the Breidè-
Fiord, on the brow of a hill, and in a very exposed situation,
but in the midst of plenty of grass. We took up our abode in
the church, while the inhabitants of the adjoining house cheerfully supplied us with curds and milk, upon which supply we
now almost entirely depended. We remained here during the
whole of the 11th, in order that the horses might recruit their
strength, for a long and fatiguing day's journey which they
were next to encounter. About Snoksdalr the country begins again to be mountainous, and towards the north and east
nothing is to be seen but long ranges of dreary hills stretching
across the narrow isthmus, which here separates the Breidè-
Fiord from the sea on the northern shores of Iceland. The
breadth of this isthmus, from Snoksdalr to the nearest inlet
on the northern coast, does not exceed forty miles. The cold
north-east wind continued to blow, and to render our abode

here by no means comfortable. The church too was in bad repair, and exceedingly damp; and at no time, during the day we passed in it, did the thermometer stand above 45°, even in the building. The latitude of Snoksdalr is about 65° 10′.

At five o'clock on the morning of the 12th, we recommenced our journey. After crossing a rugged hill behind Snoksdalr, we entered an extensive valley, well clothed with grass, and studded with a number of cottages, to each of which a portion of the pasture belongs. We were told that there were about thirty such divisions in the valley, and that the greater part of the land in it is kept for pasture, no hay being made excepting round the farm-houses. From the number of such valleys contained in this district, the Dalè Syssel, in which we were now travelling, derives its name.

From this valley, which is called Middalur, we passed into another, which gradually contracted for two or three miles, till we got to the foot of a steep and lofty ridge of mountains, called Brautarbrekkar. The ascent of this was long and laborious; but we were gratified, on arriving at the top, with a view of some fine mountains, stretching towards the south as far as the Western Skards-heidè. The descent into the valley, on the south side, was very rapid. Having advanced a little way into the valley, we dismounted, and took our breakfast by the side of a small stream. This done, the sun shining full upon us, we stretched ourselves upon the grass, and slept for an hour very comfortably, though snow lay not a hundred feet above. A lofty and singular mountain, called Baula, forms the eastern side of the valley; the direction of which is nearly north and south. After proceeding along it for some miles, we turned to the north-east, round Baula, which bears a striking resemblance to the mountain of Drapuhlid near Stikkesholm. We now were in the valley of the Norderaa, a considerable river, which we crossed several times before reaching the farm-house called Huam, where our day's journey terminated.

We were now on the border of the low country, through the centre of which the Huitaa flows. Having passed the night in the church at Huam, we recommenced our journey the following morning, attended by the farmer, who very readily undertook the office of guide. He was an oddly shaped, merry, and active little man, mounted on a miserable horse, which, by dint of constant kicking, he made to move forwards at a great rate. After crossing several rocky ridges, we descended into the valley of the Huitaa, ten or twelve miles above the place where we had formerly crossed this river. In our way we passed through some birch wood, which was the tallest we had yet met with, the trees, in general, being from six to ten feet high.

We stopped at Sidumulè, the abode of Mr Otteson, Sysselman of the Myrè and Hnappadals Syssels. This gentleman, whom we had already seen for a few moments at Stadarhraun, was greatly superior in appearance and manners to most of his brother Sysselmen whom we had met with, and his intelligence corresponded to this superiority. From his answers to the various questions we proposed to him, he seemed to be fully master of the duties of his office; in which, however, he had been only a short time installed. In his library we found Danish translations of Sir Charles Grandison, of Addison's Cato, and other English works. If we might judge from the appearance of Mr Otteson's dairy, in which we saw twenty-four large dishes full of milk, his farm, and his management of it, must be very good.

Having been informed by Mr Otteson that we could not get across the river Huitaa at this place, without great risk

of wetting our baggage, we were under the necessity of alter-
ing our plans; and instead of crossing directly to Reikholt,
we resolved to return to Hauneyrè, and from thence visit the
hot springs near the former place. Mr Otteson was going to
Huaneyrè, and offered to be our guide. The banks of the
river, almost the whole way, were swampy, and we had to
wind through the bogs in various directions, which rendered
our journey tedious and very disagreeable. We examined a
hot spring on the western bank of the river, but found nothing remarkable about it. The temperature was 165°.

About eight miles from Sidumulè, we crossed the Huitaa
at a place where it was very broad, and so deep that the
water reached our saddles. The singular colour of the water in this river was formerly mentioned. The stream, in
general, is about one hundred yards broad, and very rapid.
The left bank we found to be as boggy as the other; but,
under Mr Otteson's direction, who was very skilful in selecting the best route across these swamps, we got rapidly forward. On arriving at Huitarvellir, the place where we had
before crossed the river, we found that the hay harvest was
just begun, several peasants being engaged in cutting the
grass around the priest's house. Here we were informed that
Amtmand Stephenson had gone to Reikiavik, which to us
was a piece of bad news, as we had relied upon the use of his
horses to go to Reikholt next day. But our good fortune did
not forsake us. Not only the Amtmand, but our good friend
Mr Fell, arrived at Huaneyrè from Reikiavik, almost at the
same moment with ourselves; and this unexpected meeting gave uncommon pleasure to the whole party. Mr Fell
had come over into the Borgarfiord Syssel, to make some
inquiry respecting the salmon fishery in this district. We
also found here Mr Magnuson, who is Sysselman of the

Dalè Syssel.   This large party, added to a very large family, occasioned some consultation on the manner in which we were all to be disposed of during the night.   We were not suffered to wait for our baggage horses; the good people set themselves to work, and, by means of chairs, and mattrasses filled with Eider-down, soon made up a sufficient number of beds.   I was honoured with the bed in the room adjoining the sitting room; but I should have been more comfortable, suspended in my little portable bed, wrapped up in a blanket. From the noise over our heads, it was probable that the whole family, males and females, were crammed together in the loft. It is not uncommon in Iceland, as it must be in all countries under similar circumstances of poverty, for people of all ranks, ages, and sexes, to sleep in the same apartment. Their notions of decency are unavoidably not very refined; but we had sufficient proof that the instances of this which we witnessed proceeded from ignorance, and expressed nothing but perfect innocence.

My friend Mr Bright being somewhat indisposed, Mr Holland and I occupied the following day in visiting the valley of Reikholt, leaving him behind with great regret.   We were accompanied part of the way by the Amtmand, Mr Magnuson, and Mr Fell.   The two former were going to hold a judicial court at a place called Huitaar, in the Myrè Syssel. The eldest son of the Amtmand was our guide, a youth about sixteen years of age, and of the most promising talents. The fluency and elegance with which he spoke the Latin language, and the progress he had made in the English under his uncle the Chief Justice, were far less surprising to us than the shrewdness of his remarks on every subject which occurred in conversation.   His father intends to send him to Copenhagen.   If life shall be granted to him, and proper op-

2 b 2

portunities of prosecuting his studies, I venture to prophesy, that this young man will prove an honour to his native country, which may derive much advantage from his public services.

After traversing a great extent of swampy ground, and encountering many difficulties in our progress through it, we at length reached the entrance of the valley; the natural curiosities of which greatly exceeded the expectation we had formed of them.

The hot springs in the valley of Reikholt, or Reikiadal,* though not the most magnificent, are not the least curious among the numerous phenomena of this sort that are found in Iceland. Some of them, indeed, excite a greater degree of interest than the Geyser, though they possess none of the terrible grandeur of that celebrated fountain; and are well calculated to exercise the ingenuity of natural philosophers. On entering the valley, we saw numerous columns of vapour ascending from different parts of it. The first springs we visited, issued from a number of apertures in a sort of platform of rock, covered by a thin coating of calcareous incrustations. I could not procure any good specimens, but from those we broke off, the rock appeared to be greenstone. From several of the apertures the water rose with great force, and was thrown two or three feet into the air. On plunging the thermometer into such of them as we could approach with safety, we found that it stood at 212°.

A little farther up the valley, there is a rock in the middle of the river, about ten feet high, twelve yards long, and six or eight feet in breadth. From the highest part of this rock

---

* Reikholt, means smoky hill; Reikiadal, smoky valley; Reikiadals-aa, the river of the smoky valley.

a jet of boiling water proceeded with violence. The water was dashed up to the height of several feet. Near the middle, and not more than two feet from the edge of the rock, there is a hole, about two feet in diameter, full of water, boiling strongly. There is a third hole near the other end of the rock, in which water also boils briskly. At the time we saw these springs, there happened to be less water in the river than usual, and a bank of gravel was left dry a little higher up than the rock. From this bank a considerable quantity of boiling water issued.

About two miles farther up the valley, on the opposite side of the river, whose windings rendered it necesary for us to cross it several times, are the church of Reikholt, and the

Sketched by Sr G.M.   E. Mitchell sculpt.

BOILING SPRINGS IN THE RIVER REIKIADALSAA.

minister's house. We went thither for the purpose of examining a bath which was built nearly 600 years ago by the
celebrated Snorro Sturleson. The bath is a circular bason,
constructed of stones, apparently without any cement, but
nicely fitted together. It is about fourteen feet in diameter,
and altogether about six feet deep, the water being allowed
to fill it to the depth of about four feet. The hot water is
brought from a spring about 100 yards distant, by means of
a covered conduit, which has been somewhat injured by an
earthquake. We were told that cold water had been brought
to it, so that, by mixing the hot and cold together, any desired temperature might be obtained. All round the inside,
a little way under the surface of the water, was a row of
projecting stones, placed apparently to serve the purpose of
steps. Steps were constructed as an entrance to the bath,
close to the orifice by which the hot water entered. At present it is not much used, and the bottom is covered with vegetable matter and soil.

In the absence of the minister, we were politely received
by his wife, who gave us some excellent cream ; a good proof
of the quality of the pastures of this valley.

Proceeding down the valley on the side opposite to that on
which we entered it, we came to a groupe of cottages, situated
close to some hot springs. In the water of one of them we saw
some pots, containing milk and curds. There is a sort of natural dome, several feet in diameter, formed over part of this
spring, of clay and stones. It intermits at short, and pretty
regular, intervals. Having sat down near an orifice in the
dome from which steam was rushing, we observed that the
noise suddenly ceased, and the water, when it was visible,
sunk down amongst the stones in its channel, leaving them
dry. After a short interval, the noise recommenced, steam

rushed forth, and boiling water followed. We observed many repetitions of this phenomenon; and the intervals were scarcely two minutes. It may be easily explained in the same manner as that of ordinary intermittent springs, connecting such an apparatus as is supposed to belong to them, with one in which steam may be brought into action in order to force the water upwards. Upon part of the mound or dome mentioned above, and extending a little way beyond, a hut was constructed, the entrance to which was by a long, narrow, and low passage. The heat of the earth occasioned by the hot water was here confined, so that the temperature of the air was 73°. No use was made of this hut except for the drying of clothes. It is singular that the people have not contrived the means of heating their apartments by the hot springs that are steady in their operations. One would think, that the great scarcity of fuel, and the difficulty of procuring it, would have suggested this long ago. The fear of danger does not exist, for the habitations are close to the springs; and near the place where boiling water is thrown out with the most terrible violence, and which will afterwards be described, the natives quietly repose. Their not having taken advantage of this natural source of comfort, must proceed from that want of enterprise, which is so conspicuous in the character of the Icelanders.

About a mile farther down, at the foot of the valley, is the Tunga-hver, an assemblage of springs the most extraordinary, perhaps, in the whole world. A rock *(wacke?)* rises from the bog, about twenty feet, and is about fifty yards in length, the breadth not being considerable. This seems formerly to have been a hillock, one side of which remains covered with grass, while the other has been worn away, or perhaps destroyed at the time when the hot water burst forth. Along the face of the rock are arranged no fewer than sixteen springs, all of them boiling furiously, and some of them throwing the water to a considerable height. One of them, however, deserves particular notice. On approaching this place, we observed a high jet of water, near one extremity of the rock. Suddenly this jet disappeared, and another thicker, but not so high, rose within a very short distance of it. At first we supposed that a piece of the rock had given way, and that the water had at that moment found a more convenient passage. Having left our horses, we went directly to the place where this had apparently happened; but we had scarcely reached the spot, when this new jet disappeared, and the one we had seen before was renewed. We observed that there were two irregular holes in the rock within a yard of each other; and while from one, a jet proceeded to the height of twelve or fourteen feet, the other was full of boiling water. We had scarcely made this observation, when the first jet began to subside, and the water in the other hole to rise; and as soon as the first had entirely sunk down, the other attained its greatest height, which was about five feet. In this extraordinary manner, these two jets played alternately. The smallest and highest jet continued about four minutes and a half, and the other about three minutes. We remained admiring this very remarkable phenomenon for a considerable time, during which we saw many alternations of the jets, which happened regularly at the intervals already mentioned.

I have taken the liberty to give a name to this spring, and to call it, the ALTERNATING GEYSER.

These springs have been formerly observed, though the singularity of the alternations does not seem to have been attended to as any thing remarkable. Olafson and Paulson mention, that the jets appear and disappear successively in the second, third, and fourth openings. We observed no cessation in any of the springs but in the two under consideration.

To form a theory of this regular alternation is no easy matter; and it seems to require a kind of mechanism very different from the simple apparatus usually employed by nature in ordinary intermittent or spouting springs. The prime mover in this case is evidently steam, an agent sufficiently powerful for the phenomena. The two orifices are manifestly connected; for, as the one jet sinks towards the surface, the other rises; and this in a regular and uniform manner. I observed once, that when one of the jets was sinking, and the other beginning to rise, the first rose again a little before it had quite sunk down; and when this happened, the other ceased to make any efforts to rise, and returned to its former

Sketched by Sir G.M.   E. Mitchell sculpt.

THE TUNGU-HVER AND ALTERNATING GEYSER.

state, till the first again sunk; when the second rose and played as usual. This communication must be formed in such a manner, that it is never complete, but alternately interrupted, first on one side, and then on another. To effect this without the intervention of valves seems to be impossible; and yet it is difficult to conceive the natural formation of a set of permanent valves; so that this fountain becomes one of the greatest curiosities ever presented by nature, even though, in attempting to explain the appearances it exhibits, we take every advantage that machinery can give us. If it is occasioned by natural valves, these must be of very durable materials, in order to withstand continual agitation and consequent attrition.

Not having obtained any explanation which I can consider quite satisfactory, and having been unable entirely to overcome the difficulty myself, I leave its solution to the ingenuity of those who may think the phenomenon of the Alternating Geyser worthy the exercise of their talents.

The examination of the various natural wonders in the valley of Reikholt, detained us so long, that we did not reach Huaneyrè until a late hour in the evening; and we found the rest of the party, who had left the place in the morning, already re-assembled there.

Of all the Icelanders we had hitherto met with, we agreed that the Amtmand Stephenson had most of the appearance and manners of an Englishman. He is unassuming and mild in his address, and possesses something more than good common sense. With the exception of Mr Steingrim Jonson of Bessestad, he is the only person in Iceland who understands the French language, which he speaks with considerable facility. His property and rank, as well as character, give him a high degree of respectability among his countrymen.

The Amtmand did not terminate his great civility when we quitted his house the following morning, on our return to Reikiavik. He attended us several miles on the road to Indreholm. Mr Fell having expressed his wonder at the swift pacing of the Amtmand's horse, I was requested to try it. Though I have rode many horses esteemed excellent in England, I must confess that I was never carried so rapidly and easily before. I am not a light weight, yet the poney paced with me at the rate of twelve or fourteen miles an hour, while I felt as if sitting in an easy chair. When, on dismounting, I spoke in warm terms of my admiration of the animal's performance, and the pleasure it had given me, the Amtmand, with a politeness that could not be surpassed at the most refined court, requested that I would honour him by accepting his horse. In spite of my remonstrances against his parting with an animal so valuable to him, and which it was probably impossible for me to convey to England, he pressed the matter so much that I was obliged to comply with his desire. He is himself famous in Iceland for the rapidity of his travelling. With two or three led horses, he usually accomplishes 100 English miles in twenty-four hours.

We had a charming day for viewing the stupendous precipices of the Eastern Skards-heidè, over which mountains we again passed. On our return to Leira, we found all in a bustle preparing for a wedding, which was to be celebrated in the afternoon. Not wishing to intrude, we stopped only to take a dish of coffee. The priest, and a number of people, dressed in their holiday clothes, had already arrived. Though it would have gratified our curiosity to see the marriage feast, we feared lest our presence might interrupt the enjoyment of the party.

We found the family at Indreholm just as we had left it,

and remained there only a few hours, having resolved, as the evening was favourable, to cross over to Reikiavik by sea. When formerly here, we had observed a quantity of the bones of small whales lying scattered upon the shore. We were informed that, early in the preceding winter, a shoal consisting of nearly a thousand of these whales had come on shore, and had been taken.

At ten o'clock we went to the beach; but, on getting into the boat prepared for us, it filled so fast, that we were glad to make our escape to the beach, notwithstanding the assurances of the boatmen that the leaks were of no consequence. The Chief Justice soon had another boat launched, but we could not get away till the first had been drawn upon the beach. All hands from the house were called to assist, and men and women (among the latter, the ward of the Chief Justice) jumped out of bed on the alarm being given, and came down without waiting to dress themselves. I was perfectly thunderstruck with the appearance of this motley group, which my companions viewed from the boat with equal astonishment. A pleasant sail of four hours brought us to Reikiavik at two o'clock in the morning, after an absence of one month and two days, during which we had travelled about three hundred and fifty miles.

From the 25th of June till the end of July, the Icelanders frequent Reikiavik, in order to dispose of their commodities, and to purchase such articles as they may require from the Danish merchants. They bring oil, fish, tallow, wool, butter, fox and swan skins, &c. which are taken in exchange for tobacco, spirits, meal, rye, iron and steel, linen and cotton goods, thread, &c. &c. This period of traffic is called the Handel, and while it lasts, many thousands of laden horses come to the town. The people bring tents with them, in which they live during their stay, and on their journey. During this period of activity little bustle is observable, excepting in the shops. The Handel of this year was not so good as usual. Butter was scarce all over the country, and tallow being used in its stead, very little of that article was exposed to sale.

There was a good deal of drunkenness observable at this time. The drinking of spirits is much encouraged by the merchants, both for the purpose of promoting the sale of that article, and of enabling them to over-reach the poor people who deal with them.

During our stay at Reikiavik at this time, a day was appointed for taking the salmon from the Laxaa (Salmon river), about three miles from the town. This is a sort of gala day, and not only the people interested in the capture of fish, but all the ladies and principal people attend. Sometime before the appointed period, parties were seen gallopping off towards the scene of action.

The river, which is small, divides into two branches about a mile from the sea. The channel is dammed up early in the morning, and the water forced into one branch, while the other is allowed to run almost dry, and the salmon that happen to be in the river are thus easily taken. The river is held on lease by Mr Scheele, who keeps the tavern, and pays a rent of sixty dollars. Sometimes two or three thousand salmon are taken out. At the time we witnessed the capture, there were only nine hundred taken. The fish are caught in the early part of the season, in boxes, formed like our mouse traps. The salmon fishery of Iceland appears to be an admirable object for speculation, while the rents of our British rivers are so high. From the beginning of June to the beginning of August, vast quantities might be taken in the different rivers with very little trouble.

On the river near Reikiavik, near to the place where the salmon traps are set, are the remains of a mill, which was erected many years ago for the purposes of the woollen manufacture, which did not succeed.

About this time, Amtmand Thoranson, having been appointed one of the commissioners for managing the affairs of Iceland in the absence of the Governor, came from his residence at Eyafiord, in the north, to Reikiavik. He honoured us with a visit; and we found him to be a man of plain simple manners, but sensible, and possessed of much and accurate knowledge respecting every thing connected with Iceland; and at the same time very liberal in communicating what he knew.

Sketched by R. Bright   *E. Mitchell sc.*

SUMMIT OF SNÆFELL JOKUL.

CHAP. III.

JOURNAL OF THIRD EXCURSION.

ON the 24th of July, we attempted to leave Reikiavik, in order to visit the Geysers, and Mount Hekla; but the wind blew so strongly from the east, accompanied by very heavy rain, that we were obliged to return, after having rode a few miles. Next morning the weather seemed to be improving, and, though it was rather foggy, and heavy showers appeared all round, we departed for Thingvalla.* We were attended to the Geysers by Mr Fell, to whom we were very much obliged, and in whose society we often found great relief from the fatigues we underwent; and by Mr Jorgen Floed, private secretary to Count Trampe. In this direction the hills are low, and the country comparatively flat, and it contains much good grass, interspersed with boggy ground. We passed a deep gulley by a steep winding path, and here we once more perceived the great defect arising even in the most romantic country, from the want of trees.† On approaching Mos-

---

* The double *l* is pronounced like *tl*.

† We saw no vestiges of wood in the bogs; and were informed, that where it occurs, the trees are small. Mr Hooker, however, saw one five or six feet long, and about a foot in diameter.

fell, we saw the vapour of several hot springs at the foot of the mountain, on the south side of the valley. At that place are a church, the priest's house, and some cottages. We found the good pastor busy cutting down his grass for hay, an employment not beneath his calling, and a symptom of industry extremely pleasing. In England, the most wealthy, those of highest rank, do not disdain to amuse themselves in rural labour. With the poor priest, it was a matter of necessity to handle the scythe; yet in necessity there is often much delight; and there is little doubt that this man, and all the Icelanders occupied in the same way, were reaping the gifts of their Maker with glad and thankful hearts. On our approach, the priest left his work, and conducted us into his house. We were ushered into a very good room, and treated with abundance of milk by his wife, while he went to fasten the shoes of some of our horses. Having left Mosfell, we gradually ascended, and came upon an extensive tract of lava, which has been covered in many places to a considerable depth, with sandy soil. Our ride was now dreary and tiresome, though the path was good. We halted to refresh the horses on a small spot where there was a little grass, the principal covering of the soil being dwarf willows.

Near Thingvalla, we entered a deep and frightful fissure, called Almannagiau. This has been formed, with many others of smaller dimensions, and another large one which runs parallel to it at a considerable distance, by the sinking of the ground during some of those terrible convulsions which have shaken Iceland to its foundations. The whole rock bears marks of having been affected by fire. We came suddenly upon the brink of the precipice, and were turning aside from a scene so horrible, when we were told that we must descend. Our horses seemed prepared to carry us to the bottom, and we had already proceeded a little way, when compassion for them, not any doubt or fear for our own safety while we depended on these cautious and sure footed creatures, induced us to dismount, and allow them to find their own way. Having admired this fine scene, and the caverns which were exposed to view by the disruption of the rock, we got out of the hollow by a narrow path, and after crossing a small stream that runs into the lake, we arrived at the place of Thingvalla, which is about twenty-six miles distant from Reikiavik. Here is a small, mean, and dirty church, in which, however, we contrived to sleep. The priest is a very old man; and has had his coffin prepared and placed in the church, though his appearance did not indicate a speedy dissolution. He was the only clergyman who seemed at all to dislike our occupying the church; and he did not receive us with the same cordiality we experienced from his brethren in other parts of the country. But we were treated with civility; milk, and an excellent dish of fresh trout, fried, were soon brought to us. Trouts abound in the lake, and often go up where the water gets into the fissures of the lava; so that, by throwing a line with a baited hook into a hole at a distance from the lake, where the water is not even visible, the people frequently catch them.

The scenery about Thingvalla is romantic, but the want of wood, and the effects of subterraneous heat, combine to give an impression of dreariness. The lake is a fine sheet of water, reckoned to be about ten miles long, and from three to seven in breadth. There are two pretty large islands in the lake, called Sandey, and Nesey, composed entirely of volcanic matter. The mountains at the south end are very picturesque, and the vapour ascending from hot springs on their sides, contributes to the solemnity of the whole scene, which has been created by the most dreadful commotion, and the destruction of a country that may once have been beautiful and fertile.

Near this place was the building where the courts of justice were held formerly. Reikiavik being now the seat of government, the courts are at present held there. Why Thingvalla was originally chosen as the seat of justice, does not appear; but a town being once established, and trade carried on freely, and to a greater extent than in former times, ready recourse to the law became necessary. Though not more than ten years have elapsed since the judicial courts were transferred to Reikiavik, few remains are left to mark a spot so famous in the history of Iceland. The only building was a small wooden house in which the consultations were held, and sentence pronounced by the Stiftamtmand, or Governor. The Magistrates and people assembled on the occasion lived in tents. Those culprits who were condemned to die, were beheaded on a small island in the river Oxeraa, which here flows into the lake. The females were drowned in a deep pool below the lava, a little farther up the valley.

An ecclesiastical court used to be held at Thingvalla by the Bishop of Skalholt, attended by the Provosts and two ministers from each Syssel.*

Towards the north are several ranges of mountains, which, from the account received, and the appearances we observed, are volcanic. Among these the principal seems to be Skalbreidè, a lofty Jokul, of which description of mountains others were seen at a distance.

From Thingvalla to Skalholt, a distance of twenty-four

---

* Tingwall in Shetland, and Dingwall in Ross-shire, are evidently the same name as Thingvalla in Iceland; and were probably, in ancient times, places where justice was administered.

miles, the country is low and uninteresting; except at a place
where, from an eminence, we obtained the first view of Mount
Hekla, and the stupendous mountains beyond it.　The road
lies along the north end, and part of the east side of the lake,
where there is a considerable tract of stunted birch, and willow
trees.　The depth of the lake is said to be very great, a line
of a hundred fathoms having been sunk without bottom being
found.　After many turnings, and crossing some bogs, we came
to a low hill, round which we passed, and having got safe over
another bog, which seemed to be fully as hazardous as any
we had formerly attempted, we reached the bank of a large
river called the Brueraa, which takes its rise from the Apa
Vatn.　This lake receives the water of the surrounding bogs;
and near it, in different places, we saw vapours ascending from
hot springs.　After waiting and hallooing for some time, the
boatman arrived and carried us across the river in a very good
boat, the horses being obliged to swim.　We stopped a few
minutes at the house of the ferryman, whom, with his wife
and family, we admired exceedingly on account of their cleanliness.　Their persons, house, and the utensils in which they
brought milk to us, were all neat; but this description must
be understood comparatively.　About a mile farther on is
Skalholt, which has been erroneously denominated the capital of Iceland, in most English books on geography; but formerly it might have been entitled to this appellation as well
as any other place. Till Reikiavik became of some note, there
was nothing in Iceland that could be called a town; and it is
no wonder that the seat of a Bishop should be honoured in
preference to that of the Governor.

The situation of Skalholt is beautiful.　Towards the south,
there is a view of a noble river formed by the junction of that
discharged from Apa Vatn and the Huitaa, bounded by a

finely shaped hill in the distance; another equally picturesque hill rising on the eastern bank of the latter, and facing Skalholt. Flat meadow land, gently swelling ground, and distant mountains towards the east, among which are Eyafialla Jokul and Hekla, form altogether a magnificent amphitheatre, and compose a landscape which, even without wood, was highly gratifying to the eye.

Here we found Mr Jonson, the Lector of the school of Bessestad, for which place he was preparing to depart on our arrival. He remained, however, till the following morning, and gave us fresh cause for lamenting our having had so little of his society. A daughter of the late Bishop Finnsson, very attentively furnished us with the best provisions she had, and one of her brothers offered to shew us the way to the Geysers.

On the 27th of July we set out to visit these celebrated fountains, which are about sixteen miles to the north of Skalholt,

SKALHOLT.

The country between is varied by gentle risings, and the prospect towards the north and west is bounded by mountains,
from which there appear to have been many volcanic eruptions.   All the flat ground in this quarter is swampy, but, excepting near the lakes, it is not so soft as to occasion any risk
in travelling over it.   To the eastward of Skalholt are several
hot springs, and others rise among the low hills which we left
on the right hand in going to the Geyser. We passed one farm-
house situate on a rising ground in the midst of the bogs ; and
the weather being favourable, the people were busy making
hay ; a scene which afforded a pleasing change from dreary
solitude.   The whole of this extensive district abounds in grass,
and were draining practised, might prove a very rich pasture
country.   Further on, we found some cottages at the foot of
the mountain ; round which we turned, and came in sight
of the hill, on one side of which are the Geysers.   This hill,
which does not exceed three hundred feet in height, is separated from the mountain towards the west by a narrow stripe
of flat boggy ground, connected with that which extends over
the whole valley.   Crossing this bog, and a small river which
runs through it, we came to a farm-house at the east end of
the hill, and arrived at a place where the most wonderful and
awful effects of subterraneous heat are exhibited.

On the east side of the hill there are several banks of clay,
from some of which steam arises in different places ; and in
others there are cavities in which water boils briskly. In a few
of these cavities the water, by being mixed with clay, is thick,
and varies in colour ; but it is chiefly red and grey. Below these
banks there is a gentle and uniform slope, composed of matter which, at some distant period, has been deposited by
springs that no longer exist.   The strata or beds thus formed,
seem to have been broken by the shocks of earthquakes, par-

ticularly near the place where the great Geyser is situate.
Within a space not exceeding a quarter of a mile, there are
numerous orifices in the old incrustations, from which boiling
water and steam issue, with different degrees of force; and
at the northern extremity is the great Geyser, sufficiently
distinguishable from the others by every circumstance connected with it.   On approaching this place, it appeared that
a mount had been formed of irregular, rough looking depositions, upon the ancient regular strata, whose origin has been
similar.   The slope of the latter has caused the mount to
spread more on the east side, and the recent depositions of the
water may be traced till they coincide with them.   The perpendicular height of the mount is about seven feet, measured
from the highest part of the surface of the old depositions.
From these the matter composing the mount may be readily
distinguished, on the west side, where a disruption has taken
place.   On the top of this mount is a bason, which we found
to extend fifty-six feet in one direction, and forty-six in another.

At a quarter before three o'clock in the afternoon, when
we arrived on the spot, we found the bason full of hot water,
a little of which was running over.   Having satisfied my curiosity at this time, I went with the rest of the party to examine some other places whence we saw vapour ascending.
Above the great Geyser at a short distance, is a large irregular opening, the beauties of which it is hardly possible to
describe. The water which filled it was as clear as crystal, and
perfectly still, though nearly at the boiling point. Through it
we saw white incrustations forming a variety of figures and
cavities, to a great depth; and carrying the eye into a vast and
dark abyss, over which the incrustations formed a dome of
no great thickness; a circumstance which, though not of

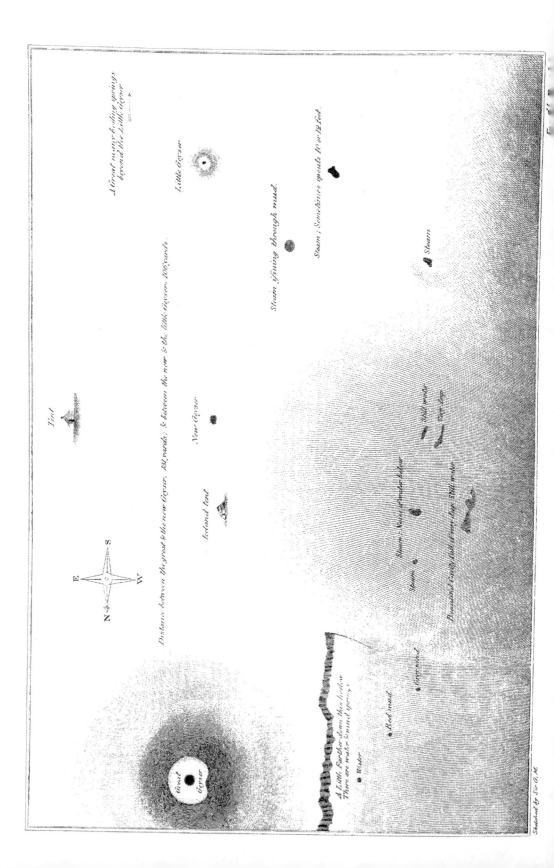

A Great water boiling springs
beyond the Little Geyser

Little Geyser

Tent

New Geyser

Distance between the great & the new Geyser, 184 yards; & between the new & the little Geyser, 100 yards.

Iceland Tent

N E W S

Steam issuing through mud.

Steam; Sometimes spouts 10 or 12 feet.

Steam

Steam

Slide mde
Dip't up

Steam · Steam of blue & White water

Steam ·

Beautiful crystalized water fro. 1501 water

Great
Geyser

A Little Farther down this hollow
There are several springs

· Water

· Red mud

· Grey mud

Sketched by Sr. G. M.

itself agreeable, contributed much to the effect of this awful scene.

Near this spot are several holes from which vapour continually rises; and from one of which a rumbling noise proceeded. This last might probably be taken for what Sir John Stanley denominates the roaring Geyser. But as the opening is not large, the beautiful cavity I have attempted to describe may have been the seat of that once furious spring.—' One of ' the most remarkable of these springs,' says Sir John, ' threw ' out a great quantity of water, and from its continual noise ' we named it the roaring Geyser. The eruptions of this foun- ' tain were incessant. The water darted out with fury every ' four or five minutes, and covered a great space of ground ' with the matter it deposited. The jets were from thirty to ' forty feet high. They were shivered into the finest particles ' of spray, and surrounded by great clouds of steam. The ' situation of this spring was eighty yards distant from the ' Geyser, on the rise of the hill.'

From the last mentioned circumstance, notwithstanding the noise emitted by the other, I am inclined to think that the first cavity I have described was the one whence these furious jets were thrown. The quantity of water that runs from it is small; and its perfect stillness at the time I saw it, formed a striking contrast with Sir J. Stanley's description. The mass of incrustations which seems to have been formed by this spring, was open in several places, and the cavities were full of water. It is probable that an earthquake has deranged the mechanism of this spring, or that the production of heat at the particular spot where it is situate, has ceased to be sufficient to produce the striking phenomena it formerly exhibited.

Having examined several other cavities, I returned to the Geyser in order to collect specimens of the incrustations on the mount.   I selected a fine mass close to the water on the brink of the bason, and had not struck many blows with my hammer, when I heard a sound like the distant discharge of a piece of ordnance, and the ground shook under me.   The sound was repeated irregularly, and rapidly ; and I had just given the alarm to my companions, who were at a little distance, when the water, after heaving several times, suddenly rose in a large column, accompanied by clouds of steam, from the middle of the bason, to the height of ten or twelve feet.   The column seemed as if it burst, and sinking down it produced a wave which caused the water to overflow the bason in considerable quantity.   The water having reached my feet, I was under the necessity of retreating, but I kept my eye fixed on what was going on.   After the first propulsion, the water was thrown up again to the height of about fifteen feet.   There was now a succession of jets to the number of eighteen, none of which appeared to me to exceed fifty feet in height ; they lasted about five minutes.   Though the wind blew strongly, yet the clouds of vapour were so dense, that after the first two jets, I could only see the highest part of the spray, and some of it that was occasionally thrown out sideways.   After the last jet, which was the most furious, the water suddenly left the bason, and sunk into a pipe in the centre.   The heat of the bottom of the bason soon made it dry, and the wind blew aside the vapour almost immediately after the spouting ceased.   We lost no time in entering the bason to examine the pipe, into which the water had sunk about ten feet, and appeared to be rising slowly.   The diameter of the pipe, or rather pit, is ten feet, but near the top it widens to sixteen feet.   The section, which is taken across the longest diameter of the bason, gives a distinct idea of the whole structure of the external part of this wonderful apparatus.   The perpendicular depth of the bason is three feet ; that of the pipe being somewhat more than sixty feet, though there may be some inaccessible hollows which extend to a much greater depth.

After the water had descended into the pipe, there was no appearance of any vapour issuing from it, till it had reached the mouth, when a little was visible. Even when the bason was full, the quantity of vapour was far from being so great as might have been expected to proceed from so large a surface of hot water. At five minutes before six o'clock it boiled a little, and continued to do so at intervals. Having thrown

Plan & Section of the bason of the Great Geyser.

a stone into the water while it was perfectly still, I observed that an ebullition immediately took place till the stone reached the bottom.　I then requested all the party to provide themselves with large stones, and to throw them into the pipe, on a signal I should give, when the water was still.　When the stones were thrown in, a violent ebullition instantly followed; and this escape of steam on agitation, may serve to assist a theory of the phenomena.

At twenty-nine minutes past six o'clock the pipe was full; and the water being within reach, its temperature was found to be 209°.　At twenty minutes before seven I looked into the bason, and it was then hardly one-fourth full.　The water was gently moved; and in some little hollows of the bottom of the bason it had the appearance of ebbing and flowing. About five minutes after, while I was collecting specimens on the edge of the bason, and expecting nothing, three jets took place, none of which exceeded thirty feet in height.　In the same manner, at a quarter past eight o'clock, jets were thrown up repeatedly during the space of three minutes, one of which was about forty feet high.　After these casual jets the water did not sink, but remained, filling the bason about three-fourths.　Sir John Stanley mentions his having been surprised by similar unexpected jets; and I must take this opportunity of advising travellers who may wish to see the Geyser, not to be rash in going into the bason while the water is rising, as an opportunity of safely gratifying their curiosity will always occur immediately after every great exertion of the fountain.

We pitched our tent at the distance of about one hundred yards from the Geyser, and having arranged matters so that a regular watch might be kept during the night, I went to my station at eleven o'clock, and my companions lay down to sleep.　About ten minutes before twelve, I heard subterraneous discharges, and waked my friends. The water in the bason was greatly agitated, and flowed over; but there was no jet. The same occurred at half past two. At five minutes past four on Saturday morning an alarm was given by Mr Bright. As I lay next the door of the tent, I instantly drew aside the canvas, when at a distance of little more than fifty yards, a most extraordinary and magnificent appearance presented itself. From a place we had not before noticed, we saw water thrown up, and steam issuing with a tremendous noise. There was little water; but the force with which the steam escaped, produced a white column of spray and vapour at least sixty feet high. We enjoyed this astonishing and beautiful sight till seven o'clock, when it gradually disappeared. This fountain we immediately conjectured to be what has been called, by Sir John Stanley, the New Geyser.

We were occupied this morning in examining the environs of the Geysers; and at every step received some new gratification. Following the channel which has been formed by the water escaping from the great bason during the eruptions, we found some beautiful and delicate petrifactions. The leaves of birch and willow were seen converted into white stone, and in the most perfect state of preservation; every minute fibre being entire. Grass and rushes were in the same state, and also masses of peat. In order to preserve specimens so rare and elegant, we brought away large masses, and broke them up after our return to Britain; by which means we have formed very rich collections; though many fine specimens were destroyed in carrying them to Reikiavik. On the outside of the mount of the Geyser, the depositions, owing to the splashing of the water, are rough, and have been justly compared to the heads of cauliflowers. They are of a yellowish brown colour, and are arranged round the mount somewhat like a circular flight of steps. The inside of the bason is comparatively smooth; and the matter forming it is more compact and dense than the exterior crust; and, when polished, is not devoid of beauty, being of a grey colour, mottled with black and white spots and streaks. The white incrustation formed by the water of the beautiful cavity before described, had taken a very curious form at the edge of the water, very much resembling the capital of a Gothic column. We were so rapacious here, that I believe we did not leave a single specimen which we could reach; and even scalded our fingers in our eagerness to obtain them. We found the process of petrifaction in all its stages; and procured some specimens in which the grass was yet alive and fresh, while the deposition of the silicious matter was going on around it. These were found in places at a little distance from the cavity, where the water running from it had become cold.

About a hundred yards from the Great Geyser towards the north, in the cleft where the disruption already mentioned had taken place, and which has probably been formed by an earthquake, are banks of clay, in which there are several small basons full of boiling mud. The mud is thin, and tastes strongly of sulphate of alumina, of which we observed many films attached to the clay, which seems to have been forced up from below, through fissures in the ancient incrustations. The clay contains also iron pyrites; the decomposition of which has given it very rich colours. Almost directly above this place, under the rock at the top of the hill, are several orifices, from which steam rushes; and there are some slight appearances of sulphur. Almost the whole of this side of the hill is composed of incrustations and clay.

The depositions of the present and former springs are visible to a great extent, about half a mile in every direction ; and, from their great thickness in many places, it is probable that they are spread under the surface now covered with grass and water, to a very considerable distance. About half a mile up the rivulet, in the direction of Haukardál, where there is a church, another hot spring appears, which deposits silicious matter. From thence we obtained one of the most curious specimens we collected ; it almost perfectly resembles opal. I mention the situation of this spring to shew the probability that the extent of the matter, which may for ages have been collecting, is very great ; and its depth, from what is seen in the cleft near the Geyser, where it is visible to the thickness of ten or twelve feet, is probably also very considerable.

It is somewhat curious, that no particular notice has been taken by the early Icelandic authors of this, the most remarkable spot in all the island. Though hot springs are without number, and occur in every part of the country, and may be regarded with indifference, yet the Geysers must have been remarkable at all times ; for the extent of the old incrustations shews them to have been deposited by springs of no ordinary dimensions. They are, it is true, on the verge of that vast district of uninhabited and desolate country which forms the interior of Iceland. In looking around as we approached the place, nothing was seen but rugged mountains, far extended swamps, and frightful Jokuls rearing their frozen summits to the sky. Nothing in this direction seemed to invite the curiosity or enterprise of people, already accustomed to the horrors of volcanic eruptions, and fully aware that their only sure subsistence was to be derived from the sea. The indifferent and casual manner in which the Geysers are mentioned by Arngrim

Jonas, shews this want of curiosity even among the learned of the Icelanders. He speaks of some great springs near Haukardal, to the north of Skalholt, which he had never himself seen, but of which he had heard that they deposited incrustations, and changed vegetable matter into stone. * At the present day, the number of the natives who have visited these springs is comparatively very small; and, by those who live near them, their extraordinary operations constantly going on, are regarded with the same eye as the most common and indifferent appearances of nature. Towards the northeast, and east, the country is low; the only elevated ground that appears towards the south-east being the summits of Hekla, and Eyafialla Jokul. Several Jokuls break the view towards the north; and we remarked one mountain which had several rugged and peaked summits soaring to a great elevation.

However strongly the feelings excited by the productions of the springs, and by the appearance of the surrounding country, were impressed upon us, we often turned anxiously towards the Geysers, longing for a repetition of their wonderful operations. To them all our wishes and hopes were directed; and we felt as if our eyes could never tire of beholding, nor our minds weary of contemplating them. The descriptions we had read, and the ideas we had formed of their grandeur, were all lost in the amazement excited on their being actually before us; and, though I may perhaps raise their attributes in the estimation of the reader, I am satisfied that I cannot convey the slight-

---

* Brev. Com. de Island. Hakluyt's Voyages, London 1809, vol. 1. p. 596.— Saxo Grammaticus, in his preface to the history of Denmark, slightly notices the Geysers; and this is the earliest account we have. It proves them to have existed at least six hundred years.

est idea of the mingled raptures of wonder, admiration, and terror, with which our breasts were filled; nor do I fear that any conception which may arise of the astonishing effect of the Geysers, will leave the traveller disappointed, who trusts himself to the tempestuous ocean, and braves fatigue, in order to visit what must be reckoned among the greatest wonders of the world.

After yielding a little to impatience, we were gratified by symptoms of commotion in the Great Geyser. At three minutes before two o'clock, we again heard subterraneous discharges, and the water flowed over the edge of the bason; but no jet took place. The same happened at twenty-five minutes past five o'clock, and at five minutes before seven. At thirty-five minutes past eight, it boiled over again, and immediately the new Geyser began to play, and continued till a quarter past nine. This Geyser gives no warning before it spouts, and it is therefore necessary to be cautious in looking down the pipe, unless it is known what time has elapsed since the preceding jet. While the spray and vapour are rushing out, one may approach with perfect safety, and stand quite close to the very brink of the pipe on the windward side. The pipe is nine feet in diameter, not perfectly round, and rough and uneven within.

Having been busily engaged in packing our specimens, and being somewhat tired, we went to sleep a little earlier than usual. We lay with our clothes on, separated from the ground by sheep-skins and a rug, in order that we might start up at a moment's notice. Mr Fell and Mr Floed had left us to return to Reikiavik; and we had soon cause to regret that they had departed before the next eruption of the Great Geyser took place. On lying down, we could not sleep more than a minute or two at a time; our anxiety causing us of-

ten to raise our heads to listen. At last the joyful sound
struck my ears; and I started up with a shout, at the same
moment when our guides, who were sleeping in their Iceland
tent at a short distance opposite to us, jumped up in their
shirts, and hallooed to us. In an instant we were within
sight of the Geyser; the discharges continuing, being more
frequent and louder than before, and resembling the distant
firing of artillery from a ship at sea. This happened at half
past eleven o'clock; at which time, though the sky was cloudy,
the light was more than sufficient for shewing the Geyser; but
it was of that degree of faintness which rendered a gloomy country still more dismal. Such a midnight scene as was now before
us, can seldom be witnessed. Here description fails altogether. The Geyser did not disappoint us, and seemed as if it
was exerting itself to exhibit all its glory on the eve of our
departure. It raged furiously, and threw up a succession of
magnificent jets, the highest of which was at least ninety feet.
At this time I took the sketch from which the engraving is
made: but no drawing, no engraving, can possibly convey
any idea of the noise and velocity of the jets, nor of the swift
rolling of the clouds of vapour, which were hurled, one over
another, with amazing rapidity.

After this great exertion, the water, as before, sunk into
the pipe, leaving the bason empty. At seven minutes before
seven o'clock on Sunday morning, the Geyser boiled over;
and again at twenty minutes past nine; and this was the last
time we saw it in motion.*

* In Olafson's and Paulson's travels, we have a description of the Geyser, in
which the height of the jet is stated at three hundred and sixty feet. This, making
every allowance for deception, is certainly an exaggeration, since in subsequent
observations made at distant periods, we find a striking uniformity. The heights
observed at the time Sir Joseph Banks visited Iceland in the year 1772, are stated

Sketched by Sir G. M.   Engr⁴ by E.Mitchell

ERUPTION
of the
GREAT GEYSER

NEW GEYSER

Published by A. Constable & C.º Edinburgh.

At thirty-two minutes past nine, the new Geyser began its operations by throwing the water out of the pipe at three or four short jets, and then some longer ones. As soon as the bulk of the water was thrown out, the steam rushed up with amazing force, and a loud thundering noise, tossing the water frequently to a height of at least seventy feet. So very great was the force of the steam, that although a brisk gale of wind was blowing against it, the column of vapour remained as perpendicular as it is represented in the engraving. It proceeded in this magnificent play for more than half an hour, during which time I had an opportunity of taking a correct sketch of this beautiful Geyser. A light shower fell from the vapour, which has been attempted to be expressed; but the imitation is very far short of the fine effect it produced. Sir John Stanley saw it throw up water to the height of one hundred and thirty-two feet. When stones are dropped into the pipe while the steam is rushing out, they are immediately thrown up, and are commonly broken into fragments, some of which are projected to an astonishing height.

This Geyser, we were told, had formerly been a comparatively insignificant spring like many which we saw around. There is no bason round the pipe, but there are some remains of incrustations on its brink, similar to those round several of the smaller springs. The water constantly boils violently, about

by Von Troil, to have been from six to ninety-two feet. Sir John Stanley mentions the highest jet to have been ninety-six feet. He visited the Geyser in the year 1789. I have stated the heights as varying from ten to at least ninety feet. From these observations it appears that the great Geyser has not failed in magnificence after the lapse of thirty-eight years. Sir John Stanley mentions that, as the jets rose out of the bason, they reflected by their density the most brilliant blue; and that in certain shades the colour was green. We did not observe any thing of this kind, which probably depended on the position of the spectator, and the brightness of the sun, which hardly shone while we were near the springs.

twenty feet below the mouth of the pipe ; but no subterraneous discharges take place to announce its operations ; and this
circumstance seems to render a different theory from that of
the great Geyser, necessary for explaining the phenomena.*

Each spring seems to have its own reservoirs, and its own
mechanism distinct from the others. There is a small Geyser
about a hundred yards distant from the new one, as it was
called by Sir John Stanley, the phenomena of which I think
worthy of being described, though after viewing the great Geyser, there is nothing wonderful in them. The description, however, may serve to shew what a singular range of cavities and
pipes, must exist under a small extent of surface, in order to produce the extraordinary effects I have detailed. This little Geyser, for so I shall call it, first attracted my attention while watch-

* This Geyser seems to have undergone a considerable change since the time
of the expedition to Iceland undertaken by Sir John Stanley. ' Its pipe,' says Sir
John, ' is formed with equal regularity as that of the great Geyser, and is six feet
' ten inches in diameter. It does not open into a bason, but it is nearly surround-
' ed by a rim or wall two feet high. After each eruption the pipe is emptied, and
' the water returns gradually into it, as into that of the old Geyser. During three
' hours nearly that the pipe is filling, the partial eruptions happen seldom, and do
' not rise very high ; but the water boils the whole time, and often with great vio-
' lence.' Sir John further informs us in a note, that before the month of June
1789, the year he visited Iceland, ' this spring had not played with any great
' degree of violence, at least for a considerable time. (Indeed the formation of the
' pipe will not allow us to suppose, that its eruptions had at no former period
' been violent.) But in the month of June, this quarter of Iceland had suffered
' some very severe shocks of an earthquake ; and it is not unlikely, that many of
' the cavities communicating with the bottom of the pipe had been then enlarged,
' and new sources of water opened into them.'

Our author also says, that the eruptions of the new Geyser resembled those of
the great one, consisting of several jets succeeding each other rapidly. It will be
seen from the theory I have formed of the phenomena, that the change has been
occasioned by the supply of water to the pipe having become less, while the great
reservoir of water, subject to sudden productions of heat, remains the same.

ing the great one on the night of the 27th. At that time I was
not at liberty to go to it, but I afterwards had an opportunity of
noting down its movements. There are several openings near
it, from one of which water occasionally spouts to the height
of ten or twelve feet, and a number of holes whence steam
rushes. The little Geyser does not throw its water above four
or five feet high, but its phenomena are similar to those of the
great one, whose grandeur it may one day rival. The pipe of
this Geyser, at the depth of eight feet, is of an irregular shape,
three feet by two in width, and opens like a funnel into a shal-
low bason about ten feet in diameter. When I went to examine it, the water was sinking into the pipe, and was at the
same time very much agitated. After it had reached its greatest depth, which was about seven feet, it remained quiet for a
considerable time, and I took the opportunity of going to examine some springs near it, which boil constantly. On my return I found all quiet, but in a very short time the water became
disturbed, and before it began to rise I counted nineteen distinct bursts of steam. It then rose gently, a burst of steam taking place at intervals, and at each burst the water rose a few
inches : sometimes it rose almost a foot, and then sunk again.
Thus it proceeded for some time, till at last the ebullition became more constant, and the water rose faster. At length there
were violent bursts of steam, and the water rapidly approached,
being thrown up in short jets; and at this time I felt a distinct
trembling of the ground, but heard no subterraneous noise.
The bason was now filled, and I left the spring throwing up jets
at short intervals. The little Geyser is pretty regular, and
continues its operations about an hour. It is probable that
the water of all these springs is of the same nature with that
of the great Geyser; those, however, that are muddy, are
different in respect to the ingredients contained in the water.

From all the circumstances I have mentioned, it is evident, that a vast variety of cavities exists in the space from which the water issues in so many different ways. In forming a theory of the phenomena, therefore, any kind of mechanism may be supposed, that, by means of steam, is capable of producing such effects. I shall submit to the reader a theory, which I formed on the spot while the phenomena were before me.

Were the appearances regular in duration, and the intervals between the jets always equal, it would not be difficult to construct an apparatus which would exhibit them with precision; but in both respects, as well as in the degree of violence, there is great irregularity. From whatever source the heat proceeds, whether from the combustion of beds of coal, the decomposition of pyrites, or any other cause, there can be no hesitation in granting the possibility of a greater quantity of heat being evolved at one time than at another; or of the heat remaining steady at intervals. It is not only possible, but very probable, that the wonders of the Geysers are caused by sudden productions of heat. By such a supposition they may easily be explained with the help of an extremely simple apparatus; but without it, a very complicated system of pipes and cavities, and perhaps, too, of valves, will be necessary.

A column of water is suspended in a pipe, by the expansive force of steam confined in cavities under the surface. An additional quantity of steam can only be produced by more heat being evolved. When heat is suddenly evolved, and elastic vapour suddenly produced, we can at once account for explosions accompanied by noises. The accumulation of steam will cause agitation in the column of water, and a farther production of vapour. The pressure of the column will be overcome, and the steam escaping, will force the water upwards along with it. Let us suppose a cavity C communicating with the pipe PQ, filled with boiling water to the height AB, and that the steam above this line is confined, so that it sustains the water to the height P. If we suppose a sudden addition of heat to be applied under the cavity C, a quantity of steam will be produced, which, owing to the great pressure, will be evolved in starts, causing the noises like discharges of artillery, and the shaking of the ground. The pressure being now greatly increased, the water must rise out of the pipe; an oscillation is produced; the water is pressed downwards from A to Q, and the steam having now room to escape, darts upwards, breaking through the column, and carrying

THEORY OF THE GEYSERS

along with it a great part of the water. As long as the extraordinary supply of steam continues, these oscillations and jets will go on. But at every jet some of the water is thrown over the bason, and a considerable quantity runs out of it. The pressure is thus diminished; the steam plays more and more powerfully, till at last a forcible jet takes place, a prodigious quantity of steam escapes, and the remaining water sinks into the pipe. This explanation, however, is not quite complete, as it requires the production of the extraordinary quantity of heat to cease the moment after the last jet, which is in general the most violent. For though we may suppose the whole of the water to have been expelled, which it is not, unless the accumulation of heat was stopped at the very instant of the last and strongest jet, we should find steam rushing from the pipe. But it uniformly happens, that after the last jet all becomes perfectly quiet; and this uniformity we know has continued since the time that Sir Joseph Banks saw the Geyser. It may be as allowable to infer a sudden cessation, as a sudden production of heat. But it is a very curious circumstance, that the heat should continue to produce steam, just as long as the pressure of the water continued considerable, and that it should cease the instant that the pressure is removed. I think that this last fact may be explained by the diminution of temperature occasioned by the escape of the vast body of vapour which accompanies the last effort.

The same configuration of a cavity will explain the phenomena of the new Geyser satisfactorily. We have only to suppose that there is a smaller supply of water, and that instead of a column reaching to P, pressing against the steam in the reservoir, the water reaches only a little way, if at all, above the level of that within the cavity. Things being thus adjusted, a sudden evolution of heat causes no explosive escape of steam, as there is but little pressure to overcome. The instant that an extraordinary supply of vapour is brought into action, part of it passes through the water, and carries some up with it. This is repeated, more and more water being thrown out, till at last there is no interruption, and the steam rushes forth with fury and noise, till, the heat abating, the force of the jet is gradually weakened, at last exhausted, and the phenomena cease.

Another way of accounting for the operations of these extraordinary fountains, which appears equally plausible with what I have just stated, has been suggested. It requires the existence of a strongly heated surface free from water; and also that of a small subterraneous fountain, operating like the little Geyser I have described, expelling its water occasionally, so that it flows over the heated surface, by which means an additional quantity of steam may be temporarily produced. But this explanation is perhaps more deficient than the other; for if we suppose the water which is to be suddenly converted into elastic vapour to be furnished from a small subterraneous fountain, the operations of that fountain must be explained, and the same difficulties that remain to be overcome in the case of the Geyser, meet us in this; as they must also do in whatever mode we may suppose water to be supplied.

About a mile from a place called Husavik, in the north of Iceland, is the Uxahver, (ox spring,*) which is more regular, and is said nearly to equal the Geyser in the magnificence of its operations.

We returned to Skalholt on the 29th. This place, during many centuries, was the residence of one of the Bishops.

---

* It is said that this name was given to it from the circumstance of an ox having accidentally fallen into it, and been boiled alive.

On the death of the last Bishop of Skalholt, the learned John
Finnsson (son to Finnur Jonson, the author of the Ecclesiastical History of Iceland), and that of the Bishop of Hoolum,
which occurred soon afterwards, application was made to the
court of Denmark to sanction a union of the two sees. This
was granted ; and the title of Bishop of Iceland was first conferred on our friend Geir Vidalin, who still holds that dignity.

The church of Skalholt is a neat small building of wood,
erected on the site of the former one, which was taken down
about six years ago. That, the Bishop's house, and a few
cottages, constituted the supposed capital of Iceland. There
is a very good picture of the late Bishop, painted at Copenhagen, hanging up in the church ; and, on the floor of the space
before the altar, is a beautiful white marble slab, inscribed to
his memory by the present Chief Justice. The font, and pulpit
of the old church, which are curiously carved and painted, are
in the present building. Near the door of the church are some
epitaphs carved on stones ; none, however, of an old date.
The following is a specimen of them :

      ' Priscis nobilibus creatus olim,
      ' Virtutisque patrum beatus hæres,
      ' Dilecti genitoris ipsa imago,
      ' Et desiderium piæ parentis,
      ' Communisque amor omnium bonorum,
      ' Quos secum sociavit alma fides,
      ' Et candor sibi nescius fraudis ;
      ' Eheu! precipiti nimis ruina,
      ' Mortis vulnifico peremptus œstro,
      ' Post vitæ decies duos Decembre,
      ' Mæstæ Thorstenides domus levamen
      ' Eggertus jacet hac sepultus in urnâ ;
        ' Amoris ergo fecit, Johannes Vidalinus.'

The easiest route from the southern to the northern parts of Iceland, is by the way of Skalholt. To Skagastrand, the nearest road is by Thingvalla and Kalmanstunga; which last place is situate a little to the north-east of Reikholt. There is another route through the eastern part of the Borgarfiord Syssel. The northern division of the island is usually called the Nordland; the others, Östland, Sudland, and Vestland. We had no time for exploring the Nordland: indeed, to travel through it would take up a whole summer; which, probably, might be spent in that quarter by a naturalist, with much profit and pleasure. Respecting that part of the country, we obtained some information from Amtmand Thoranson, which I may take the present opportunity of communicating to the reader.

A journey to the northern part of the island, from Skalholt or Thingvalla, generally occupies three or four days. The interior of Iceland, an extent of perhaps not less than forty thousand square miles, is a dreary, inhospitable waste, without a single human habitation, and almost entirely unknown to the natives themselves. Through more or less of this desert, a traveller going to the northern coast, or coming towards the south, must necessarily pass; and it is no wonder that it has become customary to travel through it night and day without stopping.

The greatest proportion of the Nordland is the property of the farmers who occupy it. Some of it belongs to the church; and part to the crown. The lands which belonged to the school of Hoolum, were sold to the farmers of the district at the time when the school was removed.*

The population is confined to the shores of the Fiords;

* For an account of the mode in which the land in general is possessed, see the chapter on Rural Affairs.

along which, and up the valleys, an extent, in many places, of above twenty miles, is occupied. In the four Syssels composing the Nordland, viz. Hunavatn, Hegranes, Vadlè, and Thingöes Syssels, there are about 12,000 inhabitants. The Vadlè, in which Eyafiord is situate, is the most populous in proportion to its extent, containing about 3000 people.

The harbour of Eyafiord is the best on the northern coast. At this place there are three wooden dwellinghouses, and four storehouses. Before the war, three ships used to be laden every year at this port, with tallow, wool, woollen goods, salted mutton, sheep-skins, &c.; the particulars of which will be found in the tables in the chapter relating to the state of commerce.

Except during the month of June, and the beginning of July, and in September and October, there are no cod-fish nor haddocks found in the Fiord; and it is only at some distance out at sea that the fish are taken at these times. The months of April and May are chiefly occupied in taking the Houkal, the same species of shark we saw at Stappen, which I believe to be the *Squalus Maximus,* or Basking Shark; known in some places by the name of Sun, or Sail-fish.* The shark fishery is principally carried on at Siglifiord, a place situate on a small bay, about fifty miles north-west from Eyafiord.

At the last mentioned place, herrings appear in vast shoals during the months of June and July; and are taken by means of Seine nets at the upper extremity of the Fiord. We heard, as no uncommon occurrence, that one hundred and fifty barrels of herrings are taken at a single hawl of a net. The fish are sold to the farmers in the neighbourhood for one rixdollar a barrel.

* See page 176.

Several rivers of considerable size run into the Eyafiord; but the courses of the rivers in this part of the island have never been traced to any great distance up the country.

Hofsos, and Skagastrand, are the next most considerable places of trade on the northern coast of Iceland. The former, situate on the western side of the Skagafiord, is a very bad harbour, and only one merchant has settled there. This Fiord receives the waters of two rivers; one of which is as large as the Huitaa of Borgarfiord, and is called Kolbeinsdalsaa. At the head of the Skagafiord, not far from Hofsos, is Hoolum, which, until the close of the last century, was the seat of one of the Bishops of Iceland. A public school was also established there; but now the place consists only of a few cottages; and, in its present state, contains nothing particularly worthy of notice. Skagastrand is situate on the western side of the large promontory which bounds the Skagafiord. It is a bad harbour; and, towards the end of September, is particularly unsafe, on account of its being exposed to the north wind, and floating ice. This place used to furnish a cargo for one vessel every year; but, since the commencement of the war, I believe the arrival of even one ship has not been regular.

Husavik is the only commercial station which remains to be mentioned. It lies to the north-east of Eyafiord, on the Skialfandèfiord, which receives the waters of a large and rapid river called the Skialfandèfliot, and also a river called Laxaa, which flows from the lake Myvatn. There is no good fishing at Husavik; but a great many seals are caught during the winter. Eider-ducks are very abundant on the coast.

Throughout the whole of the northern districts, the pasture is very good, though not so rich as that of Borgarfiord,

and some other parts of the Sudland.   It is better calculated for sheep than cows; but it is always necessary to feed the
sheep with hay during the greatest part of the winter. Labourers being scarce, and the summer short, numbers of people
go from other parts of the country, particularly from the
Guldbringè Syssel, to assist in securing the hay crop. The
Nordland is the only part of the country where goats are kept.

The Fiords on the north coast are frozen over every winter; but the open sea only in the most severe seasons. Float-
ing ice frequently comes upon the coast, both during winter
and summer.   Very little ice is ever seen on the western side
of the island, notwithstanding its proximity to Greenland;
but, on the eastern shores, it comes often farther south than
Berufiord, which it completely shut up about the middle of
May this year.

The prevailing wind proceeds from the north.   Snow generally begins to fall in large quantities about the end of September, and remains on the ground till the middle of May,
and sometimes much later.   The greatest degree of cold which
Amtmand Thoranson recollected to have observed, was about
minus 35°.   Last winter, the thermometer, about the end of
January, stood at *minus* 30°, for several weeks. The greatest heat of summer he had observed, was about 70°.   These
degrees are those of Fahrenheit: the thermometer used by
the Amtmand was that of Reaumur.

We left Skalholt on the 30th, in order to visit mount Hekla.
On approaching this mountain from the westward, it does not
appear remarkable; and has nothing to distinguish it among
the surrounding mountains, some of which are much higher,
and more picturesque.   It has three distinct summits; but
they are not much elevated above the body of the mountain.   After passing some dangerous bogs, we came to the

noble river Huitaa, which derives its name from the same cause that gives that appellation to the river of Borgarfiord. It is not, however, equally white, being more muddy, and somewhat of the colour of the Thames as it passes through London. We crossed this river in a boat, having made the horses swim over before us. We now travelled over a flat country, sometimes through bogs, sometimes among sand-banks, and occasionally on good dry turf, till we approached a farm-house, called Reikum, when we came upon lava. Indeed, we had been travelling over a particular species of this substance almost all the way; as shall be more particularly explained in the chapter on mineralogy. This place of Reikum, derives its name from a hot spring near it, which made its first appearance during an earthquake in the year 1789. Here we were informed that there was no boat at the usual place of crossing the Thiorsaa; and that we must go down the river to a place called Eyalstadir. The farmer having agreed to be our guide, he mounted his horse, and we proceeded over an extensive flat, in some parts boggy, and in others rough with lava. This is part of an extensive plain, the opening of which, from the sea, reaches from Eyarback to the Markarfliot, a distance of about thirty-six miles; and it extends a great way to the north, a number of low hills and ridges rising in it here and there. On many parts of this great flat, there has been a large deposition of loose sand, the spreading of which by the wind has done considerable mischief, and is still continuing to be injurious. This district is by far the richest in pasture that we saw during our stay in Iceland.

On our arrival at Eyalstadir, after a tedious ride, we found the Thiorsaa to be a very large muddy river; and we had some fears lest our horses should not be able to swim across, as at this place the river is about a quarter of a mile broad. The ferry boat which was destined for us was large enough for three persons, and that number was sufficient to keep the gunwale close to the water, so that the slightest motion to the right or left must have filled it in a moment; but although the wind was blowing strong against the current, we were carried in safety to the other side. In this little bark the ferryman sat with the utmost composure, and rowed across with two horses tied to the stern. The animals seemed to be perfectly aware, that if they did not swim steadily, their fate and that of the boat would be the same. We were told, that it was no uncommon thing to see this man cross in his little boat with four or five horses at a time; and that, when the people are going to Reikiavik with their goods, he frequently ferries over several hundred horses in one day, and several thousands in the course of the season. The fare for crossing the ferry was a mere trifle, a few of the small coins, called skillings, which are equivalent to our halfpence ; but I gave the man a dollar, telling him that I greatly admired his skill, which made him very happy. From the opposite bank the place of Kalfholt is about two miles distant; perhaps not so much in a direct line, but a bog intervenes which occasions the route to be circuitous. We there found a church, which we entered and occupied without much ceremony.

The Priest presented himself in a dress through which we could not possibly descry the slightest tokens of the clerical character. On his head was a greasy woollen cap that had probably once been white, elevated like a sugar loaf. A short jacket and breeches of the same stuff adorned his body, and his legs were covered with coarse black stockings reaching above the knee. His father was dressed in the same mode at our arrival; but he had a small white beard which rendered his figure somewhat more venerable than that of his son. Both of them, after some time, put on their best clothes. The Priest accompanied us, next day, a part of the road to Storuvellir, where the Provost of the district lives. The road lay through low hills and bogs; and on our way, we saw some young Ptarmigans in a place not suited to the general habits of that bird in our own country, where they frequent stony places on the tops of mountains.

We now came into the plain from which Hekla rises; but we had no view of the mountain as we approached, as it was covered with clouds. We passed through lava which had been exposed to view by the blowing of the sand that covers so great an extent of this country. Storuvellir is situate in the midst of this tract; and round it there is a great deal of excellent grass. The Provost had a large stock of old hay, which, without any report in his favour as a good rural economist, would have been a sufficient proof of his merit. He received us with great kindness, but annoyed us a little by the excess of his attention or curiosity; so that we were at length obliged to barricade against him that part of his own church which we occupied. The Provost is reputed rich; and it is said that he has made his fortune entirely by his good management of his farm, on which we saw a considerable flock of sheep, and some cows. The winter provision of stock fish kept in the church, was no advantage to its atmosphere, which can undergo little purification; for the windows of the churches, in general, did not seem to be made to open.

The weather being still foggy, we could not see Hekla as we approached it. On the 1st August, we passed through lava of the same description as that pervading every part of this flat district we had travelled over. We crossed the river called Wester Rangaa, the water of which is perfectly transpa-

rent, and flows along the foot of Hekla, on the west side.
The bed of this river is very remarkable, being formed of rugged masses of lava, which being here and there elevated in
peaks, cause great rapidity in the stream. Owing to the clefts
in the lava, it is very dangerous to attempt crossing the river
at this place without a guide. The Provost was very obliging,
and gave us instructions in what manner to follow him across;
and as soon as he saw us safe, he took leave and returned to Storuvellir. On the end of a long ridge, running nearly north and
south, close to the base of Hekla, is a small farm, called Naifurholt. Here we halted; and the grass having been recently
mowed, we found an admirable station for our tent. The
cottager, Jon Brandtson, whom we found to be the most
obliging and active Icelander we had met with, was not long
in making his appearance, and administering to our wants.
He told us that he could guide us to a place where there was
a great quantity of Iceland agate, or obsidian; a piece of
information the most welcome we had for a long time received.
That substance was one of the chief objects of our mineralogical researches; and not having before met with it, we had
given up all thoughts of seeing it in its place; when honest
Brandtson, observing us employed with the minerals we had
already collected, brought a mass of obsidian to us, and relieved us from a most severe mortification. He told us that
the place where he had seen great quantities of that substance,
was situate near the Torfa Jokul, and distant a long day's
journey from Naifurholt. Our time was now limited; but
we had no hesitation in making up our minds to endure
considerable fatigue, in order to visit a spot so interesting to
us; and even, in case of need, to relinquish the project of
ascending the far-famed Hekla; and, accordingly, we resolved to undertake this expedition next day, as the weather

did not appear favourable for the ascent of the mountain. Having made preparations for both adventures, we went early to bed.

Finding, at two o'clock in the morning, that Hekla was entirely obscured by fog, we mounted our horses ; and each taking a spare one, we departed, Brandtson leading the way. In the course of our journey, as the clouds dispersed, we had different views of the mountain, which is completely covered with slags. Few streams of lava seem to have taken their course on the west and north sides ; indeed, we saw distinctly only one. Hekla, like Snæfell Jokul, terminates a long group of comparatively low hills. Viewed from the westward, when Eyafialla, Tinfialla, and other Jokuls beyond it are in sight, the mountain makes no great figure ; but, from the east and south, it appears to rise out of the hills surrounding it, and is very conspicuous.

Our road towards the obsidian lay between the Rangaa and the Thiorsaa, the course of which is nearly from north-east to south-west. This last mentioned river here rolls its large turbid stream over rugged masses of lava rising abruptly from its bed ; and in its efforts to overcome the obstruction thus occasioned, dashes among the rocks, forming impetuous rapids and falls. Great quantities of alluvial sand appeared disposed in strata in different parts of the country through which we passed ; and in other places there were extensive accumulations of volcanic sand composed of pumice and cinders.

Having recrossed the Rangaa, we entered a wide plain, bounded by Hekla and the adjacent mountains on one side, and by a lofty, precipitous, and broken ridge on the other, the surface being completely covered with lava, sand, or minute fragments of scoriæ and pumice. The lava which has

flowed over the plain, the termination of which we could not
see, appears to have been remarkably rough, from the numerous sharp pointed masses rising out of the loose sand and
slags, the accumulation of which could alone have rendered
it passable. We travelled about fourteen miles, judging of
the distance by the time our journey occupied, and then halted at the foot of a large mass of lava, and changed our horses;
stopping no longer than was necessary for shifting our saddles. The subsequent part of our route, though still through
an extremely desolate country, was rendered more easy by
the absence of lava, and somewhat less forbidding by the appearance of thinly scattered vegetation on the valleys, and on
the sides of some of the hills. Ere long we found ourselves
inclosed in a hollow among the mountains, from which there
was no apparent outlet; but following the steps of our guide,
we pursued a winding course, passing through a number of
rivulets of very thick muddy water which proceeded from under the snow on the mountains.*

As we went along we observed several craters in low situations, from which flame and ejected matter had proceeded
during the convulsions to which this part of the island has
been particularly subjected. After having advanced about
fifteen miles farther, and traversed a part of that immense
waste which forms the interior of Iceland, and is partially
known only to those who go in search of strayed sheep, we
descended by a dangerous path into a small valley, having a
small lake in one corner, and the extremity opposite to us
bounded by a perpendicular face of rock resembling a stream
of lava in its broken and rugged appearance. While we ad-

* Extensive masses of clay are not uncommon in volcanic districts, especially
where there are hot springs, or where such have existed; and this may account for
the peculiar colour of most of the large rivers of Iceland.

vanced, the sun suddenly broke through the clouds, and the brilliant reflection of his beams from different parts of this supposed lava, as if from a surface of glass, delighted us by an instantaneous conviction, that we had now attained one of the principal objects connected with the plan of our expedition to Iceland. We hastened to the spot, and all our wishes were fully accomplished in the examination of an object which greatly exceeded the expectations we had formed. The mineralogical facts which here presented themselves to our notice, will be described in a subsequent chapter.

On ascending one of the abrupt pinnacles which rose out of this extraordinary mass of rock, we beheld a region, the desolation of which can scarcely be paralleled. Fantastic groups of hills, craters, and lava, leading the eye to distant snow-crowned Jokuls; the mist rising from a waterfall; lakes embosomed among bare, bleak mountains; an awful profound silence; lowering clouds; marks all around of the furious action of the most destructive of elements; all combined to impress the soul with sensations of dread and wonder. The longer we contemplated this scene, horrible as it was, the more unable we were to turn our eyes from it; and a considerable time elapsed, before we could bring ourselves to attend to the business which tempted us to enter so frightful a district of the country. Our discovery of obsidian afforded us very great pleasure, which can only be understood by zealous geologists; and we traversed an immense and rugged mass of that curious substance, with a high degree of satisfaction; though various circumstances prevented our tracing it so fully as we wished.

Towards the east, at the distance of three or four miles, we observed a very large circular hollow, the sides of which were chiefly of a bright red colour; from which circumstance, and its general appearance, we concluded that it was the crater of an extinct volcano. The waterfall, the noise of which we distinctly heard, though at the distance of several miles, was formed by the Tunaa, a large river, which takes its course in this part of the country, and joins the Thiorsaa.

Brandtson told us that he had never been farther in this direction; and pointed out to us the Sprangè Sands, a vast plain, consisting of volcanic matter, which are stretched over a great part of those inhospitable regions already mentioned. Numerous obstacles present themselves to any person who may think of entering this dreadful country, among which the want of food for horses is the principal. The rivers, lakes, streams of lava, all the horrors of nature combined, oppose every desire to penetrate into these unknown districts; and the superstitious dread in which they are held by the natives is readily excused, the instant they are even remotely beheld. We saw the lake called Fiske Vatn, and the summits of several Jokuls, in the distance, which will be more particularly noticed afterwards, as we observed them more distinctly from another station.

Before we had satisfied our curiosity, rain fell in torrents, and continued to do so for an hour or two. We had not proceeded far on our return to Naifurholt, when it ceased, and was succeeded by a very thick fog, through which Brandtson guided us safely, and we reached our tent soon after twelve o'clock at night, having been absent twenty-two hours, during seventeen of which we were on horseback.

After the fatigue we had undergone in our excursion towards the Torfa Jokul in search of obsidian, we did not expect to find ourselves sufficiently refreshed to attempt ascending Mount Hekla on the following day; but, as we had been long in the constant habit of enduring daily hardships, we rose at an early hour on the third of August, quite alert; and, on seeing the whole of the mountain free from clouds, we were soon ready to finish our labours, by ascending Hekla, and attaining the summit of a mountain whose fame has spread to every quarter of the world.   At ten o'clock, we were ready; and Brandtson having collected our horses, we mounted them, and began our expedition under circumstances as favourable as we could wish.   We rode through sand and lava about three miles, when the surface became too rugged and steep for horses.   Our guide proposed leaving the poor animals standing till we returned; but though they would not have stirred from the spot, we sent them back, not chusing

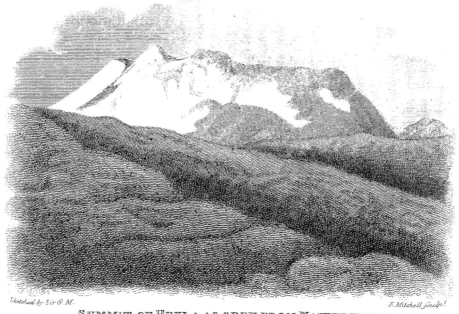

Sketched by Sir G. M.   E. Mitchell sculpt.

SUMMIT OF HEKLA AS SEEN FROM NAIFURHOLT.

that such valuable and steady servants should remain a whole day without food. We now proceeded a considerable way along the edge of a stream of lava, and then crossed it where it was not very broad, and gained the foot of the south end of the mountain. From this place we saw several mounts and hollows from which the streams of lava below appeared to have flowed. While we had to pass over rugged lava, we experienced no great difficulty in advancing; but when we arrived at the steepest part of the mountain, which was covered with loose slags, we sometimes lost at one step by their yielding, a space that had been gained by several. In some places we saw collections of black sand, which, had there been any wind, might have proved extremely troublesome. The ascent now became very steep, but the roughness of the surface greatly assisted us.

Before we had reached the first summit, clouds surrounded us, and prevented our seeing farther than a few yards. Placing implicit confidence in our guide, we proceeded, and having attained what we thought was the nearest of the three summits, we sat down to refresh ourselves, when Brandtson told us that he had never been higher up the mountain. The clouds occasionally dividing, we saw that we had not yet reached the southern summit. After having passed a number of fissures, by leaping across some, and stepping along masses of slags that lay over others, we at last got to the top of the first peak. The clouds now became so thick, that we began to despair of being able to proceed any farther. Indeed it was dangerous even to move ; for the peak consists of a very narrow ridge of slags, not more than two feet broad, having a precipice on each side many hundred feet high. One of these precipices forms the side of a vast hollow which seems to have been one of the craters. At length the sky cleared a little,

and enabled us to discover a ridge below, that seemed to connect the peak we were on with the middle one.   We lost no
time in availing ourselves of this opportunity, and by balancing ourselves like rope-dancers, we succeeded in passing
along a ridge of slags so narrow that there was hardly room
for our feet.   After a short, but very steep ascent, we gained
the highest point of this celebrated mountain.

We now found that our usual good fortune had not yet
forsaken us; for we had scarcely begun to ascend the middle
peak, when the sky became clear, and we had a full view of
the surrounding country.   Towards the north it is low, except where a Jokul here and there towers into the regions of
perpetual snow.   Several large lakes appeared in different
places, and among them the Fiske Vatn was the most conspicuous.   In this direction we saw nearly two-thirds across the
island.   The Blæfell, and the Lange Jokuls, stretched themselves in the distance to a great extent, presenting the appearance of enormous masses of snow heaped up on the plains.
The Skaptaa Jokul, whence the great eruption that took place
in the year 1783 broke forth, bounded the view towards the
north-east.   It is a large, extensive, and lofty mountain, and
appeared to be covered with snow to the very base.   On the
side next to us, though at a distance of about forty miles,
we plainly discerned a black conical hill, which very probably
may be the crater that was formed during the eruption.   The
Torfa, Tinfialla, and Eyafialla Jokuls, limit the view of the
eastern part of the country.   Towards the south, the great
plain we had passed through seemed as if stretched under our
feet, and was bounded by the sea.   The same valley was terminated towards the west by a range of curiously peaked
mountains, those in the neighbourhood of Thingvalla, and to
the north and west of the Geysers.

The middle peak of Hekla forms one side of a hollow, which contains a large mass of snow at the bottom; and is evidently another crater. The whole summit of the mountain is a ridge of slags, and the hollows on each side appear to have been so many different vents from which the eruptions have from time to time issued. We saw no indications that lava had flowed from the upper part of the mountain; but our examination, from the frequent recurrence of fog, was unavoidably confined.

After we had satisfied ourselves in surveying the surrounding country, we began to collect specimens of the slags, and perceived some of them to be warm. On removing some from the surface, we found those below were too hot to be handled; and on placing a thermometer amongst them, it rose to 144°. The vapour of water ascended from several parts of the peak. It had been remarked to us by many of the inhabitants, that there was less snow on Hekla at this time than had been observed for many years. We supposed, therefore, that the heat now noticed might be the recommencement of activity in the volcano, rather than the remaining effects of the last eruption which took place in the year 1766. Letters from Iceland dated on the 24th of October, mention, that on the morning of that day a shock of an earthquake was felt. This, however, does not always indicate an approaching eruption.

The crater, of which the highest peak forms a part, does not much exceed a hundred feet in depth. The bottom is filled by a large mass of snow, in which various caverns had been formed by its partial melting. In these the snow had become solid and transparent, reflecting a bluish tinge; and their whole appearance was extremely beautiful, reminding us of the description of magic palaces in eastern tales.

At the foot of the mountain, the thermometer at half past

nine o'clock stood at 59°. At eleven, it was at 55°, and at four, on the top, at 39°.

Our descent was greatly retarded by thick fog; and we found it much more hazardous than the ascent. We missed our way, and were under the necessity of crossing the lava we had passed in our way up, at a place where it had spread to a much greater breadth, and, from the rapidity of the slope along which it had flowed, had become frightfully rugged.

Mount Hekla has acquired a degree of distinction among volcanoes, to which it does not seem to be entitled. It is far behind Etna and Vesuvius, both in the frequency and magnitude of its eruptions. We could not distinguish more than four streams of lava; three of which have descended on the south and one on the north side: but there may be some streams on the east side, which we did not see. The early eruptions of this mountain do not seem to have been regularly recorded. Olafson and Paulson say, that after careful research they found that the number of eruptions amounted to twenty-two; and none are recorded as having happened before the year 1004. There were eruptions in the years 1137, 1222, 1300, 1341, 1362, 1389, 1538, 1619, 1636, and 1693. Flames appeared in the neighbourhood in 1728. In 1554, there were eruptions from the mountains to the eastward; and, in 1754, flames burst out to the westward. From the mountain itself, no eruption took place between the years 1693 and 1766, an interval of seventy-three years; and during this last period of activity, no lava was thrown out. The following year, flames broke out afresh, and the mountain was not perfectly quiet in the year 1768: since that time, it has remained inactive. We had no opportunity of measuring the height of Mount Hekla; and, owing to its heat, we cannot estimate it by the snow line: But, by comparing it, when viewed at a distance, with Eyafialla Jokul, which has been measured, I think it can hardly exceed 4000 feet, which I believe to be nearly the height ascertained by Mr Bain, who accompanied Sir John Stanley. It is probable that it will be found not quite so high, when accurately measured.

In the year 1755, a terrible eruption proceeded from the mountain called Kattlagiau Jokul, which is situate to the eastward of Eyafialla. From the accounts of this eruption, it does not appear that any lava flowed; but immense torrents of water carried destruction before them through the neighbouring country; and ignited stones and ashes were thrown in all directions. The electrical phenomena that accompanied this eruption, seem to have been very tremendous; several people and cattle having been killed by the lightning. The mountain continued in a state of violent activity during a whole month; and, indeed, it may be said to have been so during a whole year; for, between January and September of the year 1756, five different eruptions took place. We heard a report, that early in the summer the inhabitants in the neighbourhood had some reason to apprehend an impending eruption; but we could get no distinct account of the symptoms that had been observed. The earliest eruption of Kattlagiau, appears to have happened about the year 900; and, since that period, to the great one in 1755, only five have occurred.

It is mentioned, in the History of Iceland during the 18th century, that an eruption took place from Eyrefa Jokul, in the south-east part of the island, in the year 1720; and one from the lake of Grimsvatn in the year 1716.*

* I am uncertain of the position of this lake. The only one of that name which I could find marked on the maps of Iceland, I have noted on the small map; but, from the circumstance of all the most recent eruptions having proceed-

In the north-east quarter of Iceland, near a large lake called Myvatn, is Mount Krabla, which became remarkable by dreadful eruptions of lava that proceeded from it between the years 1724 and 1730. Some of the streams of lava flowed into the lake, destroyed the fish, and almost dried it up. There were eruptions also from the mountains round Krabla; and an extensive district of inhabited country was laid waste. At a place called Reikiahlid, near Krabla, sulphur is found in the same circumstances as that at Krisuvik, but in larger quantities. It was from the former place that most of the sulphur brought from Iceland was exported.

In the year 1000, an eruption took place in the Guldbringè Syssel. Another broke out near Reikianes in the year 1340; and one is said to have been seen at a great distance in the sea, in the year 1583, similar to that which was observed preceding the great eruption of 1783.

The total number of recorded eruptions appears to be the following:

From Hekla, since the year 1004, inclusive,		- - -	22
From Kattlagiau Jokul	900,	- - - -	7
From Krabla,	1724,	- - - -	4
In different parts of the Guld-			
bringè Syssel	1000,	- - - -	3
At sea,	1583,	- - - -	2
From the lake Grimsvatn, in 1716,		- - - -	1
From Eyafialla Jokul* in 1717,		- - - -	1

ed from the southern part of the island, I suspect that there may be a lake of the same name somewhere to the eastward of the Markarfliot.

   * This mountain is often called Öster Jokul, (eastern Jokul) in contradistinction to Snæfell Jokul, which is called western. The eruption of the eastern Jokul

| From Eyrefa Jokul | in 1720, | - - - - 1 |
| From Skaptaa Jokul, | in 1783, | - - - - 1 |

                                              42

In chronological order, the different eruptions mentioned by Icelandic authors stand recorded thus: In the years 900, 1000, 1104, 1137, 1222, 1300, 1340, 1341, 1362, 1389, 1422, 1538 (Vesuvius erupted the same year), 1554 (Etna), 1583, 1619, 1636 (Etna), 1693 (Vesuvius, 1692; Etna, 1694), 1716, 1717 (Vesuvius), 1720, 1724, 1728, 1730 (Vesuvius), 1754 (Vesuvius), 1755 (Etna), 1756, 1766 (Etna and Vesuvius), 1771, and 1772, flames seen on Hekla; 1783. Thus it appears, that many of the eruptions that are known to have taken place since Iceland was inhabited, have not been particularly noticed; and it is very probable, that numerous eruptions of less note have been passed over. We may reckon active all those mountains which have burned within the last century. Of these there are six;—Hekla, Krabla, Kattlagiau, Eyafialla, Eyrefa, and Skaptaa, Jokuls.

The most recent eruption that took place in Iceland seems also to have been the most awful. It proceeded from the Skaptaa Jokul in the year 1783. No account, which can entirely be depended on, of this destructive event has been given to the public. Mr Stephenson of Indreholm was ordered by the King of Denmark to proceed from Copenhagen, where he happened to be during the eruption, and to visit the district, that his Majesty might be enabled to alleviate the distresses occasioned by the eruption. That gentleman has published

in 1717, and the one from Grimsvatn, are recorded by Mr Stephenson in his history of Iceland during the eighteenth century.

a laboured account of the whole; but, although there is no doubt of the eruption having been one of the most terrible in the annals of volcanoes, he seems to have depended too much on reports and informations, which appear to be exaggerated. He himself told us, that the lava was so hot at the time he approached it, which was about a year after the eruption, that it could not be examined, and that it has never been traced to its source. Another account has been written, which, from what we heard in Iceland, might be depended on. It is to be lamented that the present state of the Icelandic press, prevents its being given to the public.

The whole tract between Hekla and Krabla is a desert quite impassable and unknown; and there is still subsisting a ridiculous notion that it is inhabited by a tribe of robbers. Did such people really exist, and did they know the dread which they inspire, they might easily procure more comfortable quarters.

No single volcanic mountain which we saw, appeared to have thrown out much lava. Probably this has been owing to the vast number of apertures which have given vent to the rage of subterraneous heat. In other countries, where it has for ages continued to explode from one or two mountains, the lava is confined to one place, and is abundant.

There is no country in the known world where volcanic eruptions have been so numerous as in Iceland, or have been spread over so large a surface. No part of the island is wholly free from the marks of volcanic agency; and it may be truly called the abode of subterraneous heat. Various volcanic mountains and streams of lava, are mentioned as existing in the eastern and northern districts, by Eggert Olafson, in his *Enarrationes Historicæ de Natura et Constitutione Islandiæ.* In the north-west quarter, in the district of

Isafiord, there is a volcanic mountain called Glama, which he describes as rivalling the magnitude of the Snæfell Jokul. Thus it appears, that the force of subterraneous fire has been exerted upon every part of this extensive island; and when we consider the eruptions that have been seen at a distance in the sea, we are safe in estimating, that, in this part of the earth, one continued surface of not less than 60,000 square miles has been subjected to that engine of destruction.

On the 4th of August we took leave of Hekla, and our excellent guide Brandtson, whose great activity, and obliging disposition, must recommend him to all travellers who may have occasion for his services. After passing the different streams of lava on the south side of the mountain, we left the plain, and ascended a ridge from which we had a view of the Westmann Islands and Eyafialla Jokul. Having descended towards the valley of the Markarfliot, we arrived at Hlide-

Sketched by Sir G. M.   E. Mitchell scalp.

MOUNT HEKLA FROM THE SOUTH.

Plate II.

EYAFIALLA IOKUL from HLIDERENDE.

rendè, the house of Sysselman Thoranson, brother of the Amtmand. The Markarfliot is a large and remarkable river. Its course to the sea is short, and it is formed by numerous streams, which descend precipitately from the Jokuls, bringing down a quantity of clay which gives it a white colour, and a fetid smell, especially perceptible in autumn. It divides into numerous branches, and very frequently changes its course, sometimes keeping close to the mountains, and at other times flowing ten or fifteen miles to the westward.

We were received very hospitably at Hliderendè; and had we been less welcome, we should have very readily excused any deficiency of attention, when we saw how very busy the people were in getting home the hay, during the fine weather which had prevailed for some time. The Sysselman has a large farm; and such of the pastures as we saw were excellent. A considerable quantity of angelica grows here, which is used as an article of food in many parts of the island. The carraway grows abundantly in the meadows at this place, but is not indigenous. A small quantity of seed was brought from Copenhagen by some person, and in a short time it disseminated itself.

Our supper consisted of baked mutton and melted grease. In the morning we had coffee; and for breakfast the same viands which were presented to us at supper; and our dinner was a repetition of the breakfast. There was no kind of bread in the house; and the only liquor presented to us was corn brandy. It was with much difficulty we could obtain leave to drink water; and we were afraid lest our entertainers should think our asking for it as great a piece of rudeness, as they esteem offering such liquor to their guests.

From the Sysselman's lady I purchased the dress which

has been already described.   We were shewn how the figured
stuff used for saddle-cushions, and with which the cloaks are
ornamented, is manufactured.   It is first made as our country-
women make coverings for footstools; only instead of working on canvas, the Icelandic women use a small frame, on
which threads are stretched.   These are crossed with others,
and worsted loops are wove in, which being afterwards cut, the
stuff resembles a very coarse velvet.   To form the figures on
trimming, part of the rough nap is cut out with scissars.   By
using different needles and different coloured worsteds, very
neat figures are worked in this manner; and the piece we
saw in the little loom was really very pretty.

While I was engaged in examining these things, my friends
went to a neighbouring church, attracted by the report of the
singular character of the priest, and desirous of seeing the
religious ceremonials in a country church.   They were accompanied by a student, a sensible young man, who officiated
as a sort of secretary to the Sysselman.   On their way, they
found the scenery very interesting and picturesque.   Many
fine streams precipitate their waters over the lofty cliffs forming the western boundary of the Markarfliot.   One of these
streams falls upon a ledge in the rock, and has worn a
pipe through it, having four apertures at different heights.
When there is little water in the river, it falls directly
down the pipe, and issues only from the lowest; but
when swelled by rain, it rushes from all the apertures,
forming a very curious and magnificent cascade.   Some
of the rocks composing these cliffs, consist of very fine ranges
of lofty columns.   Eyafialla Jokul soars above the eastern
side of the valley.   It is covered with perpetual snow for
nearly two thirds of its height, which has been ascertained
by the Danish officers now engaged in surveying the coast

to be about 5,500 feet. On some parts of the sides of this mountain, magnificent glaciers have been formed, the snow having descended in several places almost to the valley, and become frozen into solid masses, which assume a great variety of singular and abrupt forms.

The church is at a place called Hyindarmulè. The people were assembled, all dressed in their best suits, and were waiting for their pastor at the time my friends reached the place. On entering the church, a little while after the service had begun, the priest, without interrupting his devotions, beckoned to them to take seats near him. He soon handed to them his snuff-box, and, still going on with the service, he invited them by signs and gestures to partake of its contents. A dram bottle stood upon the altar, to which he made frequent applications. Before the sermon, he left the church, and said that he was very desirous to accompany them to Hliderendè; that he had already shortened the service, and would abridge his sermon as much as possible, in order that they might not be detained. They then returned to the church, and sat quietly till their new friend had vociferated a sermon which occupied half an hour. The sacrament was then administered, and the service concluded. They had advanced a short space on the road to Hliderendè, when the priest drew from a side-pocket a fresh bottle of spirits, which he offered to the party. On their declining his civility, he consoled himself by taking their share as well as his own, and, by the time they reached their place of destination, the bottle was almost empty. This poor wretch is, as far as we could learn, a solitary instance of habitual drunkenness among the people of Iceland.

The appearance of the Westmann Islands from Hliderendè is very picturesque; and we were told that they consist entirely

of lava. The nearest to the coast is about twelve miles distant; and the most remote about twice that distance. Only
one of them, called Heimaey, is inhabited; and the people
are by no means respected by their neighbours on the mainland, who represent them as being remarkably indolent,
and depraved in their habits. Their food consists chiefly of
fulmars and puffins, *(procellaria Glacialis, et Alca Arctica of
Linnæus)*, which are slightly salted and barrelled. This is
the principal aliment of the people of St Kilda, the most
remote of the western islands of Scotland, which I visited in the year 1800; and a peculiar and fatal disease
which attacks children, is common in both places, and
may probably be occasioned by the mode of living. These
islands produce a great quantity of feathers; and, until
the great eruption in 1783 took place, there was abundance
of fish around them; but since that period, the fishing on this
coast is reported to have been much less productive. There
is a church in the island of Heimaey, said to be one of the
best in Iceland; but it does not appear to be of much use
in improving the characters of those for whose benefit it
is intended. In the same island, there is a small creek, which
forms a tolerable harbour, but it is equally difficult to enter
or to leave it, on account of the strong currents, and the
heavy sea, which generally rolls around the islands when the
wind is a little more than moderately high.

A journey to the eastern part of Iceland, along the southern coast, is rendered very difficult by the rivers; and is not
often attempted, even by the natives of the country. From
Kattlagiau Jokul, the road is chiefly along the shore; but
when the rivers are swollen, it is usual to cross some of the
mountains to avoid them. During two days of the journey
to Berufiord, which is the most southerly station on the east-

ern coast, no habitations are met with; and an express cannot reach that place from Reikiavik in less than fourteen
days. Berufiord, Rodefiord, and Vapnafiord, are the only
ports on this side of the island.

We had intended to proceed farther to the eastward, in
order to examine Kattlagiau Jokul, and the lava of 1783; but
being aware, that should the Elbe return, even at this time, we
should necessarily be detained till a late period of the year,
and probably the greatest part of the winter; and having been
informed that the brig Flora, which we had left at Stromness,
after failing to procure a cargo on the east and north coasts,
was daily expected at Reikiavik, and would remain there only a few days, we resolved to avail ourselves of this opportunity of returning to Britain. We therefore proceeded, on
the evening of the 5th, down the valley of the Markarfliot,
towards Reikiavik, having been furnished with fresh horses
by Sysselman Thoranson, who attended us to Oddè. The
Sysselman is famed in the country as a good horseman, and
for being possessed of an excellent stud. All the Icelanders
shew a great regard for their riding-horses, and emulate each
other in breaking them to pace rapidly. They also pay great
attention to their cushions, and other trappings, which they
arrange and put on with much pains, and afterwards tie up
the horses tails into a knot, to prevent their being spoiled.

The pasture in this district, and the appearance of inclo-
sures, are far superior to any thing we had seen before; and if
ever an attempt shall be made to raise corn, potatoes, or turnips,
this is the district which, in respect to soil and situation, seems
to hold out the greatest temptations to agricultural experiment. The vicinity of Eyafialla, and the other lofty Jokuls
to the eastward, render the climate perhaps too unsteady for
extensive cultivation. We passed the church of Breidè-bol-

stadr, which is the richest living in Iceland; the stipend being 182 dollars.

We found Oddè under some small green hills, not far from
the bank of the Rangaa. The church is one of the best structure, very like that of Skalholt; and the minister's house is
large and commodious. The house was at this time occupied by the widow of the late minister, whose successor, the
Lector Theologiæ, Steingrim Jonson, had not yet taken possession of it. The widow and her family were very hospitable; and her son, a young man of superior manners and understanding, was very assiduous in his attentions. Although
we arrived at a late hour, and were not much disposed for
eating, it was thought necessary to prepare a repast for us;
and accordingly, a little before twelve o'clock, baked mutton, and a dish of rice boiled in milk, were set before us.

Early in the morning of the 6th, we were preparing for
our departure, when a substantial breakfast of mutton, coffee,
and chocolate, detained us; and, in the course of conversation, we discovered that the late minister had left a number
of books, some of which we might purchase. Accordingly,
sundry chests and other receptacles were opened, and a variety of books were released from dust and cobwebs. We
made several purchases; but the most curious, and perhaps
the most valuable, was a superb Icelandic Bible, which fell
to the lot of Mr Bright. The way to Eyarback from Oddè,
is long and tiresome. We first passed through the Rangaa
by a deep and difficult ford; and, after scrambling among
bogs, we crossed the Thiorsaa at a place where it was very
broad. Owing to the shallowness of the stream, the boat
grounded when not much more than half way over, and we
had to mount our horses, and ride about a hundred and fifty
yards through a very dangerous quicksand.

Sketch'd by Sir Geo. Mackenzie.

MOUNT HEKLA from ODDE.

Pub.ᵈ by Constable & Cᵒ Edinburgh. 1811.

J. Clark direxit

We now went towards the coast, where the plain becomes more sandy. On approaching Eyarback, we saw, at the distance of a few miles, the vapour ascending from the hot springs of Reikum, which, had our time not been much limited (an express having been sent to Oddè by Mr Fell, acquainting us of the arrival of the Flora, and wishing us to hasten our return, if we desired to seize this opportunity of embarking), and had we not been quite familiarized with similar phenomena, we should have visited. Our not having done so is not to be regretted, as Sir John Stanley has given an excellent description of Reikum, in a letter to the late Dr Black, which is printed in the Transactions of the Royal Society of Edinburgh ; from which I shall now transcribe it.

‘ The valley is in this place fertile, and nearly half a mile
‘ in breadth. It becomes more narrow towards the north ;
‘ and it is there rendered barren by heaps of crumbled lava,
‘ or other rubbish, brought down from the hills by the wa-
‘ ters. These have the appearance of artificial mounds, and
‘ a great number of springs are continually boiling through
‘ them. Below the surface, a general decomposition seems
‘ taking place : for almost wherever the ground is turned
‘ up, a strong heat is felt ; and the loose earth and stones
‘ are changing gradually into a clay, or bole of various co-
‘ lours, and beautifully veined, resembling a variegated jas-
‘ per. The heat may possibly proceed from a fermentation
‘ of the materials composing these mounds ; but more pro-
‘ bably (I should conjecture) from the springs and steam
‘ forced up through them. The springs must have acquired
‘ their heat at some greater depth, from some constant, steady
‘ cause (however difficult to explain), adequate to the length
‘ of time they have been known to exist, with the same un-
‘ varied force and temperature.

' Springs do not boil on or near these banks only. They
' rise in every part of the valley; and, within the circumfe-
' rence of a mile and an half, more than an hundred might
' easily be counted. Most of them are very small, and may
' be just perceived simmering in the hole from whence the
' steam is issuing. This, trailing on the ground, deposits, in
' some places, a thin coat of sulphur. The proportion va-
' ries, for, near some of these small springs, scarce any is
' perceptible, whilst the channels by which the water escapes
' from others, are entirely lined with it for several yards.
' Neither the water nor the steam from the larger springs,
' ever appear to deposit the smallest proportion of sulphur;
' nor can the sulphureous vapour they contain be discovered,
' otherwise than by the taste of what has been boiled in them
' for a long time.

' Many springs boil in great caldrons or basons, of two,
' three, or four feet diameter. The water in these is agitated
' with a violent ebullition, and vast clouds of steam fly off
' from its surface. Several little streams are formed by the
' water which escapes from the basons; and as these retain
' their heat for a considerable way, no little caution is re-
' quired to walk among them with safety.

' The thermometer constantly rose in these springs to the
' 212th degree; and in one small opening, from whence a
' quantity of steam issued with great impetuosity, Dr Wright
' observed the mercury rise, in two successive trials, to the
' 213th degree.

' I have already said, that the ground through which many
' of the springs were boiling, was reduced to a clay of vari-
' ous colours. In some, the water is quite turbid; and, ac-
' cording to the colour of the clay through which it has pass-
' ed, is red, yellow, or grey.

'The springs, however, from whence the water overflows
'in any great quantity, are, to appearance, perfectly pure.
'The most remarkable of these was about fifty or sixty yards
'from our station, and was distinguished by the people of
'the neighbourhood, by the name of the Little Geyser. The
'water of it boiled with a loud and rumbling noise in a well
'of an irregular form, of about six feet in its greatest dia-
'meter; from thence it burst forth into the air, and sub-
'sided again nearly every minute. The jets were dashed in-
'to spray as they rose, and were from twenty to thirty feet
'high. Volumes of steam or vapour ascended with them,
'and produced a most magnificent effect, particularly if the
'dark hills, which almost hung over the fountain, formed a
'back-ground to the picture. The jets are forced, in rising,
'to an oblique direction, by two or three large stones, which
'lay on the edge of the bason. Between these and the hill,
'the ground (to a distance of eight or nine feet) is remark-
'ably hot, and entirely bare of vegetation.- If the earth is
'stirred, a steam instantly rises; and in some places it was
'covered with a thin coat of sulphur, or rather, I should
'say, some loose stones only were covered with flakes of it.
'In one place, there was a slight efflorescence on the surface
'of the soil, which, by the taste, seemed to be alum.

'The spray fell towards the valley, and in that direction
'covered the ground with a thick incrustation of matter,
'which it deposited. Close to this, and in one spot, very
'near the well itself, the grass grows with great luxuriance.

'There is another fountain in the valley, not much infe-
'rior in beauty to that which I have described. It breaks
'out from under one of the mounds, close to the river. Its
'eruptions are, I think, in some respects, more beautiful than
'those of the former. They rise nearly to the same height;

' and the quantity of water thrown up at one time is great-
' er, and not so much scattered into spray. The jets conti-
' nue seldom longer than a minute; and the intervals be-
' tween them are from five to six minutes. They are forced
' to bend forwards from the well, by the shelving of the bank,
' or probably their height would be very considerable; for
' they appear to be thrown up with great force. We never
' dared approach near enough to look deep into the well;
' but we could perceive the water boiling near its surface,
' from time to time, with much violence. The ground in
' front of it was covered with a white incrustation, of a more
' beautiful appearance than the deposition near any other
' spring in this place. By a trial of it with acids, it seemed
' almost entirely calcareous.

  ' I have now described to you the two most remarkable
' fountains in the valley of Reikum, the only two which throw
' up water to a considerable height with any regularity. There
' are some from whence, in the course of every hour, or half
' hour, beautiful jets burst out unexpectedly; but their erup-
' tions continue only a few seconds, and between them the
' water boils in the same manner as in the other basons.

  ' Towards the upper end of the valley, there was a very
' curious hole, which attracted much of our attention. It
' seemed to have served at some former period as the well of
' a fountain. It was of an irregular form, and from four to
' five feet in diameter. It was divided into different hollows
' or cavities at the depth of a few feet, into which we could
' not see a great way, on account of their direction. A quan-
' tity of steam issued from these recesses, which prevented us
' from examining them very closely. We were stunned while
' standing near this cavern, and in some measure alarmed, by
' an amazing loud and continued noise which came from the

' bottom. It was as loud as the blast of air forced into the
' furnace from the four great cylinders at the Carron iron-
' works.

' We could discover no water in any of the cavities; but
' we found near the place many beautiful petrifactions of
' leaves and mosses. They were formed with extreme deli-
' cacy, but were brittle, and would not bear much handling;
' their substance seemed chiefly argillaceous.

' We perceived smoke issuing from the ground in many
' places in the higher parts of the valley, much further than
' we extended our walks. I am sorry to say we left many
' things in this wonderful country unexamined; but we were
' checked in our journey by many circumstances, which al-
' lowed us neither the leisure nor the opportunity for explor-
' ing every part of it as we could have wished. The sub-
' stances deposited near the different springs seemed to me,
' in general, a mixture of calcareous and argillaceous earths;
' but near one spring, not far from our tents, there seemed
' to be a slight deposition of silicious matter. To the eye it
' resembled calcedony; but with its transparency, it had
' not the same hardness, and, if pressed, would break to
' pieces. The water you have analysed came from this
' spring, and we were obliged to take some care in filling
' the bottles; for though gradually heated, they would break
' when the water was poured into them, if it had not been
' previously exposed to the air for some minutes in an open
' vessel.

' The water of this spring boiled, as in most of the
' others, in a cauldron four or five feet broad. I do not
' recollect to have seen any of it ever thrown up above a
' foot, and some meat we dressed in it tasted very strong of
' sulphur.

2 L

' Mr Baine, by a measurement of the depth, the breadth,
' and the velocity of the stream flowing from the Little Gey-
' ser, found the quantity of water thrown up every minute
' by it to be 59,064 wine gallons, or 78.96 cubic feet. Mr
' Wright and myself followed the stream, to observe how
' far any matter continued to be deposited by the water.
' We found some little still deposited where it joined the
' river, a quarter of a mile at least from its source. At
' that place, it retained the heat of 83° by Fahrenheit's ther-
' mometer.'

Eyarback is situate near the mouth of the river Elvas, which
is formed by the rivers which join near Skalholt, and that
which flows from the lake of Thingvalla. Here a large quantity of the *fucus palmatus* (called in Scotland, *dulse*) is prepared by drying. It is packed in casks, and in a short time
gives out a white and somewhat saccharine powder. In this
state it is eaten by the natives, either raw with butter, or
boiled in milk. The merchant who is settled here is a Mr
Lambasson, who was at this time in Denmark. During his
absence, the business is carried on by his wife, and his agent
Mr Peterson, who received us very civilly. The harbour is
by no means a safe one ; and, in bad weather, exceedingly
dangerous of approach. Finding that, owing to the state of
the tide, we could not cross the river (which is here about half
a mile broad) till midnight, we refreshed ourselves by sleeping
a few hours. We had some trouble in getting the ferrymen
out of their beds, but we were carried over in an excellent brat,
our horses swimming after us. Many thousand horses cros at
this place during the summer, and we were told that so ie-
times as many as 900 crossed in one day.

Having proceeded a mile or two along the river, we began
to ascend the ridge of hills which extends almost without in-

Plate 13.

Sketch'd by H.Holland.

EYAFIALLA IOKUL, MOUNT HEKLA, & the RIVER ELVAS, from the Westward.

I. Clark direxit

terruption from the lake of Thingvalla to the extremity of the
Guldbringè Syssel. We had a fine view from the heights, of
Hekla, and the whole country we had passed through, towards Eyafialla Jokul, which bounded the scene. We now
encountered lava, many streams of which we passed; and
after travelling through twenty-five miles of a district entirely laid waste by fire, we arrived at Reikiavik about noon,
on the 7th of August, considerably fatigued; our progress
having been very slow on account of the extreme ruggedness of the country. The length of the journey we had now
accomplished, was about 280 miles. Thus our travels terminated; and we immediately commenced our preparations
for leaving a country, whose inhabitants and natural curiosities we had surveyed with singular interest and gratification.

We found the vegetables, the seeds of which we had sown
on our arrival in Iceland, tolerably well advanced. The white
turnips were of a good size; the peas were just out of blossom; the radishes, cress, and mustard, had mostly gone to
seed; but the cabbages had not made much progress. In
Mr Frydensberg's garden, the potatoes were very good, and
we partook of a dish of them, and of well-grown kol-rhabie.
Nasturtium, lupines, and some other hardy annual flowers,
had advanced to perfection in Mr Frydensberg's dining-room,
which they greatly contributed to ornament.

The attentions of all our friends were renewed on our arrival, and the Bishop repeated a present of a sheep for the
third time; a mark of his hospitality and friendship which
we shall never forget. That our wants might be supplied, we
had only to express them to Mr Frydensberg or Mr Simonson.

On Sunday the 12th, we went to an entertainment to which
we had been invited by the merchants, who desired to pay us
a compliment before we departed. In the morning they all

displayed their flags from their houses, and at two o'clock we sat down to a very good dinner.

Having, a short time before our departure, addressed a letter to the Deputy-governors, expressive of our gratitude for all the attention which had been bestowed upon us, and requesting them to communicate our thanks to the people of the various districts through which we had travelled, for their hospitality and kindness, I received the following letter in reply :—

 ' Perillustri, nobilissimo Domino Georgio Mackenzie,
  ' salutem plurimam.

 ' Literas tuas, vir nobilissime, humanissimas accepimus,
' et nostri officii erit, incolas regionum, per quas iter fecisti,
' de singulari tuâ, et itineris sociorum, humanitate, benigni-
' tate, et grata erga eos animo certiores facere. Te et socios
' vicissim excusare rogamus, si antiqua illa hospitalitas, quæ
' incolis hujus insulæ a primis usque temporibus propria fuit,
' vobis alicubi defuerit, et ut hoc verecundiæ, paupertati, ac
' linguæ vestræ, ut et vitæ consuetudinis, imperitiæ, non in-
' humanitati gentis tribuatur, enixè postulamus.

 ' Accipe denique, vir nobilissime ! nostras integerrimas
' grates, pro tuâ et sociorum dulci, amicâ, et urbanâ conver-
' satione, quam gratâ diu colemus memoriâ. Prosperrima te,
' tuosque itineris socios, nostra et insulæ hujus incolarum
' prosequuntur vota !

          ' THORANSON.
          ' EINARSON.
          ' FRYDENSBERG.

 ' Reikiavicæ, d. 13 Augusti 1810.'

I must not omit mentioning, that I also received a letter,

written in tolerably good English, from the Chief Justice Stephenson, in which he expressed himself in terms highly flattering to us. He made me a present of several books, and of his picture; and of these, as well as many other marks of his regard and hospitality, I shall ever retain a grateful remembrance.

We were now about to take leave of a people whose situation had often excited our pity. Being of quiet and harmless dispositions; having nothing to rouse them into a state of activity, but the necessity of providing means of subsistence for the winter season; nothing to inspire emulation; no object of ambition; the Icelanders may be said merely to live. But they possess innate good qualities, which, independently of the consciousness of their former importance, have preserved their general character as an amiable community. They have indeed become negligent with respect to the cleanliness of their persons and dwellings; but they deserve a high place in the scale of morality and religion. The example of the Danes has done very material injury to the moral character of those with whom they have constant intercourse; but beyond the precincts of Reikiavik, the people are found possessed of their pristine worth and simplicity. To religious duties they are strictly attentive; and though the clergy are not in general raised above the level of the peasantry, in any respect but in their sacred office, yet they have been able to preserve the regard due to those who are considered as peculiarly the servants of the Supreme Being.

To say that crimes are rare, is perhaps a slight compliment to people who have few temptations to commit them. Except at Reikiavik, vice is hardly known; and even there, when we reflect on the loose lives of the Danes, it is astonishing how little progress it has made among the natives.

To the laws of hospitality they are particularly attentive. If they give little, it is because they have little to give. To measure their disposition by their power of bestowing, would be a very unjust estimate.

The history of the Icelanders points out sufficient reasons for the decline of activity and enterprize. In pronouncing upon their character, therefore, some caution is necessary. Travellers, when they find themselves obliged to submit to privations before unknown to them, when they experience a deficiency of alacrity in supplying their wants, and a great degree of indifference in the behaviour of the people among whom they sojourn, are too apt to form a hasty and partial judgment of their character. Some of the occurrences I experienced in Iceland might have entitled me to speak unfavourably of the inhabitants, had I been disposed to judge of them inconsiderately. But when I recollected what Icelanders once were; when I saw the depressed state of this poor, but highly respectable people; and perceived that they still retained that mild spirit (once, too, an independent and an enterprising one) which taught them to regulate their affairs with prudence, and to live together in the utmost harmony I could not help admiring their patience and contentment.

> Yet still, e'en here, content can spread a charm,
> Redress the clime, and all its rage disarm.
> Tho' poor the peasant's hut, his feasts tho' small,
> He sees his little lot, the lot of all;
> Sees no contiguous palace rear its head,
> To shame the meanness of his humble shed;
> No costly lord the sumptuous banquet deal,
> To make him loathe his *hard-earn'd scanty* meal;
> But calm, and bred in ignorance and toil,
> Each wish contracting, fits him to the soil.

I trust, that in these pages enough will be found to excite compassion in every British breast, for the calamitous situation of an innocent and amiable people, at that critical period when oppression or neglect may overwhelm them in misery. The distracted state of Europe will not, I trust, be considered as a reason that Britain should disregard their wants, or withhold relief. Iceland requires no sacrifice of blood nor treasure. Though very rarely a complaint was uttered, I sometimes heard the wishes of the people expressed in the relation of an ancient prophecy delivered in these terms,—
' When the Danes shall have stripped off our shirts, the
' English will clothe us anew.'

The possession of Iceland would not be burdensome to England. An exuberant and inexhaustible supply of fish from the sea, and the rivers, would alone repay the charitable action of restoring freedom to the inhabitants, who, under the fostering care of a benevolent government, might soon improve their soil and their own condition. I must not be understood as intending to convey any insinuation against the government of Denmark, which has done every thing that was possible to encourage the trade with Iceland. But in doing so, with the best intentions, the people have been neglected, and the Danish merchants alone regarded. Whatever good the regulations of Denmark might have been calculated to effect, the prohibition against trading with other nations has left the Icelanders nearly in the same state they were in, when subjected to the monopoly of a company.

On the 19th of August we set sail for England, and, after a tempestuous voyage of fourteen days, during which Captain Butterwick of the Flora paid us every attention in his power, we landed at Stromness. The only occurrence, during the voyage, worth mentioning, was that of a flock of wagtails and two sparrow-hawks perching on the rigging of the ship one morning, during a heavy gale of wind, when we were at least 100 miles from any land. On our coming in sight of the rocks called Barra and Rona, off the north of the island of Lewis, the wagtails left us. The sailors had deprived one of the hawks which was caught, of the means of flying, by clipping its wings. We saw, at different times, a considerable number of large whales; and were struck with the resemblance between the forcible jets of water which they threw from their nostrils, and some of the fountains we had seen in Iceland. We remained in Orkney two days, and went to see the town of Kirkwall. Although this country did not strike us as by any means beautiful when we first saw it, we now derived very high gratification from the sight of corn fields and gardens; and were particularly delighted with the few small trees which grow about Kirkwall. Having hired a boat, we intended to proceed by water to Inverness; but bad weather forced us into the harbour of Wick in Caithness-shire. From thence we proceeded to Dunrobin, and crossed the Dornoch Frith to Ross-shire, pursuing our journey by land to Edinburgh, where we arrived in health and safety, after an absence of nearly five months.

CHAP. IV

RURAL AFFAIRS.

THE terms on which a tenant holds a farm in Iceland, are similar to what is called *steelbow* in Scotland. The rent is paid in two parts. First, there is a land rent, or Land-skuld as it is called, which is a fixed sum rated according to an old valuation; secondly, there is a certain rent paid for a permanent stock of cattle and sheep, which is transferred from tenant to tenant, every succeeding one being obliged to take it on certain conditions, and to leave the same number on his quitting the farm. The tenant, however, is at liberty to keep as much stock as he can support, without paying any additional rent. The Land-skuld is paid in various ways; in money, wool, tallow, &c. &c. That for the permanent stock chiefly in butter.

Leases for a term of years are not common in any part of the island. The same tenant continues to possess the land, unless the proprietor can prove that the farm has been neglected, or that the farmer has misconducted himself. The law is effectual in preventing abuses in the dismissal of te-

2 M

nants; for if a farmer can prove by a survey of the Hrepsti-
orè, or two respectable persons of his own profession, that
his farm has not been neglected, he cannot be removed;
but he may quit his farm whenever he pleases. The practice
of letting farms from year to year is not uncommon; six
months notice being necessary for the tenant to quit.

A farm, the disposeable value of which is about 200 rix-
dollars, pays a Land-skuld of from four to six. The nominal
price of land has, in many instances, doubled within the last
forty years; not, however, in consequence of any improvement, but of the depreciation of the government paper. The
rix-dollar, which is paper, is worth four shillings English,
when at par. A guinea in Iceland, at the time we left the
island, was worth fifteen paper dollars; and since my arrival
in Scotland I have been offered twenty for a guinea. The increase of rent has taken place chiefly on the permanent stock
of the farm.

Besides the rent payable to the proprietor, a farmer is
obliged to pay a proportion to the parish priest, according to
the rent of his farm; and to keep a lamb for him during the
winter season, taking it in October, and returning it in good
condition about the middle of May.

A general description of a farm-house was given in the
account of our first excursion. I shall now describe one of
the best sort, that of the Provost at Storuvellir, of which an
accurate plan was taken.

    a. The entrance passage, 40 feet long, and 4 feet wide.

    b. The kitchen.

    c. Fuel-room. *

---

    * Bad turf, dried cow and sheep dung, and fish-bones, are the articles used as
fuel. In the Westmann islands, they use dried sea-birds. Fuel is very scarce;

d. Bunn, or store-room.

e. Bed-room, 40 feet long by 8 feet wide, with a recess 10 feet by 8.

f. A wainscotted room, with bedsteads. This is an appendage only to some of the principal dwellings, and is usually crowded with saddles, harness, and implements of various kinds. It has frequently a small window in the end.

g. Dairy.

h. Out-house.

i. Smithy.

and in the houses of the Icelanders, there is only one fire in the kitchen, which is placed on the floor; stoves being seldom seen even in the houses of the better sort of farmers.

Sketched by R. Bright   *E. Mitchell sc*

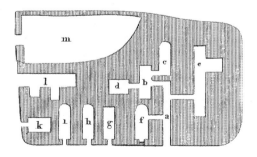

k. Out-house.

l. Cattle-house.

m. An enclosure for hay and turf, to which there is a path, often over the top of the house.

The servants are generally orphans, or the children of very poor farmers. As they are considered nearly on a level with their master's children, it is not uncommon for marriages to take place between them; and a poor farmer sends his son or daughter to serve in the house of one in more affluent circumstances, in hopes of such a connection being formed.

The wages given to servants, male and female, amount to from four to six dollars a-year, sometimes more, besides food and clothes. By these, and the other members of the family, every thing that is necessary for subsistence and clothing is prepared, and all business performed. During the winter season, the family rises about six or seven o'clock in the morning. One is sent out to look after the sheep; another attends the cattle; some are employed in making ropes of wool or horse hair; one is in the smithy making horse shoes and other articles. Spinning is performed with a spindle and distaff, and sometimes with a wheel; some, both men and women, knit and weave, and others prepare sheep-skins for fishing dresses. While so many are thus occupied, one generally reads aloud, in a singing tone, different tales and histories. Most farm-houses are supplied with books containing such tales; and the people exchange books with each other for the sake of variety. The only opportunity they have of making this exchange is when they meet at church, where, even during the most inclement part of the season, a few always contrive to be present. The people sometimes amuse themselves with a game somewhat like drafts; with cards; and many play chess extremely well.

The Icelanders divide the day and night into nine periods. From midnight to three o'clock in the morning they call Otta; from three to six, Midurmorgun; from six to nine, Dagmal; from nine to twelve, Hactei: the first hour and a half after noon, Midmunda; from half past one to three o'clock, Noon; from three to six, Miduraftur; from six to nine, Nattmal; from nine to twelve, Midnat. There are but few clocks in the island, and they are not very good.

We saw in different places, particularly at Huaneyrè, pieces of very good cloth which had been manufactured in the country. The sort called wadmal differs from cloth, in being what is called in this country, tweeled. Blue and black are the most common colours. One piece of cloth which we saw was a mixed black and white. Different shades of yellow are used, and not unfrequently for stockings. The processes of dyeing are very simple. The leaves of the *arbutus uva ursi*, the *lycopodium alpinum*, the *lichen Islandicus*, and some others, are employed. Stockings are filled with the lichen Islandicus, and boiled. When cloth is to be dyed, the vegetable substances are chopped small, and spread over the cloth, which is then rolled up and boiled. Black is obtained by strewing a rich black earth, found in some of the bogs, over the cloth, after it has been boiled with the *arbutus uva ursi*, when it is again rolled up and boiled. We saw none of this earth, but probably it contains a considerable proportion of iron, which, with the astringent matter of the plant, affords the black colour. Indigo is used for dyeing blue.

The skins of horses and cows, after having been steeped for some time in urine, are frequently put into the liquor which has been used for dyeing black; by which means they undergo a slight degree of tanning. Sheep-skins are prepared by being soaked in water till the wool loosens, which is removed; and then the skins are drawn over a ram's horn fastened to the roof by its ends.

Farm-houses are for the most part built on dry knolls, and the ground immediately around them is allotted for hay. The extent is greater or less according to circumstances; and though hay is by far the most important article to a farmer in Iceland, I do, not recollect to have seen any signs of exertion to improve a hay field by draining, or otherwise. All the manure is bestowed upon the little hillocks, which surround the houses like graves, into which the hay ground is generally partitioned. The people believe that a greater quantity of grass can grow upon an extended surface of this sort; and this erroneous notion is entertained even by the higher classes. That a greater surface is procured, is true; but as every plant grows perpendicularly, or as nearly so as circumstances will admit, a greater produce cannot be obtained. The speedy evaporation of moisture, occasioned by the smallness of the hillocks, and the air circulating between them, must render the grass that does grow, less luxuriant than it would be otherwise. About the time of our arrival in Iceland, the people were busy spreading the dung; and about the end of July, the hay harvest had begun in many places. The grass is neither close, nor long, at the time it is reckoned fit for cutting. We did not observe any field in which the useless or less nutritious plants did not exceed, or at least equal in number, those that were really valuable. Every thing that grows is cut down by means of a short narrow scythe, with which the Icelanders work expeditiously and neatly, making all the little knolls perfectly bare. When cut, the grass is commonly gathered together on some even place, where it can be turned and tossed conveniently. I observed in many places that no more was cut at a time than what would employ the people on the farm to dry; and before any more was cut, the first portion was carried home. When bog-grasses are accessible, they are carefully cut and made into hay. The process of drying is the same as with us; and when carried home, the hay is made up into long and narrow stacks, often before it is perfectly dry, and consequently much of it is spoiled by heating. The hay is kept chiefly for the cows, on which the people depend for much of their subsistence. In severe weather, a little is given to the sheep and horses; but they often struggle through a hard winter without any sustenance but what they can procure for themselves.

As soon as the hay around the house is secured, the farmers give a feast, or harvest-home. This is a supper of which the chief delicacy is porridge, made of meal of some sort, and milk. When the whole hay-harvest is finished, another feast takes place, when a fat sheep is killed. Though neither dancing nor singing are called in aid, these feasts are chearful and merry.

The immense extent of the bogs and swamps of Iceland renders it obvious to any one who has attended to the subject, that the climate must be greatly deteriorated by the evaporation from them. Were the people to set about draining the bogs, they would find not only the climate improve, but the quantity of grass fit for hay to increase largely. There seems to be some prejudice against draining, which a little intercourse with Britain may probably remove. I do not know any place where draining could be more easily or more advantageously practised than Indreholm, and in the country lying between Akkrefell and the Skardsheidè.

The cattle, in point of size and appearance, are very like the largest of our highland sorts, except in one respect.

that those of Iceland are seldom seen with horns. As in other countries, we meet with finer cattle on some farms than on others; but, from every observation I could make, and information I could obtain, the Iceland farmers know nothing of the art of breeding stock. The bulls are in general ugly, and no use is made of them till after they are five years old. In rearing a bull-calf no more attention is paid to him than to others. Taking all the circumstances of management together, I had some reason to be surprised to find the cattle upon the whole so handsome. The cows in general yield a considerable quantity of milk, many of them ten or twelve quarts per day, and some a good deal more. Milk is usually made into what is called *skier*, which has been already mentioned.

Sour whey, mixed with water, is a favourite beverage of the Icelanders, and they seldom travel without a supply of it. Butter, however, is the chief article among the products of the farm, and of this the Icelanders eat a surprising quantity. They value it most after it has been barrelled, without salt, and kept several years. It is wonderful how well butter keeps in this manner; it arrives at a certain degree of rancidity, beyond which it does not pass. The smell and taste of the sour butter are very disagreeable to English palates, though Icelanders delight in it. When there is a scarcity of butter, the people eat tallow. The former was not very plentiful last summer, and consequently little tallow was brought to market; and I have seen children eating lumps of it with as much pleasure as our little ones express when sucking a piece of sugar candy. When people go to the northern districts for the purpose of cutting hay, they are paid for their work in butter, at the rate of 30lbs. per week. It is made in churns of the form most common in this country, in which the cream is agitated by the perpendicular motion of a plunger. Sometimes two are worked by one handle fastened to a cross piece of wood, to which the plungers are connected by projecting arms, the cross piece forming the angle between them and the handle, and turning on two pivots. There is not much cheese made in Iceland, and they do not begin to manufacture it till late in the season. It is of very inferior quality. The manufacture of butter and sour whey employs the farmer's wife during his absence, while he is engaged in fishing. In some parts of the country the servants or children are employed in gathering lichen and angelica root. The former is carefully dried and packed for use; and the latter is buried, and used more as an article of luxury than of subsistence.

The sheep of Iceland appear to be the same with the old Scotch highland sort, which is now nearly extinct. They are larger, however, and the wool is long and soft, but not fine. Many of them are entirely black, and a great proportion are black and white. The wool is never shorn, but pulled off. Much of it is lost before it is taken off; and what remains, after hanging for a time on the animal's back, becomes spoiled and felted by the rain. The sheep are very much infested by vermin, known in England by the names of *ticks* and *keds.* The lambs are early restrained from sucking; and the ewes are milked, and butter is made from the produce.

It is part of the employment of the women, during winter, to pick and clean the wool, and to spin it. A considerable quantity is exported; and it is so valuable an article in Denmark, that it sells in Iceland for as much as coarse wool in the north of Scotland.

About the year 1756, an attempt was made to improve the

wool in Iceland, by the introduction of Spanish rams; but, owing to negligence, it was unsuccessful. With that zeal for bettering the condition of his country which distinguishes him, Mr Stephenson of Indreholm brought a few Merino rams and ewes from Norway in the year 1808. Their wool is tolerably fine, but by no means so good as that of the Merinos in England. I saw the lambs of the first cross between them and the Iceland ewes, and they promised very well. If Mr Stephenson perseveres in his laudable exertions, and if the people can be made sensible of the advantages to be derived from improving the wool, he will have the satisfaction of having begun a most beneficial improvement.

The gathering of the sheep from the mountains before the commencement, of winter, is a very important part of the business of an Iceland farmer. As soon as the hay harvest is over, and when the Hreppstiorè, or parish officer, thinks that the farmers are ready, he informs the Sysselman of the district, who causes a notice to be given in the churches, that on a certain day the gathering of the sheep shall commence, and, at the same time, appoints a place of rendezvous. Every farmer who has a considerable part of his stock feeding on the mountains, must send one man; or, if the number of his sheep be very small, he may join with another whose case is similar, and together they send one. When the men destined for this service assemble, they chuse one who has had much experience, whom they agree to obey, and they give him the title of king, and the power of selecting two associates as counsellors. On the appointed day they meet at the place fixed upon, perhaps to the number of 200, on horseback. Having pitched their tents, and committed their horses to the care of children who have accompanied them, the king, on horseback, gives his orders, and sends the men off two and two, strictly enjoining them not to lose sight of their comrades. Having collected as many sheep as they can find, they drive them towards the tents; and then shift their quarters. Thus they go on during a week, when they take all the sheep to one of the large pens constructed for the purpose, which consist of one large enclosure, surrounded by a number of smaller ones, for the purpose of separating the sheep belonging to different persons. This business is quite a rural festival; but the merriment is often mingled with the lamentations of those who have lost some of their sheep, or the quarrels of others who have accidentally fixed upon the same mark for their property. The search for sheep is repeated about the middle, and again about the end, of October. At this last time, those only who have failed in recovering their sheep on the former occasions, are engaged. Every animal that is unproductive, or which cannot be used, must, by a law which is strictly enforced, be sent to the mountains about the end of May, in order that as much fine grass as possible may be saved for the milch cows and ewes, and for making hay.

Mention has been made in the Journal, of the excellence of the riding horses of this country. When a young horse is thought to promise well, his nostrils are slit up, the Icelanders believing, that when exercised, or ridden hard, this operation will allow him to breathe more freely. I do not suppose that the horses of Iceland could run on our roads at the great rate at which I have seen them go, for any length of time. They are accustomed to scramble slowly through the bogs and over rocks, and to dart rapidly forward whenever they come to dry and smooth ground. In travelling, a man has generally two or three horses with him, and he changes from one to another as they become tired.

The saddle for the use of women resembles an elbow-chair, in which they sit with their feet resting on a board. Some of them are highly ornamented with brass, cut into various figures. The common people all ride in the same way, with the legs astride, the women having their feet raised so high, that their knees are considerably above the back of the horse.

For grinding corn, the Icelanders use small handmills, the same with those known in Scotland by the name of quern.

Though there is little encouragement from the climate, yet there are some parts of Iceland where experiments might be made in cultivating barley, potatoes, and turnips. Along the shores, where the soil is sandy, and where sea-weeds can be procured in abundance, something in this way might be done. But nothing can be effected without the superintendance of some active and intelligent person, able to combat the prejudices, and to encourage the exertions of the natives.*

* When this was going to press, I was informed, by the best authority, that the people of Faro were in a most distressing state for want of food; and that a ship was to be sent out to inquire into the reality of the intelligence.

Names of Districts.	Farms.	Families.	Farmers.	Hirelings having grass.	Hireling without grass.	Priests.	Civil Officers.
Southern Amt.							
East Skaptafells Syssel............	53	126	88	29	—	7	2
West ditto	133	248	214	26	1	8	1
Rangaavalle & Westmann Isles..	374	664	530	104	13	14	4
Arnes Syssel............................	418	709	495	153	40	21	1
Guldbringe and Kiose Syssels.....	256	704	302	132	245	13	12
Borgarfiord Syssel..................	227	285	250	9	11	9	4
Western Amt.							
Myre & Hnappadals Syssels......	180	235	202	13	9	9	2
Snæfell Syssel........................	270	652	337	143	167	10	2
Dale Syssel............................	181	231	208	10	4	7	3
Bardestrand Syssel..................	203	374	332	12	21	10	2
Western Isafiord Syssel............	123	261	236	5	13	7	1
Northern Isafiord Syssel..........	170	305	255	20	22	7	1
Strande Syssel........................	118	150	140	3	—	6	1
Northern and Eastern Amts.							
Hunavatn Syssel.....................	375	433	405	3	6	20	2
Skagafiord Syssel....................	412	492	459	9	2	20	3
Eyafiord Syssel.......................	448	535	499	6	13	17	3
Norder Syssel.........................	387	451	420	1	7	23	1
North Mule Syssel..................	217	267	222	32	3	11	1
Southern Mule Syssel..............	211	279	231	25	13	12	—

Males.	Females.	Total Inhabitants.
400	511	911
678	861	1539
1876	2311	4187
2053	2572	4625
1949	2066	4015
854	1028	1882
694	784	1478
1627	1914	3541
699	893	1592
1161	1332	2493
884	966	1850
961	1076	2037
438	544	982
1288	1592	2880
1385	1756	3141
1572	1881	3453
1337	1665	3002
781	1981	1762
839	998	1837
21476	25731	47207

This table, exhibiting the state of the population of Iceland in the year 1801, is taken from the register of the Bishop, to whom returns are regularly made by the priests. The original table contains also a detailed statement of the ages of the whole population of the island, of which the following is the general result:—

	Males.		Females.
Under 10 years of age	6231	……	6300
From 11 to 20	3207	……	3299
21 … 30	3385	……	4060
31 … 40	3190	……	3901
41 … 50	1838	……	2460
51 … 60	1707	……	2453
61 … 70	1162	……	1842
71 … 80	592	……	1096
81 … 90	158	……	285
91 … 100	6	……	35
Total	21476		25731

aving been unable to arrange them before our departure, could only furnish us with the increase.

rable; and that the average longevity of the women is greater than that of the men.

amounted to 50,444
.... 46,201
.... 47,287
.... 47,207
.... 46,349
.... 48,063

[To face page 284.]

Names of Districts.	Population.	Working Men.	Hired Servants.	Cows.	Heifers with one Calf.	Bulls and Oxen.	Calves.	Milch Ewes.	Rams and Wethers above one year old.
East Skaptafell......	853	94	11	391	34	29	56	2241	408
West Skaptafell	1499	224	17	747	102	150	116	3230	371
Westmann Islands..	157	42	5	39	8	1	—	268	88
Rangaavalle.........	3854	358	48	2074	251	174	325	8079	744
Aarnes	4398	505	105	2371	336	176	375	8550	700
Guldbringe	2866	915	739	490	36	6	50	1022	140
Kiose.....................	1088	135	37	444	31	110	109	1449	244
Borgarfiord...........	1796	260	130	764	69	94	120	3870	426
Myre & Hnappadals	1924	276	41	701	94	55	113	3898	372
Snæfellsnes.............	3541	346	598	671	55	16	56	1995	339
Dale........................	1506	128	41	408	50	18	60	1662	218
Bardestrand............	2225	448	123	665	61	31	76	3374	610
Isafiord...................	3887	418	60	965	62	46	84	6182	1911
Strande	895	89	26	201	4	4	16	2489	35
Hunavatn................	2873	260	79	894	87	52	86	11310	479
Skagafiord..............	3080	209	29	1086	114	29	52	9598	641
Eyafiord..................	3234	189	128	1012	74	22	116	9115	263
Norder	2928	246	43	677	29	28	74	10013	616
Northern Mule......	1845	285	30	475	28	52	75	8039	1116
Southern Mule......	1900	365	59	520	31	39	83	5921	1082
Total.........	46,319	5,792	2,349	15,595	1,556	1,132	2,042	102,305	10,803

		Occupied Farms.		Cattle.		Sheep.		Horses.
* 1703.........	50,444	4039		35,860	A. D.	279,812	A. D.	26,909
1770.........	46,201	3647		30,096	1760...	491,934	1770...	32,689
1783.........	47,287	3567	A. D.	21,457	1770...	112,809	1783...	36,408
1801.........	47,207	4661	1784...	†9,986	1784...	42,243	1784...	8,395
1804.........	46,349	—		20,325		218,818		26,524

* In the year 1707, the small-pox destroyed 16,000 people.   Between 1759
† In the year 1783, the great eruption from Skaptaa Jokul destroyed a gre

Rams, Ewe-Lambs, and Wethers, of one year old.	Lambs.	Horses & Mares broken in.	Horses & Mares not broken in.	Colts and Fillies.	Boats with 8 and 10 oars.	Boats with 4 and 6 oars.	Small Boats.	Number of Gardens in use.
1154	2024	615	97	78	—	19	—	5
1989	3153	1209	323	135	4	1	3	13
165	225	39	1	—	2	11	3	1
3750	7479	2889	949	468	9	8	2	23
2944	7050	2886	551	255	35	30	21	17
500	974	745	38	30	9	225	252	139
979	1487	469	203	94	2	16	31	13
1841	3250	900	249	125	3	25	88	9
1697	2929	1112	218	105	1	19	17	9
214	1671	670	92	27	67	118	50	8
867	1471	543	173	58	10	31	5	1
1809	2443	550	45	25	4	115	90	17
4048	5582	493	54	39	41	119	129	16
255	942	324	7	8	—	30	67	—
2620	4370	1885	294	137	—	42	3	1
2299	1322	1462	432	165	8	40	17	2
1842	6069	1241	177	109	10	56	20	9
3106	3217	945	38	32	2	68	55	—
3735	6549	733	71	73	1	40	11	2
2910	4779	663	97	79	—	55	23	8
38,724	66,986	20,373	4,109	2,042	208	1,068	887	293

A. D. { 1770 | 408 | 811 | 650 | 1,869 Total.
{ 1804 According to the above table.................. 2,163

The population of Reikiavik in the year 1806 (including 27 prisoners in the Toght - Huus) was 446. Of this number 134 were working men. Belonging to the town there were 35 cows, 2 sheep, 92 horses, 3 large boats, 26 of a middle size, 9 small ones, and 17 gardens.

53 and 1759, famine carried off 10,000, and vast numbers of cattle perished.
eat number of cattle; but the loss in this table appears to be exaggerated.

CHAP. V.

PRESENT STATE OF EDUCATION AND LITERATURE.

In the Dissertation prefixed to this volume, an attempt has been made to explain the circumstances in which the literature of the Icelanders originated, and to trace its progress through the successive periods of the history of the island. It will be the object of this chapter to complete the view of Icelandic literature, by exhibiting the present state of mental cultivation among the people; their institutions for the promotion of learning; and the modes of education among different classes of the community. From the more minute description to which they lead, these circumstances could not with propriety form a part in the general history of the country; though, as a sequel to it, they may possibly be interesting to the reader.

The picture of the present state of literature in Iceland is much less imposing than that of its early condition and growth. The changes, however, which the lapse of time has effected, are rather relative than absolute in their nature; and though the glory of the Icelanders is now for ever sunk, and their name almost lost among nations, yet in their own island they still keep alive much of that spirit of literary pursuit by which the character of their ancestors was so greatly distinguished. A few of the names which adorn the modern history of the country have already been mentioned. At the present time, there are many individuals, living on this remote spot, and from their situation exposed to innumerable privations, whose talents and acquirements would grace the most refined circles of civilized society. The business of education is systematically carried on among all ranks of the inhabitants; and the degree of information existing, even among the lower classes, is probably greater than in almost any part of continental Europe.

This state of mental culture will appear more wonderful, when it is considered that the circumstances of the country do not allow of any extended scheme of public education, and that the transmission of knowledge can take place only through the private and domestic habits of the people. In the existence among the Icelanders of habits which are fitted to this end, we contemplate a feature which is justly entitled to admiration and esteem.

At the present time, the school of Bessestad is actually the only establishment for education in Iceland. About the middle of the 16th century, when the reformation of religion took place in the island, two schools were founded; one at Skalholt, the other at Hoolum in the northern province; and a landed property was attached to these institutions, sufficient for the support of between twenty and thirty scholars at each place. Towards the close of the last century, the two schools were united into one, and transferred to Reikiavik; while in lieu of the school lands, which were appropriated by the crown, an annual sum from the public money was allotted to the support of the establishment. A few years ago, the school was again transferred to its present situation at Bessestad; the building being vacant which was formerly the abode of the Governors of Iceland. This edifice, though by no means in good repair, is from its size better adapted than any other in the country for the purposes to which it is now applied; and, but for the intervention of the war between England and Denmark, would have been further improved by the completion of some additional buildings, which are yet in an unfinished state.

The establishment at Bessestad consists at present of three masters, and twenty-three or twenty-four scholars; the funds of the school not allowing the reception of a greater number. The head master, or Lector Theologiæ, has an annual salary of 600 rix-dollars. It is his office to superintend the general concerns of the school, and to conduct more especially the theological department, and the study of the Hebrew language. At the time of our arrival in Iceland, the person who held this situation was Mr Steingrim Jonson; a man apparently not more than thirty-five years of age, but possessed of talents and learning which well fitted him for the discharge of its important duties. For several years he was the pupil and secretary of the late Bishop Finsson at Skalholt, after whose death he studied some time at Copenhagen; where, as a classical scholar, he acquired very great credit. His knowledge of the Greek and Hebrew languages is said to be accurate and extensive; and to theological studies he has given a very minute attention, being intimately acquainted with the writings of the most eminent of the German theologians. This gentleman, during our stay in Iceland, was removed from Bessestad to the church of Oddè, in Rangaavallè Syssel, one of the most valuable livings in the island. He was succeeded by another person, of the same name, who is likewise reputed to be a man of learning and acquirements.

The two inferior masters of the school have salaries of 300 rix-dollars each. The office of the second master comprehends the instruction of the scholars in Latin, history, geography, and arithmetic; while the third is occupied in teaching the Greek, Danish, and Icelandic languages. It is a singular circumstance in the regulations of the school, that each scholar, whether intended for the pastoral office or not, is obliged to study the elements of Hebrew, and to undergo some examination in this language. By far the greater number, however, of those who attend the school, are preparing themselves for this future situation in life; and in the admission of scholars, a preference is always given to the children of priests. A youth is not allowed to enter until he has been confirmed; and a certificate of his talents and dispositions is required from the minister of the parish in which he has resided. The period of annual study extends from the beginning of October to the end of May; the summer being made the season of vacation, to accommodate the rural occupations, in which all ranks among the Icelanders are obliged to partake. It is a part of the office of the Bishop to visit the school at the commencement and close of each session; and at the latter time, to superintend the examinations of the scholars, which then take place. These examinations continue during several days, with a prescribed form of proceeding, of which a sketch has already been given in the narrative.

After a certain degree of progress in the studies allotted to him, each scholar becomes what is termed a *demissus;* leaving the school, and pursuing his future studies at home. No particular period is fixed for a *demission.* This is determined solely by the proficiency of the student, as ascertained by an examination; for which it is required that he should be able to read and write Latin with accuracy, that he should have some knowledge of Greek and Hebrew, and of the rules for interpreting the Old and New Testaments; and that he should be acquainted with the Danish language, with history, arithmetic, and geography. The knowledge of Greek and Hebrew, though officially required, is, however, in the practice of these examinations, by no means very rigorously exacted. Where the students are preparing for the priesthood, as is generally the case, they are farther questioned upon the Bible and ecclesiastical history, upon the doctrines of the Lutheran church, &c. If a youth has continued seven years without attaining the qualifications which entitle him to become a *demissus*, the Lector writes to his family, representing the matter to them, and he is not allowed to remain longer at the school.

A library is attached to the establishment at Bessestad, containing probably twelve or fourteen hundred volumes; among which are a few good editions of the classics. The greater part of the library consists of Icelandic and Danish works; beside which there are a considerable number of volumes in the German language, and a few in the English and French. The number of manuscripts is very inconsiderable, and they appear to be of little value. The private library of the Lector Theologiæ, though smaller, is more select, and contains the works of Mosheim, Heinzius, Reinhard, Lowth, Griesbach, Michaelis, and numerous other authors of minor note, on ecclesiastical history and doctrine. It is the best theological collection in the island.

Among the young men educated at this school, there are some who afterwards go to Copenhagen, with the view of prosecuting their studies at the university there; this advantage being occasionally afforded to the children of those who hold civil offices, or possess landed property, and to the sons of some of the wealthier among the clergy of the country. The number of students, however, who enjoy such opportunities is very limited ; and the remainder, oppressed by poverty and the necessities of their situation, are generally compelled to take up their abode for life in solitary spots, where their intercourse, even with each other, is almost wholly suspended, and where any future progress in knowledge can only be effected by their independent and unaided exertions. This is the condition of all the country priests in the island, and of many of the more respectable of the proprietors and farmers. Deprived, as they thus appear to be, both of the means and motives for mental cultivation, it could scarcely be expected that instances should occur, where the ardour of literary pursuit is still maintained, and the acquisitions of former study not only preserved, but even increased and improved. The occurrence, however, and even the frequency of examples of this kind may render necessary some explanation of a fact so extraordinary. Among the more obvious of the causes which present themselves, is the long period of leisure which the Icelanders enjoy, during the protracted winters of their northern region. This leisure, those who have acquired in their youth the habits of literary pursuit, will naturally devote to a continuance in occupations, which are so well adapted to relieve the weariness of the passing time. Their means of study are indeed very limited, and the enjoyments of participation almost wholly denied ; but these comparative disadvantages are in some measure overcome by the habits of perseverance, which necessity creates, and which are maintained from an experimental sense of their value. Nor is the great name of their ancestors without its influence upon the present generation of Icelanders. There are few amongst them who cannot refer back to the times, when those, born on the same soil with themselves, were raised to honours and renown in foreign lands: and never is this appeal made without an animated feeling of patriotic pride and satisfaction.

Among the class of priests, another motive to mental cultivation is the desire of maintaining in their office, an influence, which cannot be derived from any difference of external circumstances. The pastor must undergo the same labours and hardships as the meanest of his flock : he enjoys few additional comforts or refinements of life; and but for the superiority of his intellectual attainments, would speedily lose that station in society, which it is so necessary he should retain. It forms, too, an important part of his duty, to superintend the business of domestic education in the families placed under his pastoral care. This office is not, indeed, strictly required by the ecclesiastical statutes of the country; but it is founded upon usage, and ultimately upon a sense of the necessity for such a superintendance, where the public means of education are so greatly limited by the poverty of the people, and the dispersion of their numbers. An interesting example of the attention with which this duty is sometimes exercised, has been given in the Journal, (p. 143); and the instance of the parish priest of Saurbar is by no means singular among the ministers of religion in Iceland. Their poverty, indeed, rather increases than lessens the influence of these exertions, by associating them more intimately with their parishioners, and promoting that free and unreserved communication, which a more refined state of society has so much tendency to preclude.

By this superintendance of the priests, and the long established habits of the people, a regular system of domestic education is maintained; in the benefits of which even the lowest ranks of the community partake. With the exception
of those who inhabit the coast, in the vicinity of the great
fishing stations, it is a rare thing to meet with an Icelander
who is unable to read and write, or who does not possess considerable intelligence on all subjects to which his situation al-
lows him access. The instruction of his children forms one
of his stated occupations; and, while the little earthen hut
which he inhabits is almost buried by the snows of winter,
and darkness and desolation are spread universally around,
the light of an oil lamp illumines the page, from which he
reads to his family the lessons of knowledge, religion, and
virtue. The importance of these domestic habits has been
well understood by the Icelanders themselves. In the ecclesiastical code of the country, an article is extant, singular
perhaps in its nature, but admirable in its design, which gives
to the Bishop, or even the inferior clergy, the power of preventing any marriage where the female is unable to read.
This law, which provides so powerful a pledge for the instruction of the rising generation, is still occasionally acted upon,
though probably not with so much strictness as in former
times. The books in the possession of the lower classes are
chiefly of a religious nature; a great number of these works
having been printed in Iceland during the last two or three
centuries, and very generally circulated through the country.
In many parishes, there is a small collection of books belonging to the church; from which, under the superintendance of the priest, each family in the district may derive
some little addition to its means of instruction and improvement.

The historical and poetical writings which the early literature of Iceland produced, are by no means generally known
among the Icelanders of the present time ; such studies
being principally confined to the priests, and to those of
the higher classes. The calamities which oppressed the island
during the 15th century, and which entirely extinguished the
celebrity, and almost even the name of the people, interposed a sort of barrier between the ancient Icelanders and their
posterity. Learning was restored under an altered form ;
the works of former genius were only partially revived ; and
these circumstances, together with the changes progressively taking place in the language of the country, have removed
from the possession of the present race of people, all the more
striking evidences of the ancient condition of their community. A great number of manuscripts are still to be found in
the churches, and in the houses of the priests and principal
inhabitants ; but, with few exceptions, they are all of modern
date, and are merely the representatives of works which were
intended for publication, but which the poverty of their writers, or other circumstances, have unavoidably suppressed.
The greater proportion of the Icelandic manuscripts which
derive value from their antiquity, have been gradually transferred to Copenhagen, and deposited in the public or private
libraries of that metropolis.* Here they have been carefully
collated, with a view to the publication of those which were
found most remarkable or important ; and it is principally
through this channel that the earlier writings of the Icelanders are known to the present inhabitants of the country. The

---

* The library of Professor Thorkelin, which contained a valuable collection
of Icelandic works, is said to have been destroyed during the late bombardment
of Copenhagen.

valuable editions of these writings printed at Copenhagen, have come into the possession of all who bear a literary character among the Icelanders; and a few editions of the works of this period, which have been printed in the island, have given a further diffusion to this branch of knowledge among the people. It is, however, by no means general; the tales and traditions which now prevail in the country, relating for the most part to more recent times, and being in few instances derived from the Sagas and poems, in which the events of antiquity are described.

Among those individuals of the present day who have made the early literature of Iceland an object of study, the name of Finnur Magnuson may particularly be mentioned. This young man, who holds the situation of public pleader in the courts of law at Reikiavik, and is distinguished by his classical acquirements, has bestowed very great attention upon the early writings, and especially upon the ancient poetry of his country; and is considered to have a more intimate knowledge of them than any other person in the island. He has likewise been enabled, from his residence on the spot, and from his family connections with several eminent Icelanders, to collect some manuscripts of considerable value from their age and rarity. The industry and success of Professor Thorkelin in the same pursuits are more generally known; but the long absence of this gentleman from Iceland has lessened, in some degree, his connection with the modern literature of the country.

In describing the state of knowledge among the present race of Icelanders, their attainments in languages and in classical literature must particularly be noticed. This is one of the first of those circumstances which engage the attention and admiration of the stranger, in visiting the island.

He sees men whose habitations bespeak a condition little removed from the savage state; who suffer an almost entire privation of every comfort or refinement of life; and who, amid the storms of the surrounding sea, seek, in their little boats, the provision upon which alone their families can scarcely depend. Among these very men, he finds an intimate knowledge of the classical writings of antiquity; a taste formed upon the purest models of Greece and Rome; and a susceptibility to all the beauties which these models disclose. While traversing the country, he is often attended by guides who can communicate with him in Latin; and, arriving at his place of nightly rest, he not unfrequently draws forth from the labours of his little smithy, a man who addresses him in this language with the utmost fluency and elegance. This cultivation of the ancient languages has been common among the Icelanders from an early period in their history; and it will be seen from the Preliminary Dissertation, that many of the principal works which distinguish their literature, and especially those of the historical kind, have been composed in Latin. At the present time, this language forms a part of the education of the priests, and of all the principal inhabitants of the island. It is still very frequently the vehicle of their writings; and a great number of inedited Latin books, both in poetry and prose, may be found throughout the country, destined for ever to remain in the obscurity which gave them birth. Epigrams and short descriptive poems in the same language are exceedingly common; and, through this medium, the Icelanders often indulge that tendency to personal satire, which it formerly required even the operation of laws to restrain.

The study of Greek, as might be expected, is by no means equally general; but there are, notwithstanding, several very

excellent Grecian scholars, who now do credit to the literature of the island. In the first place among these is the present Bishop, Geir Vidalin; a man whose acquirements in
every department of literary pursuit, would do honour to any
country, or condition of society. To classical studies, he has
devoted peculiar attention; and, in his colloquial Latin, he
displays a facility and correctness of style, and a richness
and propriety of quotation, which evince the most intimate
acquaintance with the writers of the best ages of Rome. In
Grecian literature, his reading has been almost equally extensive; and he is said to be a very excellent Hebrew scholar. Among the other Icelanders of the present time who
have distinguished themselves in classical literature, are,
Steingrim Jonson of Bessestad; the Rector Hialmarson, who
formerly conducted the school at Hoolum; and Arnas Helge-
son, the priest of Vatnsfiord, at the most northerly extremity of the island. Few translations from the classics have been
published in the Icelandic language, though it is probable
that many manuscripts of this kind exist in the country. The
Transactions of the Icelandic Society, a work afterwards to
be mentioned, contain translations of the Idylls of Theocritus,
which possess very great merit; and, in the same work,
there is a translation of Plutarch's Paidagogia by the Assessor Einarson. The fables of Æsop, and many of the odes of
Horace, have likewise been given to the Icelanders in their
native verse.

It may be proper to notice here the great attention to the
studies of philology and criticism, which has existed among the
learned men of Iceland during the last two centuries. Many
valuable works, connected with this department of literature,
have been already published, either in the island or at Copenhagen; and numerous writings of the same kind are still

to be found in manuscript, in different parts of Iceland. A catalogue of a few of these works is given in the subjoined note. *

In the study of the modern languages, the proficiency of the Icelanders is as great as can be expected from their limited intercourse with the continental nations. With the Danish language all the higher class of inhabitants are perfectly familiar; the German is understood by many; and of late years the English has been cultivated by a few individuals with much success. All these languages, as is well known, originate from the same root; and the resemblance still retained between the Icelandic and Danish, or still more the Norwegian, is such, that the natives of each country can, without much difficulty, make themselves mutually understood. Through these different channels the Icelanders have acquired considerable information respecting the modern literature of Europe, particularly that of Germany and Denmark; and they possess not only the

---

* Lexicon Runicum *Magni Olavii.* Havniæ 1650.

Lexicon Islandicum *Gudmundi Andreæ.* Havniæ 1683.

Monosyllabica Islandica *Johannis Rugmanni.* 1676.

Præcepta Grammatica et Syntactica *Theodori Thorlacii.*

Linguæ Septentrionalis Elementa, per *Runolphum Jonæ.* Havniæ, 1652. Republished at Oxford, 1689; and in Hickes's Thesaurus, 1705.

Gustus ad Isocratem *Arngrimi Vidalini.* Havniæ 1698.

Dissertatio Philologica *Arngrimi Vidalini,* de vocis באזו (Psalm xx. v. 17.) genuina lectione et significatione. 1689.

Lexicon Juridicum *Pauli Vidalini.*

Expositio Alphabeti Hebraici *Gunnlaugi Snorronis.* Havniæ 1775.

Tractatus de Orthographia Islandica *Eggerti Olavii.*

Claves Metricæ *Thorsteini Magni, Gudmundi Olavii, Thorlaci Gudbrandi,* &c.

NOT PUBLISHED.

Lexicon Islandico-Latinum *Johan. Arnæi, Episcopi Skalholtensis.*

Lexicon Islandicum *Gudmundi Olavii.*

Lexicon Islandicum *Johan. Olavii,* &c. &c.

originals, but translations of many of the works, which have acquired reputation in these countries in later times. Their knowledge of English literature is obtained chiefly through the medium of the Danish and German; in which languages the works of Addison, Pope, Richardson, and Young, are known and admired by many individuals in the island. They possess likewise a few translations of English works into their native language. Twenty or thirty years ago, the whole of Milton's Paradise Lost was translated into Icelandic verse by Jonas Thorlakson, a priest at Backa, in the northern part of the island; of which translation two books were published in the Acts of the Icelandic Society: the remainder are yet in manuscript. The merits of the poetry in this translation are spoken of in terms of high eulogium by the Bishop; who is, however, unacquainted with the original. The same Jonas Tholakson has also translated Pope's Essay on Man, of which a considerable edition was printed at Leira in 1798, in a duo-decimo form.

The cultivation of poetry in Iceland, though by no means so general as in ancient times, still forms a striking feature in the literature of the country. Among those of the natives who enjoy the reputation of talents or learning, there are few who have not occasionally tempted the Muse; and where such efforts have been seconded by the true inspiration of genius, the poet has received his reward in the unlimited applause and admiration of his fellow-citizens. The days indeed are past, when the bard ' poured forth his unpremeditated lay' to the assembled and admiring multitude; but in exchange for these rapid and irregular effusions of fancy, a more classi-cal style has been acquired, and greater scope is given to the exercise of selection and taste in poetical compositions. A few only can be mentioned of those individuals who still adorn this branch of Icelandic literature. One of the most eminent is the Assessor Benedict Grondal, a judge in the higher court of justice, and a man of an elegant and cultivated mind. His published poems, which are regarded as the best modern specimens of the Icelandic language, are not, however, either very numerous or considerable in length; consisting chiefly of odes, epitaphs, and other detached pieces; among which are many excellent translations from Theocritus, Anacreon, and Horace. A translation of Pope's Temple of Fame, which was published some years ago, is greatly esteemed by the best judges of Icelandic poetry. He has composed also several poetical satires; in which, according to the information of the Bishop, there is much successful ridicule, after the manner of Horace, of the follies and vices of his countrymen; but these satires, in consequence of the express prohibitory article in the laws of the island, he has not ventured to publish. The general style of his poetry is described to be terse, pointed, and elegant.*   Finnur Magnuson is another of the Icelandic writers of the present day; who has acquired much credit from the facility with which he composes in the Latin and Danish languages, and for the extreme accuracy of his Icelandic style. He has translated into Danish verse, the poem of his uncle Eggert Olafson, on the rustic life of the Icelander, and published also several smaller pieces. Jonas Thorlakson, the venerable translator of Milton, is still living in a remote part of the island, and has composed many original poems of great merit; of which, however, nearly all are unpublished. Another individual, possessing some reputation, is Sigurdar Peturson of Reikiavik, who has written among other works a poem in six books, called Stella; in which, under a fictitious

---

* From his state of health, when we were in the country, it is much to be feared, that Assessor Grondal can no longer be counted among the living poets of Iceland.

form, the manners and habits of the Icelanders are minutely described. This poem is likewise unpublished, and will probably ever remain in obscurity. The poverty and other circumstances of the Icelanders, offer indeed such multiplied obstacles to their literary progress, that it is impossible not to admire the ardour and industry which in pursuits of this nature they continue to display. In the department of poetry more especially, the number of manuscript works, doomed from the situation of their authors to perpetual oblivion, is exceedingly great; yet the muse is still invoked; and the taste and feeling for such compositions are still awake in the minds of the people, though so little cherished by opportunity, or by the aspect of surrounding nature.

The religious character of the Icelanders has strongly disposed them to the cultivation of sacred poetry; and a great number of writings of this kind have appeared in the island during the last two centuries. Besides numerous collections of psalms and hymns, various parts of the Old and New Testament, as the books of the Pentateuch, the history of David, and the life of the Apostle Paul, have been published in the form of poetical paraphrase; and a few of these works may be found in the possession of almost every family in the country.

With the scanty materials for history which the Icelanders possess, it is scarcely wonderful that this branch of literature should be less cultivated now than it was in former times. Among those of the natives, who have received patronage and support at Copenhagen, many, during the last century, have well maintained by their historical writings the reputation of their country; but in Iceland itself, few considerable works of this kind have lately appeared; and the greater number of these relate merely to the events of the passing age. The Chief Justice Stephenson is undoubtedly entitled to the first place among the present historical writers of the island. This gentleman, whose zeal in the pursuit of knowledge has been seconded by better opportunities than most of his countrymen enjoy, has held during the last twenty years the most conspicuous place among the literary characters of Iceland; a situation to which his acquirements and influence would seem to justify his claim. The attainments he has made are various and extensive; a residence of several years at Copenhagen having afforded him access, not only to the literature, but also to some part of the science of modern times. To the English language he has paid particular attention, and besides speaking and writing it with facility, he is familiar with all the more eminent of our writers in the department of the belles-lettres. Mr Stephenson is himself a very voluminous author. As President of one of the Literary Societies of Iceland, he has published many books for the use of the Society; and from the catalogue of his writings, which is given in the note, it will be seen that his labours have comprehended a singular extent and variety of subjects.* The most important

* 1. Treatise on Meteors, 1783.

2. Description of the eruption of a new volcano in Skaptaa-fell Syssel, with engravings, 8vo. 1785.

3. Mournful Thoughts, and a Funeral Song, at the tomb of the celebrated and beloved Bishop Finsson, 8vo. 1796.

4. A Panegyric on Bishop Finsson, read to a meeting of the Icelandic Society, 8vo. 1797.

5. Amusements for Friends; containing useful pieces in prose and poetry, on natural history, physic, astronomy, history, morals, and divinity, 8vo. 1797.

6. A Biographical Memoir of the Lady of the late Bishop Teitson, 8vo. 1797.

7. *Minnisverd Tidindi,* or Memorable News, containing the most remarkable historical events from 1795 to 1801; in five vols. 8vo.

8. Jest and Earnest; or a collection of pieces in poetry and prose; on jurisprudence, morals, theology, medicine, and the belles-lettres, 8vo. 1798.

9. A defence for the injured Icelandic Magistrate, 8vo. 1798.

among his writings, is the History of Iceland in the 18th century ; a work which abounds in valuable information respecting the civil condition, the natural history, and the literature
of the island during this period.   In the style of the book, a
singular example of the *prosopopeia* occurs; Iceland being made
occasionally to tell her own tale, and to speak in a personal
form of the various events which have befallen her : a mode
of narrative, which though sometimes verging towards the
ludicrous, has nevertheless a simplicity congenial to the subject, and capable even of rising into the pathetic and sublime.
Another historical work, edited and chiefly written by Mr
Stephenson, is a sort of political register, of which in the period
between 1795 and 1802, a volume was published annually at the
Leira printing-office, under an octavo form. This contained a
narrative of the political events which had occurred in Europe
during the preceding year ; a separate article being allotted
to the affairs of every state. The narratives appear to be
drawn up with much care and considerable minuteness. Under
the article of England, as an example, not only are the more

10. A select collection of Hymns and Psalms for the Churches, 8vo. 1801.
11. Translations from the Danish of Select Royal Edicts : published by the
    Court General of Iceland at different periods since 1801.
12. A speech made at the establishment of the Royal Court General, 8vo. 1801.
13. Iceland in the 18th century, 8vo. 1806.
14. The same book translated into Danish, with additions, 8vo. 1808.
15. A Treatise on the Fuci ; published by the Royal Society of Denmark, 1809.
16. A Treatise on the Sheep, Cows, and Horses of Iceland; published by the
    Copenhagen Veterinarian Society, 1809.
17. Encouragement to the good people of Iceland during these times of war.
    Written and published by the King's order, 8vo. 1808.
18. Instructions for the Officers of the Police, 8vo. 1809.
19. Commentary on these Instructions, 8vo. 1810.
20. Help in Distress ; or Instructions to the Icelandic farmers in their hus-
    bandry, 8vo. 1810.

important national events described ; but the state of parties is accurately detailed ; extracts are given from the Parliamentary debates ; and notice is taken of many provincial occurrences. The information necessary to this work was almost entirely obtained through the medium of Denmark. The greater number of the volumes were written by Mr Stephenson : one by his brother the Amtmand Stephenson ; and the last which was published, by Finnur Magnuson of Reikiavik.

The vast number of works of divinity, which have appeared in Iceland since the period of the reformation of religion, testify the diligence with which such studies have been pursued by the learned men of the country ; and from catalogues which are extant, it would appear that the writings on this subject, yet inedited, are much more numerous than those already published. Many of these works are translations from the German and Danish ; with a few also from the English language. Among the latter, may be mentioned the translation of the ' Whole Duty of Man' by John Vidalin, a Bishop of Skalholt at the beginning of the last century, very eminent for his piety and learning. The original writings of the natives are for the most part either commentaries on particular parts of scripture, or collections of prayers, homilies, and sermons; the doctrinal parts of theology being less frequently the subjects of discussion. At the present time, the works of the Bishop, whose name has just been mentioned, possess great reputation in the country ; and of the collections of sermons which he published, some are to be found in almost every habitation. In all departments of literature, there is a strong disposition among the Icelanders to critical severity ; and in theological writings more especially, this severity has occasionally assumed a very rigorous form. A curious instance of this kind occurred about a hundred years ago, when an unfortunate man was publicly whipped, as a punishment for the errors he had committed in a translation of the book of Genesis.

Metaphysical studies do not engage much attention among the learned men of Iceland. A few individuals, from their residence at Copenhagen, have become intimately acquainted with the metaphysics of the German schools, and have themselves published treatises connected with the subject; but these writings are by no means numerous, nor does it appear that they possess any peculiar value.* Publications connected with practical morality are, however, very common in Iceland; and several excellent books of this kind have lately appeared in the island, adapted chiefly to the use of the farmers, or those of the middle class; in which moral instruction is judiciously blended with amusing information in various branches of knowledge. The most valuable of these writings is a work, called ' Evening Hours,' which was published by the late Bishop Finsson, a few years before his death.

While the studies of literature are thus cherished among the Icelanders of the present time, science, strictly so called, engages but few votaries; and these follow with feeble and tardy steps the rapid progress which has been made among the European nations. Even in the department of natural

---

* The following are some of the metaphysical writings of the Icelanders:

Dissertatio de Essentia Consecutiva; by Stephen Biornson, 1757.

Dissertatio de Ente; by Finnur Thoralfson, 1770.

Examen facultatis cognoscitivæ superioris et inferioris; by the same author, 1772.

Dissertatio de Infinito; by Skule Thorlacius, 1762.

Disputatio de Nihilo; by John Olafson, 1758, &c.

The Dialectics of Aristotle, and the Logic of Peter Ramus, have been translated into the Icelandic language.

history, where the situation of the people does not oppose the
same obstacles as in other scientific pursuits, there are few
individuals who have acquired more than a superficial knowledge of the subject, and few works have been published,
which possess value either from the extent or accuracy of
their information. Most of the writings of the Icelanders
upon the natural history of their own country, display indeed
a singular vagueness of description, and more of superstitious
belief than is entirely consistent with the other habits and
attainments of the people. When considered, however, the
latter circumstance will scarcely be thought surprising. In
forming the scenes which surround them, nature seems to
have deserted all her ordinary operations, and to have worked
only in combining the most terrific extremes which her powers
can command. Nor is it merely a passive and silent desolation which marks this tremendous influence. After the lapse
of ages, the fire of the volcano still bursts out among regions
of eternal snow, the earthquake still shakes the foundations
of the island, and the impetuous thundering of the Geyser
yet invades the stillness of the surrounding solitude. Living
amidst so many wonders of nature, and ignorant of natural
causes, the Icelanders are readily infected by superstition relating to these objects ; and this influence is observable in most
of the descriptions they have given of their own country. In
later times, however, such superstitions have greatly declined;
and during the last century, several works have appeared,
descriptive of the natural history of the island, in which
accurate observation is conjoined with some degree of scientific knowledge. The author of most reputation in this department is Eggert Olafson, who in 1749 printed his ' *Enarrationes Historicæ de Islandiæ Naturâ et Constitutione;*' and afterwards in conjunction with Paulson, another naturalist, pub-

lished a larger work, under the title of 'Travels in Iceland;' in which the various objects in its natural history are carefully and minutely described. The mineralogical details in this book are very ample; but owing to the want of arrangement, and of suitable nomenclature, they are not easily intelligible to the reader. In 1780, a work by Olaf Olafson, entitled, 'Economical Travels through the northern parts of Iceland,' was published in two volumes quarto; containing much valuable information upon the natural history of this district. Two smaller treatises are subjoined to the work; one on the Surturbrand, the other on the Sulphur beds of Iceland. Several descriptions have been published of the different volcanic eruptions during the last century; among which may be mentioned, the treatises of John Sæmundson on the eruptions around the lake of Myvatn, in 1724, 1725, 1727, and 1728; the treatise of Bishop Finsson on the eruption from Hecla in 1766; and that of Mr Stephenson on the great eruptions in 1783, at Cape Reikianes, and from the mountains of the Skaptaa Jokul. The person said at present to be the best naturalist in Iceland, and particularly intimate with botanical science, is Swein Paulson, one of the medical practitioners of the country, whose abode is near the volcano of Kattlegiau Jokul on the southern coast. His principal original work is on the diseases of Iceland; but he has written also several treatises on the natural history of the island, and on the rural economy of the people, which are said to possess great merit. Mr Stephenson has distinguished himself in the same department; and in many of his writings, has laboured to make his countrymen avail themselves of all the means of improving their condition, which are rendered possible by the nature of their soil and climate.

The sciences of astronomy and mathematics are by no

means generally cultivated among the Icelanders; though there are some individuals who have successfully pursued these studies, either in the island itself, or with the better opportunities which were afforded by a residence in the Danish metropolis. A few of the astronomical treatises published during the last century are noticed below.* The study of the mathematics, though prescribed by the regulations of the school at Bessestad, receives but little attention there; nor does it in general form a part of the private occupation of those, who enjoy a literary character in the island. No purely mathematical work is extant in the Icelandic language; but in a book of arithmetic, which was written some time ago by the elder Mr Stephenson, trigonometry and the elements of equations are briefly included. Stephen Biornson, formerly the master of the school at Hoolum, published in the acts of the Icelandic Society a treatise on statics, which is well spoken of. This man, whose acquirements in various branches of science were very considerable, died at Copenhagen about the beginning of the present century.

After all that has been said in the preceding pages of the poverty of the Icelanders, and of the nature of the country upon which their destiny is cast, it will not be thought wonderful that the fine arts should desert a clime so little congenial to their growth. Painting exists here only in it rudest

---

* Dissertations on the Zones of the Terrestrial Globe, and on the Phases of the Moon; by Magnus Areson. 1710.

Dissertations on the Astronomy of the Chaldeans; on the Sphere of the Fixed Stars; on the Pythagorean Harmony, &c.; by Thorleif Haltorson; 1706, 1707, 1708.

Dissertation on the Solar Year of the Ancients; by Jonas David Gam. 1733.

Dissertation on the effects of Comets; by Stephen Biornson. 1758.

Dissertation on Celestial Physics; by the same. 1760.

Introduction to Tetragonometry; by the same. 1780.

forms; the native music of the island is inharmonious and uncouth; while the art of sculpture is almost entirely unknown. In proof, however, that these deficiencies must be ascribed to the situation of the people, and not to a defect in original genius, it may be mentioned that Thorvaldson, the son of an Icelander, dwelling on the classic ground of Rome, is at the present moment second only to Canova among the statuaries of Europe.

Before closing this sketch of the literature of Iceland, it will be necessary to say something of the Icelandic Societies, to which a reference has already more than once been made. The first of these was instituted at Copenhagen in 1779, with the professed object of aiding the literature of the island, and bettering the condition of the inhabitants, by the suggestion of improvements in their rural economy. The Society, at its first establishment, was composed of only twelve members; but the number was afterwards extended to about a hundred and thirty; comprizing all the most learned and intelligent men in Iceland; and, as extraordinary members, many individuals of much eminence on the continent of Europe. Fourteen volumes of transactions have been published by the Society; the contents of which are various, comprehending many valuable papers upon the fishery, agriculture, and natural history of Iceland; poetry; historical essays, &c. The principal writers were Bishop Finsson; the elder Mr Stephenson, then Governor of the island; Mr Ericson; Mr Paulson; Mr Ketilson, Sysselman of Dalè; and many others of the literary Icelanders, whose names have before been mentioned. These transactions were entirely composed in the Icelandic language, and great attention was paid to the preservation of its purity; two censors being appointed to judge of the style of every essay which was presented for publication.   In 1790, considerable dissensions arose in the Society, in consequence of a project for transferring it to Iceland; and this circumstance, with other concurring events, had the effect of suspending all the proceedings of the institution. Though still nominally in existence, it has now for a long period been entirely dormant.

The second Icelandic Society was established in the island in 1794; chiefly under the auspices of the Chief Justice Stephenson, who had been a strenuous advocate for the removal of the former Society from Copenhagen.   The original number of members was not less than twelve hundred; most of the farmers, as well as the priests and civil officers of the country, being included.   The object of the institution was the promotion of knowledge and improvement among the people; and with this view, a fund was provided by the annual contribution of a dollar from each member, and devoted to the publication of books, to be distributed among the subscribers.   The printing establishment at Hoolum, which had fallen into decline, and another which in 1773 was instituted at Hrappsey, an island in the Breide-Fiord, were purchased by the Society; and a printing office, under their management, established at Leira, in the Borgar-Fiord Syssel. From this press have issued, for the use of the Society, fifty or sixty different works; some of them translated, but the greater number original, and comprizing a very great variety of subjects; history, poetry, divinity, law, medicine, natural history, and rural economy.   In his situation of President, Mr Stephenson has had the superintendance of the funds of the Society, the entire direction of their various publications, and the management of a small, but well chosen collection of books, which was presented to them by some literary characters of the Danish metropolis.   Though his exertions, however, for the support of the institution have been incessant, it has notwithstanding greatly declined during the last few years; and, at present, the number of members does not exceed thirty or forty. The successive occurrence of several unfavourable seasons, and the evils entailed upon the Icelanders by the war between England and Denmark, have contributed in some degree to produce this decline. It was found, too, that there was much difficulty in accomplishing the scheme for the distribution of books, among a people dispersed over so vast an extent of country, and during a great part of the year so entirely separated from each other by the barriers of nature. Some little fault has, perhaps, also existed in the general management of the institution. The office of censorship of the press, vested in one individual, has had the effect of disgusting or deterring many, whose connection would have done credit to the Society; while by giving too much to this single judgment, it has been the means of bringing forth many works, little adapted to the comprehension of those, for whose use they were designed. In the present state of the Society, it is particularly unfortunate that the only printing establishment in Iceland should be thus exclusively appropriated; and as a matter of general policy, it may be doubted whether, under any circumstances, such a corporate institution ought thus to monopolize the literary growth of the country. As guides and protectors to the progress of knowledge, institutions are good :—when they become arbitrary rulers, their influence but retards the course of mental improvement, and proves baneful to the very energies which gave them birth.

Such is the present state of mental cultivation among this singular people. The disparity of their physical and moral circumstances forms an interesting feature equally in the history of literature, and in that of the human species.   While the calamities of internal warfare, and the oppression of tyrannical governments, have clouded with ignorance and barbarity countries on which the sun of nature sheds its brightest beams,—the possession of peace, of political liberty, and well ordered laws, has given both intellectual and moral exaltation to a community, which has its abode at the very confines of the habitable world.

CHAP. VI.

ON THE GOVERNMENT, LAWS, AND RELIGION.

SINCE the period when Iceland was first annexed to a European monarchy, the progress of time has effected little change either in the physical condition or political situation of its inhabitants. The general form of government, which was established nearly six centuries ago, is still preserved; and the circumstances of the people have required few important alterations in the code of laws, which was then transmitted to them by the Norwegian monarchs. Even this form of government, and these laws, were founded upon the existing usages of the country; and we must principally look to the wisdom of those, who framed the ancient commonwealth of Iceland, for the origin of institutions, which, in this later age, preserve to their posterity all the blessings of tranquillity and social order.

The government of Iceland is committed to an officer, appointed by the crown of Denmark; who is occasionally a native of the island, but more frequently a Dane or Norwegian by birth. This supreme magistrate has the title of Stiftamtmand. It is his office to conduct the various public con-cerns of the country; to preside in the courts of law; to superintend the execution of the laws, and the collection and disposal of the public revenue; and, in conjunction with the Bishop, to regulate the schools and certain ecclesiastical con-cerns of the island. He receives a salary, as Governor, of 2,000 rix-dollars per annum; and has a house appropriated to his use.

Since the disturbances which occurred in Iceland in the summer of 1809, there has been no regular governor in the country; and it is probable that the appointment of one will not be made, till the war between England and Denmark has been brought to a termination. After Jorgensen had been deposed, and Count Trampe taken to Britain in the Talbot sloop of war, the functions of the government were under-taken by Mr Stephenson, who retained this office till the month of June 1810; when the change took place which has already been mentioned in the narrative.

Immediately subordinate to the Governor are the Amt-men, or Provincial Governors. The island is divided into four provinces; but as the jurisdiction of the northern and eastern are united, and as the Governor of Iceland assumes the office of Amtmand of the southern province, in which he resides, there are, in fact, only two officers who possess this title. Their duties are very similar to those of the Stift-amtmand, though on a more limited scale. They inspect the conduct of all the subordinate officers, and hold provincial courts, at which a report is made of all the public proceed-ings within their districts. The present Amtmen are Mr Stephenson of Huaneyrè, who has the jurisdiction of the western province; and Mr Thoranson, who holds the north-ern and eastern provinces of the island.

The Sysselmen, though of inferior rank, are, in the offices they sustain, of great importance in the country. They have the personal charge of collecting the taxes in their several districts or Syssels; they hold courts of law, and pronounce judgment in all cases; they watch over the public peace, officiate as public notaries, and maintain the rights of inheritance. The Sysselman is appointed by the Crown; and the office, on account of its importance, is always given to one of the most respectable landed proprietors within the district.

There is still a subordinate local officer, called the Hreppstiorè, whose jurisdiction is a parochial one, and whose duty it particularly is to attend to the condition and maintenance of the poor, and to assist the proceedings of the Sysselman in all that relates to the preservation of public order. Where the population of a parish exceeds four hundred persons, the office is committed to two individuals; who usually belong to the class of farmers, and are invariably chosen from a regard to their abilities and steadiness of character. By a late edict of the Danish Government, these inferior magistrates, as well as their children, are exempted from the levies for the support of the poor, and are allowed also to use a dress appropriate to their situation. In each parish, beside the Hreppstiorè, there are a certain number of individuals (Forlikunarmen), appointed for the express purpose of accommodating any disputes which may arise among the inhabitants.

The laws of Iceland, it has been already remarked, are founded principally upon the ancient code, called the Jonsbok, which was introduced into the island A. D. 1280. Certain changes have since been introduced into the substance of these laws, and several additions made to them by the edicts of the Norwegian and Danish Kings; but none such as to affect materially their general spirit or character.

The alterations, however, which have taken place in the judicial establishments of the country, have been much more considerable; and the forms of justice, in consequence of progressive changes, are now in many respects similar to those in the continental dominions of Denmark.

Of the judicial establishments of the island, the court of the Sysselman is the first to which all cases, whether criminal or civil, are referred. This court, which is called the Hierads-thing, is officially held only once in the year in each Syssel; but should the public business of the country require it, an extraordinary court may at any time be appointed, with the same jurisdiction; either by the Crown in criminal cases, or by the petition of the litigants in causes of a civil nature. Every public suit brought before this tribunal is instituted by the Amtmand of the province, as the representative of the Crown.

In criminal cases, an examination of the suspected person and of the witnesses is made immediately after the appre-hension of the former; and the results of the examination are transmitted to the Amtmand, who decides upon these grounds whether a trial is required. If it be deemed necessary, the prosecution is conducted in the Sysselman's court by a public pleader on the part of the Crown, who is opposed by another pleader appointed for the defendant. Upon the evidence and the pleadings, the decision of the court is founded; the Sysselman in his judicial capacity being assisted by four per-sons, called Meddoms-menn, (*consessores judicii*) who both register the proceedings, and give their suffrages, together with the Sysselman, in the decision upon every cause.* The

---

* A similar institution, with respect to the assistant inferior judges, exists in the inferior courts of Denmark. See Dissert. de Offic. Judicum Inferior. in Dania, (Havniæ 1801, p. 17,) by Snæbiorn Stadfeldt.

sentence proceeds upon a plurality of these suffrages. At Reikiavik the Bifoged, or Sheriff of the town, has a jurisdiction similar to that of the Sysselmen in the larger districts.

From these provincial courts, appeals are permitted, and are usually made in all important civil or criminal cases, to the superior court of justice, which has its ordinary sittings at Reikiavik six times in the course of the year.* This tribunal, in its present form, was constituted as lately as the year 1800 ; at which period the judicial assemblies, held annually at Thingvalla during so many centuries, were finally abolished, and their place supplied by the present establishment, considerably altered in its constitution and forms. The court is composed of the Stifftamtmand or governor ; of three judges ; of a secretary ; and two public pleaders. The governor officiates as president, but takes no part whatsoever in the judicial proceedings. Of the judges, one has a superior rank, with the title of Justitiarius ; which office, since the institution of the court, has been held by Mr Stephenson of Indreholm, with credit to himself, and advantage to his country. The other judges, entitled the Assessors, are Mr Grondal of Reikiavik, whose poetical celebrity has elsewhere been mentioned, and Mr Einarson of Bessestad. Though inferior in rank to the Justitiarius, they have an equal weight with him in the decisions of the court, every sentence being determined by a plurality among their three votes. In the proceedings of the tribunal, much impartiality is observed. The evidence and pleadings for each party are respectively heard ; and the sentence is pronounced by the Justitiarius, after the votes of the other judges are obtained. It does not

---

* This court is called the *Konunglegi Lands-yfur-rettur :* according to a literal translation, The Royal court of justice over the land.

appear that any vestige of the trial by jury at present exists in Iceland ; though there are evidences in some of the ancient writings of the country, that it was not unknown to the people during the early periods of their history. * Their acquaintance with this mode of trial was doubtless obtained from their Scandinavian descent; but it seems to have been resorted to only in particular instances, and was not adopted in the code of laws which was framed for the republic of Iceland.

From the superior court of justice at Reikiavik, a final appeal is still in all cases reserved to the authority of the higher courts at Copenhagen. Since the interruption, however, of the ordinary intercourse between Iceland and Denmark, by the war with England, it has been found desirable to establish a court in the island, with the authority of passing a final judgment in criminal cases; which court is constituted by the governor, the three judges, and one or two other members specially appointed by the governor in every cause. This tribunal, being created by an emergency, will doubtless be abolished as soon as the necessity for it has ceased.

Though the transference of the superior court from Thingvalla to Reikiavik has probably, on the whole, been attended with advantage, yet there are some reasons why the Icelanders, as a people, should contemplate this change with feelings of regret. The annual meeting at Thingvalla was not merely that of a tribunal of justice, but an assembly of the nation ; and though the importance of this assembly was di-

* The Eyrbyggia Saga (Havniæ, 1787, p. 47.), contains a curious narrative of the trial of a woman named Geirrida, accused of practising the arts of magic ; to judge in which case, twelve men were appointed, and put upon their oaths. This happened A. D. 981. An account of the same transaction is given in the Landnama Book, p. 82, note.

minished, and its dignity degraded, by the subjection of the
island to a foreign power, yet, on the spot where the greatest
among his ancestors had so often stood, the mind of the Icelander must ever have been awake to enthusiasm and patriotic
pride. ' Hic sacra, hic genus, hic majorum multa vestigia !'
To the eye too of the poet, every thing is lost in this change.
The Icelanders are now summoned to the public courts of
the country in a small and miserable apartment, destitute of
all ornament, and even of common furniture ; where there
is nothing present to confer external dignity upon the meetings, or to mark the character of a national establishment.
At the assemblies of Thingvalla, though artificial splendour
was wanting, yet the majesty of nature presided, and gave a
superior and more impressive solemnity to the scene.   On the
banks of the river Oxeraa, where its rapid stream enters a
lake, embosomed among dark and precipitous mountains, was
held during eight centuries, the annual convention of the people. It is a spot of singular wildness and desolation ; on every
side of which appear the most tremendous effects of ancient
convulsion and disorder ; while nature now sleeps in a death-
like silence amid the horrors she has formed.   Here the legis-
lators, the magistrates, and the people met together.   Their
little groupe of tents placed beside the stream, was sheltered
behind by a rugged precipice of lava ; and on a small grassy
spot in the midst of them was held the assembly, which provided by its deliberations for the happiness and tranquillity
of the nation.

    The study of their own laws, as well as of the principles
of law in general, has ever been a favourite pursuit among
the Icelanders ; and both in ancient and modern times, a
great number of writings, connected with this subject, have
appeared in the island.   In consequence of this minute

attention, all the laws of the country, both civil and criminal, are very distinctly defined; and even among the inferior magistrates, are so well understood, that their execution is every where conducted with fidelity and exactness.

The punishments for theft, prescribed in the criminal law, are varied by the degree of the offence. In cases where the theft is of little importance, or the crime committed for the first time, the offender is whipped, in the presence of only the judge and two witnesses. This punishment is allotted also to other trifling offences, when the poverty of the persons convicted makes it impossible for them to pay a pecuniary fine. In cases where petty thefts have been a second time committed, the criminal is usually sent to Copenhagen; in the workhouse of which city he is confined for the term of three or five years, according to the degree of his guilt. Thefts of a more serious nature, as the breaking into churches or houses, or the stealing of horses, are punished either by public whipping, or by a sentence of perpetual confinement in the Copenhagen work-house*. Where such thefts have been committed for the fourth time, or still more frequently, the punishment is confinement for life in the public prisons of Denmark. The operation of these more severe laws is, however, very seldom required; crimes of this description being by no means frequent among the natives of Iceland.

The only public prison in the island is that of Reikiavik, which was erected about fifty years ago. By a mistake, not unnatural in such a country as Iceland, this building has been rendered greatly more comfortable than the common habita-

* In the work-house at Copenhagen there are different sections, allotted to different classes of criminals. The men condemned to confinement there are kept in a part of it called the *Rasp-huus*, where they are employed in rasping dye-woods; an occupation considered very dangerous to the health.

tions of the natives ; so that, were it not for the privation of
liberty, the Icelander might well be content to exchange his
own abode, for one where his actual comforts are little inferior, and where he is exempted from many of the evils incident
to his usual mode of life. Sheep-stealing is the most common
offence for which imprisonment here is adjudged ; the term of
confinement extending from two to five years, and a certain
portion of daily labour being appointed for each prisoner.
The crime of adultery, committed for the third time, is punished by a confinement of two years. At the time we visited
Iceland, there were six people imprisoned in this place ; but
this is probably rather below the usual number.

Capital punishment, though strictly provided for by the
laws in cases of murder, &c., is scarcely ever required among
a people, gentle in all their dispositions, and possessing moral
qualities of the most excellent description. Examples of this
kind have been so very rare, that a few years ago, when a
peasant was condemned to die for the murder of his wife, no
one in the island could be induced to perform the office of
executioner, and it was necessary to send the criminal over
to Norway, that the sentence of the law might be carried into
effect. The method prescribed for inflicting death, is that of
taking off the head with an axe. In all cases where capital
punishment or perpetual imprisonment have been adjudged by
the courts, the ratification of the king of Denmark is required,
before the sentence can be acted upon.

By a law enacted a few years ago, it is provided that no
Icelander, unless under an accusation which might subject
him to capital punishment, or to imprisonment for life, shall
be kept in confinement before the time of his trial. When an
individual is accused of any inferior crime, he is admonished
by the Hreppstiorè, in the presence of witnesses, not to leave

the parish, in which he resides.　If he infringes upon this obligation, and is afterwards apprehended, he remains under strict confinement, until judgment upon his case has been pronounced.

Some of the Icelandic laws with respect to property have been mentioned in the chapter on Rural Affairs.　The law of inheritance is well defined, and acted upon with much strictness.　No entail of landed property is allowed; but upon the decease of an individual, a division is made of his lands, or of a value equivalent to them by estimate, in which an equal share is allotted to every son; with the right, however, in the case of the eldest, of chusing the farm or share of the property, which may be most agreeable to him.　The daughters have each the half of a son's portion.　If the wife survives her husband, she has half of his estate; or if she dies first, the husband retains the same proportion of the property which she brought him at her marriage.

The tributes paid by the Icelanders are by no means considerable; and do not even suffice for the support of the civil establishment of the island.　They are collected in different ways.　Some of them are strictly taxes on property; founded upon an estimate which is annually made, under the superintendance of the Hreppstiores, of the possessions of the several individuals in each parish.　This estimate is conducted in a somewhat singular way; its basis being a very ancient regulation of property, according to the number of ells of *wadmal*, the cloth of native manufacture, which each individual possessed, or was enabled to manufacture in the course of the year.　The term *hundred,* which was formerly a division derived from the number of ells, is now applied to other descriptions of property.　An Icelander is reckoned the possessor of a *hundred*, when he has two horses, a cow, a certain

number of sheep and lambs, a fishing boat furnished with nets
and lines, and forty rix-dollars in specie; and it is by this
ratio, that the amount of all possessions is ascertained, and
the tributes levied upon them. One of the tributes, called
the *Tuind's*, requires from every person possessing more than
five hundreds, the annual payment of twelve fish, or an equivalent amounting to twenty-seven skillings, or somewhat more
than a shilling of English money. This tax increases in an
uniform ratio with the increase of property; and its produce
is allotted in equal portions to the public revenue, to the
priests, to the churches, and to the maintenance of the poor.
Another tribute, called the *Skattur*, consisted in former times
of twenty ells of wadmal, but is now commuted to money, at
the rate of four skillings and a half per ell. It is paid to the
public revenue by the owners of farms, and by all whose property, estimated in hundreds, exceeds the number of individuals
composing their families. A third tax, called the *Olaf-tollur*, is
paid either in fish or money; likewise in proportion to the property of each individual. A few others are collected in the
country; but they are very inconsiderable in amount, and
devolve little burthen upon the inhabitants. The commerce
of the island has, since the year 1787, been exempted from all
duties.

The management of the taxes is entirely in the hands of
the Sysselmen, who collect them from the inhabitants at the
public meetings which they hold in their respective districts.
The payment is for the most part made in produce of various
kinds; fish, tallow, butter, fox-skins, wool, or woollen goods.
As the Sysselman is required to pay the amount of the taxes
in money to the Landfoged, or treasurer of the island, it becomes a part of his office to dispose of these articles to the
merchants; in which transaction, he is himself subject to the

chances of gain or loss that may arise from fluctuations in the market price. A third part of the produce of the taxes is retained as his own salary ; nor is this more than sufficient to compensate him for the labour and responsibility which he incurs in the discharge of his various duties.

The Icelandic laws respecting the condition and maintenance of the poor, are very strictly enforced ; and become much more burthensome to the farmers and peasants of the country, than the taxes to which they are subject. With the exception of three small buildings, for the admission of a few incurable lepers, there is no public establishment in the island, which affords a permanent abode to the aged, and destitute ; and by all such the more immediate assistance of their fellow-citizens is therefore imperiously required. The laws render it necessary for every farmer or householder to receive into his family, and to give support, to those of his relations, even in the fourth degree of kindred, who may be in a destitute condition. If he has no such calls made upon him by consanguinity, he is still required to assist in the support of the poor, either by admitting some orphan or aged person into his house, or by contributing an annual sum proportioned to the value of his property. It not unfrequently happens, that a landed proprietor, who pays little more than two rix-dollars to the public revenue, is called upon for forty, fifty, or even sixty, as his ratio towards the maintenance of the poor in the district ; when he is unwilling to receive any of these into his own habitation. The execution of the poor-laws is committed to the Hreppstiorè of each parish ; and forms the most essential part of the duties of his office.

In the preceding parts of this volume, much has been said respecting the history of religion in Iceland, the services of the Icelandic church, and the general condition of the priests throughout the country. A brief account of the nature of the religious establishment, and the mention of a few miscellaneous facts, will give the reader all the information that remains upon this subject.

The reformation of religion in Iceland took place A. D. 1551; since which period the doctrines of the Lutheran church, as it exists in the northern kingdoms of Europe, have been strictly maintained in the island. At the present time, not a single dissentient is to be found from the established religion of the country; and the only instance of the kind on record, is one which occurred about the end of the 17th century; when Helgo Eiolfidas, a man who had acquired much knowledge of German literature, espoused the Socinian doctrines, and taught them openly to his children and friends; till compelled by the judgment of the Ecclesiastical court to make a public renunciation of his belief. Doctrinal discussion is of course little known among the Icelanders; and the contests which have existed in their church, relate chiefly to external ordinances, and to the situation and rights of the clergy of the island.

The religious establishment of Iceland is formed on a more extensive scale, than might have been expected from the nature of the country and the condition of the people. The inhabited parts of the island are divided into 184 parishes; a division which gives to each parish an average population of about 260 persons. From the great extent, however, of these districts, it has in many instances been found necessary to erect more than one church in a parish; and the total number of churches in the island somewhat exceeds three hundred.

The duty of each parish devolves upon a single priest; with the permission, however, if his own circumstances do not allow the full discharge of his duties, to take an assistant from among the young men educated for the church, who have not yet obtained a permanent situation in life. The number of the officiating ministers of religion is of course various at different times, though never greatly exceeding that of the parishes. Immediately superior to the common priests are the Provosts, or Deacons, whose office it is to exercise a general superintendance over the churches in each Syssel, and who are chosen in general from a regard to their talents and respectability of character. There are nineteen of these deacons in the island; but their number is included among that of the priests, just mentioned, as they severally have parishes allotted to them, of which they discharge all the ordinary duties. A small additional stipend is attached to the office, which renders their situation somewhat superior to that of the other clergy.

During a period of seven centuries, Iceland was divided into two bishoprics; that of Skalholt comprehending the southern, that of Hoolum the northern districts of the island. The sees becoming vacant at the same time, they were united in 1797 by the order of the Danish government; and the title of Bishop of Iceland was conferred upon the learned and respectable Geir Vidalin, the present possessor of this dignity. The duties of the office are important and extensive. The Bishop superintends the general concerns of the religious establishment, and the particular affairs of each church in the island: he inspects the conduct of the priests, regulates any ecclesiastical disputes which may occur, ordains those who are entering upon the pastoral office, and watches over the education and moral conduct of the people at large. It is a part of his duty also to visit at stated periods the different districts of his diocese, for the purpose of personal inspection; and the farmers of the country are required to assist him, while making these journies, with every accommodation which their means may afford. The appointment of the Bishop is entirely vested in the Crown. While there were two Bishoprics in Iceland, the revenues of each were extremely small, and ill adapted to support the dignity, scarcely even the necessary duties of the office. In consequence of the union of the sees, a considerable augmentation was made in the revenues of the present Bishop, which now amount to about 1600 dollars per annum; derived chiefly from the public treasury of the island. Did he reside in the interior of the country, this sum would raise him to the highest rank of opulence; but making his abode in Reikiavik, he is subject to many additional expences, not only from the different mode of life among the Danes, but also from the necessity of entertaining the country priests, who come to barter their commodities with the merchants at this place. The singular hospitality and kindness of heart, which distinguish the character of Bishop Vidalin, would keep him in a state of poverty, even were his means of exercising these dispositions much greater than they actually are.

The patronage of the church in Iceland was formerly in the hands of the people and the proprietors of land; was afterwards assumed by the Bishops, as the representatives of the Papal authority; and finally, at the period of the reformation, was transferred to the crown of Denmark. The power is now, in most cases, exercised by the Governor of the island, with the assistance and advice of the Bishop. The revenues of the clergy are derived in part from the lands annexed to the churches; partly from tithes upon the landed

property of the country. These tithes are paid by the farmers, in a ratio determined, not by the quantity of produce
raised upon each farm, but by the fixed rents of the land;
from the nature of which rents, as described in the chapter
on Rural Affairs, it will appear that the value of the tithes is
subject to very little variation. More than half a century has
elapsed since the estimate was made, upon which the regulation of their value was founded; but the results of the tables
drawn up at that period, probably do not differ greatly from
the present revenues of the Icelandic church, in as far as
these are derived from tithes. To afford an idea of the extreme
scantiness of the provision which is thus made for the clergy,
it may be sufficient to state the general fact, that the whole
revenue by tithe, in 184 parishes, does not exceed the sum of
6400 specie dollars; giving an average of 34 or 35 dollars for
each parish in the island. The distribution of the stipends is
by no means equal, owing to the difference in the extent and
value of the land under cultivation in different districts. The
most valuable living in the island is that of Breidè-bolstadr,
in Rangaavallè Syssel, the stipend of which is upwards of
180 dollars : the parish contains 376 people. In the parish
of Kröss, in the same district, where there are two churches,
and a population exceeding 500, the stipend amounts only to
33 dollars. In Aarnes Syssel, the parish of Torfastadir, in
which the Geysers are situated, contains five churches; while
the salary of the priest and his assistant, amounts scarcely to
30 dollars. In numerous instances, however, the stipends
are still much smaller; and there are two or three parishes in
the island, where the annual sum of five dollars forms the
whole provision which is made by tithe for the support of the
ministers of religion. The stipends, though specified according
to their value in money, are very generally paid, like the

taxes, in different articles of produce ; which the priests either consume in their own families, or barter with the merchants for other articles which they more immediately require.

These scanty pittances would obviously be insufficient to the support of the religious establishment, were they not assisted by the value of the glebe-land, which is annexed to the church in each parish. Every priest thus becomes a farmer ; and though the land which they hold is in general of small extent, yet there are certain rights attached to it, which augment considerably the profits derived from this source. Beside the tithe upon his rent, each farmer in the parish is required to give annually to the priest, either a day's work, or an equivalent value in money ; and likewise to keep one of his lambs during the winter season ; taking it home in October, and returning it in good condition the following spring. It is customary, also, for the more wealthy of his parishioners, to make him a small offering, of the value of eightpence in English money, three times in the course of the year ; besides which, a trifling perquisite is occasionally obtained for the performance of particular services, as baptism, marriage, and burial. These are all the sources from which the Icelandic priest obtains a livelihood for his family.

In the preceding narrative of our travels, the general appearance and construction of the churches in Iceland has been minutely described. It would be difficult, indeed, to convey to one who has not visited the country, an adequate idea of the extreme wretchedness of some of the edifices which bear this name. But it must be recollected, that if a greater size, or more decoration, had been given to these places of worship, their number would have been diminished in the same proportion ; and in looking therefore at the Icelandic churches, as they now are, no feeling of contempt can have place in the mind, but rather a sentiment of admiration for the propriety and judgment with which the means of the people have been applied to the great object in view. The charge of attending to the condition of the churches is committed to the Hreppstiorè of each parish ; while to provide for any necessary repairs, a small tax is levied upon the inhabitants, and the personal labours of the peasants are occasionally required. The present war between England and Denmark, unfortunate for Iceland in so many points of view, has here also inflicted some of its evils. The accustomed supply of timber from Norway being suspended, many of the churches in the country are getting into a ruinous state ; and during the last summer, communications were made to the Bishop from different parishes, representing the impossibility of continuing public worship from this cause.

The education of the priests at the school of Bessestad, was described in the last chapter. When a young man, intended for this office, has undergone the required examinations, he leaves the school, and usually returns to his native place ; where, in assisting his family to obtain their scanty and hardly-earned provision, he submits to the same labours as the meanest of those around him. During our first journey in Iceland, we were attended by a person in this situation, who performed for us all the menial offices of a servant and guide. These young men are still called upon, however, to pursue their theological studies in as far as their limited means will allow ; and, to provide for this necessary part of discipline, the superintendance of the Bishop is still continued, who annually transmits to each candidate for the priesthood, a series of Latin questions, as a test of his diligence and proficiency. The nature of these questions will be seen from the subjoined list, which was sent to some of the students of

divinity in the summer of 1810. * The dissertations in reply
to them, are conveyed to the Bishop at Reikiavik by those
who come down to this part of the coast to fish, or to dispose
of their tallow and other commodities to the merchants. After
a certain period of probation, and a personal examination by
the Bishop on the doctrines and duties of their profession,
the candidates are received into orders, and await the occurrence of vacancies, which may afford them a place of final
settlement. It is not, however, a life of luxurious ease which
they enjoy, when their abode is thus determined. From
the scantiness of the provision which is made for them in their
public situation, the toil of their own hands is necessary to
the support of their families ; and besides the labours of the
little farm which is attached to his church, the priest may
often be seen conducting a train of loaded horses from the
fishing station to his distant home ; a journey not unfrequently
of many days ; and through a country wild and desolate beyond description. Their habitations are constructed merely
of wood and turf, like those of the farmers of the country,
and are equally destitute of all internal comforts. A stove,
or place for containing fire, is scarcely ever to be found in
them : often there is only one apartment in the house to which

* Examen Theologicum Candidato ...... solvendum.

1. Quanam cautione opus est in prophetiis Veteris Testamenti explicandis ?
2. Quid libri Veteris Testamenti docent de resurrectione mortuorum ?
3. An mali genii homines ad peccandum solicitant ?
4. In quo consistit venia peccatorum nobis per Jesum parta ?
5. Æternitas pœnarum post hanc vitam quibus argumentis probatur, et quo-
   modo cum benignitate Summi Numinis concilianda est ?
6. Explicentur Matt. xv. 4, 5, 6 ; et i. Cor. iii. 15, 16.
7. Qualis fuit status religionis in patria nostra ante reformationem ?
8. Cur Deus hominibus salutem æternam, tantum conditione vitæ emendandæ,
   pollicetur ?

the light of the sun has free access, or where there is any flooring but the naked earth ; and the furniture of this room seldom comprehends more than a bed, a broken table, one or two chairs, and a few boxes, in which the clothes of the family are preserved. Such is the situation during life of the Icelandic priests ; and amidst all this wretchedness and these privations, genius, learning, and moral excellence, are but too frequently entombed.

The ordinary service of the churches in Iceland consists of prayer, psalms, a sermon, and readings from the Scriptures. The prayers and readings are rather chaunted than spoken by the priest, who performs this part of the service at the altar of the church. The sermons appear in general to be previously composed, and are delivered from notes. Of the style and character of these compositions we had not the means of forming an accurate judgment; but in those instances where we attended the public worship of the country, it seemed, from the warm and empassioned manner of their delivery, and from the frequent use of the figure of interrogation, that a powerful appeal was made to the feelings, as well as to the understanding, of the audience. In the conduct of the religious service much decorum is generally maintained. One striking instance to the contrary occurred indeed to our observation ; but the case was a singular one, and must be received merely in the light of an exception to a general statement.

The moral and religious habits of the people at large may be spoken of in terms of the most exalted commendation. In his domestic capacity, the Icelander performs all the duties which his situation requires, or renders possible ; and while by the severe labour of his hands, he obtains a provision of food for his children, it is not less his care to convey to their minds the inheritance of knowledge and virtue. In his intercourse with those around him, his character displays the stamp of honour and integrity. His religious duties are performed with cheerfulness and punctuality; and this even amidst the numerous obstacles, which are afforded by the nature of the country, and the climate under which he lives. The Sabbath scene at an Icelandic church is indeed one of the most singular and interesting kind. The little edifice, constructed of wood and turf, is situated perhaps amid the rugged ruins of a stream of lava, or beneath mountains which are covered with never melting snows; in a spot where the mind almost sinks under the silence and desolation of surrounding nature. Here the Icelanders assemble to perform the duties of their religion. A groupe of male and female peasants may be seen gathered about the church, waiting the arrival of their pastor; all habited in their best attire, after the manner of the country; their children with them; and the horses, which brought them from their respective homes, grazing quietly around the little assembly. The arrival of a new-comer is welcomed by every one with the kiss of salutation; and the pleasures of social intercourse, so rarely enjoyed by the Icelanders, are happily connected with the occasion which summons them to the discharge of their religious duties. The priest makes his appearance among them as a friend: he salutes individually each member of his flock, and stoops down to give his almost parental kiss to the little ones, who are to grow up under his pastoral charge. These offices of kindness performed, they all go together into the house of prayer.

There are two versions of the Bible in the Icelandic language; the first of which was translated by Gudbrand Thorlakson, Bishop of Hoolum, from the German Bible of Martin

Luther, and published in 1584; the second was executed chiefly by Bishop Skulasson, in conformity with the Danish version of Resenius, and appeared about sixty years afterwards, under the more immediate patronage of the King of Denmark. The latter of these versions is preferable to the former, merely from the division of the text into verses; which division the edition of Bishop Thorlakson did not supply. At present, owing to the length of time which has elapsed since any edition appeared, there is a great deficiency of Bibles in every part of Iceland; an evil which, from the depressed state of the printing establishment of the island, it is scarcely possible that the unaided efforts of the people should be enabled to remove.

CHAP. VII.

STATE OF COMMERCE.

FROM the beginning of the 17th century till the year 1776, the trade of Iceland was in the hands of a chartered company, during the existence of whose monopoly the Icelanders were greatly oppressed. The Iceland trade however did not continue to hold out its original temptations ; and at length an unwillingness to risk capital in prosecuting it, became apparent. These circumstances induced the Danish government to adopt a system, the liberality of which deserves the highest praise. Before the trade was declared free, it was nominally vested in the King for a period of ten years, and was carried on with a fund, amounting to 4,000,000 dollars, provided by the government, and of which the King was director. At the end of ten years, when the vessels and stock were sold at greatly reduced prices, it was found that the capital had diminished 600,000 dollars. The remainder of the fund was placed under the management of commissioners, who were empowered to lend money at four per cent. interest to those who embarked in the trade to Iceland. The merchandize

being now freed from imposts of every kind, the encourage-
ments held forth did not fail to take effect. The present state
of the fund is not known in Iceland; but the events of the
war in retarding, and indeed putting a stop to regular communication, render it probable that considerable loss has been
sustained. The freedom from impost was proclaimed to continue for twenty years, at the end of which period, in the year
1807, it was further prolonged for five years. It is impossible
however, that, in so short a time, the trade can recover from
the severe shock it has recently received. In the years 1797,
1798, and 1799, a very considerable traffic in fish, was carried
on to Spain and the Mediterranean; and this period was certainly one of the most favourable for the commerce of Iceland that has ever occurred. Mr Thorlacius, a native merchant residing at Bildal, in the north-western part of the
island, speculated largely at that time, and made a considerable fortune. At present he is esteemed the most wealthy
man in Iceland.

Before the war, about fifty vessels, chiefly galliots of 100
to 250 tons, were employed in the trade. Last year, not
more than ten ships arrived in Iceland; and while we remained there, not more than seven, including three from Britain,
and one galliot laden with salt from Liverpool, on account
of Messrs Phelps and Company of London.

The nature of the trade with Iceland will be seen in the
following tables, taken from Mr Stephenson's history of Iceland during the 18th century; and, to render them intelligible, the weights and measures are annexed.

Current Prices of Icelandic Produce,
in the Year 1810.

1 Pair of Mittens, ... 4 to 6 Skillings.
1 do. of Stockings, ... 12 to 18 do.
1 do. do. fine, ... 64 Sk. to 1 Rixdollar.
1 Woollen Jacket, ... 40 to 64 Skillings.
1 do. fine, ... 2 to 3 Rixdollars.
1 Pund Wool, ... 12 to 20 Skillings.
1 do. Eider-down, ... 2 Rixd. 48 Sk. to 3 Rixd.
1 do. Feathers, ... 16 to 20 Skillings.
1 do. Tallow, ... 16 to 22 do.
1 do. Butter, ... 10 to 28 do.
1 Skippund Stock-fish, . . 12 to 20 Rixdollars.
1 do. Salted Fish, ... 15 to 30 do.
1 Barrel of Oil, ... 12 to 20 do.
1 White Fox-skin, ... 80 Skillings to 3 do.
1 Black do. . . . . 5 to 8 do.
100 Swan-quills, ... 2 Rixd. 48 Sk. to 3 Rixd.
A Horse, ... 6 to 40 do.
A Cow, ... 16 to 24 do.
A Ewe with Lamb, . . . 2 to 2½ do.
A Wether, ... 2 to 5 do.
A Lamb, ... 80 Sk. to 1 Rixd. 32 Sk.

The circulating medium of Iceland is the same as that of
Denmark ; the coins being 10 sk., 8 sk., 4 sk., 2 sk., and 1
skilling pieces. A considerable number of specie dollars are
in the island, but are seldom seen ; the natives being in the
practice of hoarding them. The paper rixdollar is used in
all transactions.

In the preceding tables, it will be seen that the island is divided into four commercial districts; viz. Reikiavik, Eskefiord, Eyafiord, and Isafiord. Formerly, there were six districts. This division is not merely nominal, regulations being established for merchant ships visiting each district, and the settlement of Danes or Icelanders as merchants. A vessel arriving from Denmark in any district, is allowed to visit all the ports included in it; but is not allowed to go into any of the harbours of the other districts.* Every person desirous of settling in Iceland as a merchant, must become a burgher, or freeman, of the district in which he wishes to establish himself. A settlement is obtained without difficulty, most commonly through the favour of the Governor or the Sysselmen; and with no other expence than that of a few dollars for writings.

The districts of Reikiavik and Isafiord supply the greatest quantity of salted and dried fish; and from the latter the greatest exportation of oil takes place, on account of the productiveness of the cod and shark fisheries. The northern and eastern coasts furnish the greatest quantity of tallow, salted mutton, wool, and woollen goods. The large quantity of tallow and woollen stuffs exported from the Reikiavik district, is not owing to the number of sheep and cattle kept in this part of Iceland, but to these articles being brought by the

* REIKIAVIK includes,	ESKEFIORD,	EYAFIORD,	ISAFIORD,
Reikiavik.	Eskefiord.	Eyafiord.	Isafiord.
Havnefiord.	Rodefiord.	Husavik.	Patrixfiord
Kieblivik.	Berufiord.	Siglefiord.	Bildal.
Eyaı back.	Vapnafiord.	Hofsos.	Olafsvik.
Westmann Islands.		Skagastrand	Gronnefiord.
			Stikkesholm.
			Stappen.
			Buderstad.

Districts.	Rye Meal.	Rye.	Barley.	Oats.	Peas.	Pearl Barley.	Barley Groats.	Buck Wheat.
	Barrels.	*Barrels.*	*Barrels.*	*Barrels.*	*Barrels.*	*Barrels.*	*Barrels.*	*Barrels.*
Reikiavik............	3502	2265	27	72	777	531	128	7
Eskefiord.............	563	575	....	4	310	145	109	....
Eyafiord............	978	2965	32	6	856	190	80	5½
Isafiord.............	1017	701	26	2	136	161½	56½	2
Total......	6140	6506	85	84	2079	1027½	373½	14;
In the year 1630...	4501	....	....	....	17	83	....	
1743...	8038	....	....	....	52	135	....	
1779...	10,665	475	1138	98	133	367	....	
	Rye and Barley.							

Districts.	Vinegar.	Mead.	Beer.	Malt.	Coffee.	Sugar.	Treacle.	Tobacco.
	Hhds.	*Barrels.*	*Barrels.*	*Barrels.*	*Skpd.L.lb*	*Skpd.L.lb*	*Skpd.L.lb*	*Skpd.L.l*
Reikiavik............	5	24	7	10	13 15	24 17	13 14	96 1
Eskefiord............	1	9	....	11	4 10	10 0	4 6	52
Eyafiord............	2	17	34½	18	2 7	6 13	5 10	44
Isafiord...	2¼	3	11	....	6 6	7 5	1 13	44 1
Total......	10¼	53	52½	39	26 18	48 15	25 3	238
In the year 1693...	....	216	....	38	....	....	....	
1743...	....	67	....	12	....	....	....	20 1
1779...	....	67	....	213	10 7½	27 0	....	256 1

CELAND *in the Year* 1806.

Oat roats.	Rice.	Wheat Flour.	Rye Bread.	Biscuit.	Wheaten Bread.	Brandy.			Rum.	Wine.
						Danish.	French.	Grape.		
Barrels.	Lispd. lb.	Lispd. lb.	Sk.pd.Llb.	Sk.pd.Llb.	Skpd.L.lb.	Barrels.	Barrels.	Barrels.	Barrels.	Hhds.
8	3 10	66 0	90 15	226 4	47 4	255	43	149	28	60
4	.....		51 12	21 4	.....	59	98	6	4	8
4¼	1 0	28 5	35 14	19 17	0 13	105	52	25¼	12½	6½
4	30 13	16 12	53 6	3 5	2 15	150	33½	32	3	12
20¼	35 7	111 1	231 7	270 10	50 12	569	226½	212¼	47½	86½
			Barrels	Barrels.						Pipes.
.....	.....	.....	352	93	.....	262	.....	.....	.....	13¾
.....	.....	.....	1239	684	.....	722	14	12½	.....	57⅓
.....	.....	.....	986	422	.....	1160	36⅔	.....	18	71

Paper.	Soap.	Salt.	Iron.	Tar.	Coal.	Hemp.		Fishing Lines.	Cable.	Twine.
						Hackled.	Unhackled.			
Reams.	Firkins.	Barrels.	Skpd.L.lb.	Barrels.	Lasts Bar.	Skpd.L.lb.	Skpd.L.lb.	Pieces.	Pieces.	Skpd.L.lb.
63	46	842	170 13	159	26 4½	27 0	17 6	7117	1122	4 8
25	1	325	5 4	31	9	.....	.....	817	.....	0 3
65	7	549	2 13	13	2 17	1 16	1 16	907	.....	0 1
4	1	862	21 14	116	3 0	3	0 3	3630	23	2 0
157	55	2378	200 4	319	32 12½	27 4	19 5	12471		6 12
.....	.....	834	781 0	61	.....	.....	.....	34412	.....	9 1
.....	.....	1864	272 0	147	.....	.....	.....	.....	.....	.....
	lbs.									
218	2054	2954	310 0	291	22 0	15 0	.....	12890	.....	9 5½

[*To face page* 337.]

Districts.	Fish.	Dried Fish. (Stock Fish.)	Salted cod in barrels.	Salted cod in bulk.	Oil. Cod.	Oil. Shark.	Oil. Seal.	Fish Liver.	Tallow.
	Sklb. L.lb.	Sklb. L.lb.	Barrels.	Sklb. L.lb.	Barrels.	Barrels.	Barrels.	Barrels.	Sklb. L.lb.
Reikiavik.............	1606 3	1809 2	66	2 8	495½	110	10	12	149 15
Eskefiord*.........	....	....	28	....	36	79½	....	....	151 10
Eyafiord............	30 18	....	36	....	17	561	....	....	278 6
Isafiord............	364 5	524 16	20	7 13	259	913	14	....	19 8
Total.........	2001 6	233 18	150	10 1	807½	1663½	24	12	598 19
									Barrels.
In the year 1624..	843 0	5817	444½	....	930	....			337
1630..	207 0	2823	142	....	1445¼	....			133¾
									Sklb. L.lb.
1743..	392 0	5380	658	....	471	....			475 6
1779..	3612 0	4901	1905	....	1402	....			609 8

Districts.	Frocks or Jackets.	Mittens.	Wadmal.	Lamb Skins.	Salted Sheep Skins.	Small Shark Skins.	Fox Skins.	Swan Skins.	Goat Skins.
		Pairs.	Pieces.						
Reikiavik.............	130	77,203	3	2442	190	233	63	52	....
Eskefiord............	345	6,737	8	642	23,516	....	32	....	....
Eyafiord............	5790	57,798	....	3723	9,091	.. ..	15	....	115
Isafiord............	17	141,338	....	620	6	1335	35	3	....
Total..........	6,282	283,076	11	7427	32,803	1568	145	55	115
			Ells.						
In the year 1624..	....	12,232	12,251	....					
1630..	....	13,004	4,042	....					
1743..	1211	110,507	876	....					
1779..	884	186,624	521	20,722	....	....	406	98	....

A volume was published, in the year 1787, at Copenhagen, by Royal authority, entit

the trade was declared free ; and the volume contains Tables of the Exports of Iceland f

* Eskefiord is situate on the

ICELAND in the Year 1806.

[Salted] mon. Barrels	Wool White Sklb. L.lb.	Wool Mixed Sklb. L.lb.	Woollen Yarn Skl.l. L.lb.	Stockings Pairs.
28½	327 14	58 14	9	19,567
......	92 12	56 4	4 3	26,186
......	166 17	11 5	24 11	79,900
......	92 5	7 18	......	56,023
28½	679 8	134 1	29 3	181,676
5½	......	......	......	......
5	......	......	......	......
3	265 0	......	......	......
16½	23 0	......	......	......

Eider Down lb. L.lb	Feathers Sklb. L.lb.	Iceland Moss Barrels.	Rein Deer Horns.
11	14 12	4	......
) 19	......	......	......
0 19	......	153	......
3 7½	12 1	¼	......
6 16½	26 13	4¼	153
......	......	......	......
......	......	......	......
6 0	2 9	......	......

Weights and Measures used in the Island.

Liquid Measure

1 Pipe contains	3 Ame,	or 120 Gallons.
1 Oxhoved,(Hogshead), ...		60
1 Ame.....................	4 Ankers, or	40
1 Tonde ·...................	3 Ankers, or	30
1 Anker....................	5 Kutting,or	10
1 Kutting................	4 Kander, or	2
1 Kande	2 Potter, or	0½
1 Pot.....................	4 Pæle, or	2 Pints.
1 Pæle............... or		0½ Pint.

Corn Measure.

1 Tonde, (Barrel), = 8 Skepper, or 4 English Bushels.

1 Skepper, (½ Bushel), contains 18 Potter or Quarts.

Cloth Measure.

1 Alen, or Yard, = 25 English inches, or two-thirds of a Yard, and is divided into Quarters.

Weights.

1 Skippund, = 20 Lisepund, or 3 cwt. 22lb. English.

1 Lisepund, = 16 Pund.........or 17 lb.

1 Pund... ... = 16 Onzer........or 1 lb.

The Danish Pound is 12 per cent heavier than the English.

ed, ' Regulations for the Trade and Navigation of Iceland.' This was at the time when om 1764 to 1784; but I have not had an opportunity of copying them.

ame bay as Rodefiord.

people from different parts of the island, for the purpose of bartering with the merchants of this place.*

There is perhaps no part of the world where the cod fishery can be carried on so extensively, so easily, or so safely, as in Iceland. When the distance of Newfoundland, and the stormy weather which prevails in that quarter, are comparatively considered, together with the expence of our establishments there, Iceland offers the most important advantages as a fishing station. The facility with which the fishing is carried on by the natives, is really astonishing. In the morning they go out in small skiffs, to the distance of a few miles from the shore, and in the afternoon return with as many fine fish as their boats can contain. Even in the very harbours, as was the case at Reikiavik soon after we left it, abundance of cod are sometimes taken. The rivers are frequented by vast numbers of salmon, an article in great demand both in this country and in the West Indies. But they are neglected; no means being employed for a regular capture, except in the small river near Reikiavik; and the salmon from that river are almost all consumed in the country.

Fish and oil are the chief articles of export, which could be extended to an indefinite amount. Wool is an article not required in Britain; but, by improving the land, the stock of cattle and sheep might be greatly increased, so that the quantity of hides and tallow would become considerable. Whether Iceland will ever be thought an object of importance by the English government, remains to be known. The disposition to be humane towards that miserable country, cannot be doubted; and to the feelings which dictated the Order in Council, dated in February 1810, much credit is due. That

* See Journal, page 204.

compassionate and well-meant Order was not attended with any advantage to the Icelanders. Though permission was given to trade with the natives, the duties on the goods brought home still remained prohibitory; and several ships belonging to Iceland, which came to this country on the faith of the Order, were, on account of the duties, long detained at Leith, till strong representations being made, they were permitted to sail to Denmark with their cargoes. Sir Joseph Banks, with that humanity which distinguishes him, has made many applications in behalf of Iceland; and I have not failed to represent the case of that country, as I found it, to those under whose management our foreign relations and trade are placed. I do not know that any attention has been paid to these representations; though, indeed, I learned from Mr Parke, the gentleman appointed Consul for Iceland, that, in consequence of my statements, some alterations were made in the licences to trade with the island. These were made, however, for preventing some abuses, not to benefit the natives. I am happy to find that Mr Hooker, in his narrative of the revolution in 1809, entertains the same sentiments with me in respect to Iceland. He concludes his narrative with strongly expressing the propriety of its being taken possession of, and considered as a dependency of Britain; this being the only effectual mode of relieving the inhabitants.

CHAP. VIII.

ZOOLOGY AND BOTANY.

In a general outline of the zoological productions of Iceland, it is by no means necessary to be minute ; nor, indeed, would the few observations we were enabled to make, authorize such an undertaking.

Iceland does not present many of those species of animals which are strictly confined to the land ; but of those which require land only as a resting place, while the sea supplies their other wants, many have found in this country every requisite for support. We will proceed, however, to take a cursory survey of all the tribes of animated nature which exist there under any circumstances.

The catalogue of mammiferous animals inhabiting Iceland, is nearly confined to the following :—The dog, the fox, the cat, the rat, the mouse, the rein-deer, the goat, the sheep, the ox, and the horse ; together with seals and whales, and a few Polar bears which annually make their appearance. Bears cannot be considered as inhabitants ; they are merely visitors, brought on detached masses of ice. They are chiefly landed on the north coast; and twelve or thirteen appears to be the greatest number ever seen in one year. They are not suffered long to enjoy themselves on land; for, hungry and voracious after their voyage, they commit great devastations among the flocks. The people take the alarm; and, with whatever weapons they can command, generally with musquets, they attack, and soon destroy them.

The dogs which are generally seen in Iceland, bear a strong resemblance to those of Greenland. Like them they are covered with long hair, forming about their necks a kind of ruff. Their noses are sharp, their ears pointed, and their tails bushy, and curled over their backs. Their predominant colour is white; yet they vary considerably; and some are entirely brown or black. Very few of them can be induced to go into the water; and though some are of service in guarding the cottages and flocks, and preventing the horses from eating the grass intended for hay, yet the greater number appear very useless. Scarcely any family, however, is without one or two of them.

Two distinct varieties of the fox present themselves in Iceland: the arctic, or white fox, (Canis Lagopus), and one which is termed the blue fox, (Canis Fuliginosus), and varies considerably in the shades of its fur, from a light brownish or blueish grey, to a colour nearly approaching to black. It is a more gracefully formed animal than the white fox; has longer legs, and a more pointed nose. Horrebow mentions a dark red coloured fox, in the existence of which we had no reason to believe. He likewise says, that the black fox is sometimes brought over on the ice. Frequently at night, in travelling through the country, you hear the discordant cries of the two former varieties. But if we may judge from the quantity of skins exported, the number of foxes in Iceland, though considerable, cannot be great. The inhabitants do their utmost to destroy them; being induced not only to prevent the great devastation which they commit among the young lambs, but to obtain the reward given by government, and to profit by the furs, which is an advantageous article of traffic. There is no particular ingenuity, however, displayed in the methods by which they are taken; they are shot, caught in gins, or forced from their holes by smoke.

Rats in considerable numbers, and mice, are met with, particularly at the Danish factories; but, as far as our observation went, there was nothing to render them particularly worthy of attention.

The hog, which has from time to time been imported from Denmark, has, from the scarcity of proper food, been found so expensive to keep, that it has never been much propagated; and it is doubtful whether, independently of two or three sows and pigs which were taken from England during the last summer, a single animal of the species exists in the country.

The rein-deer has been introduced into the island, and has increased rapidly. Out of thirteen which were exported from Norway in 1770, three only reached Iceland. They were sent into the mountains of the Guldbringè Syssel; and they have since multiplied so considerably, that it is now no uncommon thing for those who pass often through the mountains in various parts of the island, to meet with herds, consisting of from forty to sixty, or a hundred. They are very little molested, the Icelanders satisfying themselves with complaining that the deer eat their lichen; and though, sometimes, for the sake of amusement, the Danes go out in pursuit of them, very few are destroyed. They live almost entirely among the mountains, and are very shy; but sometimes, in the depth of winter, come down into the plains, particularly about Thing-valla, to feed on the moss which abounds in that quarter.

Goats were at one time more numerous in Iceland than they now are. At present, they seem to have been completely expelled from the southern part, because vegetation being very scanty, they were constantly injuring the roofs of the houses by climbing on them in search of food. There are still a few in the north, where farmers keep flocks of thirty or forty.

The cow, the horse, and the sheep, afford the principal source of wealth, comfort, and subsistence to the Icelanders. Milk is almost their only summer beverage. Whey becomes a wholesome, and to them a pleasant drink in winter. Even fish itself, their primary article of food, is scarcely palatable to an Icelander without butter; and curds, eaten fresh in summer, and kept through the winter, yield the most precious change of diet, both for health and pleasure, which he enjoys. A cow on the farm of the Amtmand Stephenson, we were assured, gave regularly every day twenty-one quarts of milk. Their value is well known and appreciated by the Icelanders, who take the greatest care of them through the winter, and seem to shake off their habitual listlessness, while employed in gathering in the hay that is to support them through the inclemencies of that season.

If the horse be less useful in Iceland than the cow, the care which is devoted to him is proportionally less: still, however, the assistance which he affords is by no means to be overlooked. But it will be unnecessary, after what has been stated respecting the frequent intercourse between different parts of the island, and the extreme roughness of the country, to say any thing farther of the utility of this animal. The Iceland horse is about thirteen hands in height, stoutly made,

and frequently evincing much spirit.   These animals are in
very considerable numbers throughout all the inhabited parts
of the island; no farmer being able to carry on the necessary
affairs of life without their assistance; and many of the Icelanders, particularly those who, from their avocations as
judges or magistrates, are obliged to take long journies, are
at great pains in the breeding and rearing of them.   But by
the inhabitants in general, they are let loose to provide themselves with food and shelter; in consequence of which, a
great number are annually carried off by the severity of the
winter.

The sheep furnish much milk and butter; and besides affording, when smoked or salted, a part of the winter food of the
inhabitants, form a considerable article of export.   Almost
every part of the Icelandic dress is manufactured from wool:
and of the sheep-skins, without much preparation, they make
their fishing-dresses, which they smear repeatedly with oil,
for the purpose of rendering them impervious to the water.

Of the seal, three or four species (Phoca vitulina, Lepor-
rina, Barbata, and Grœnlandica) frequent the shores.   Their
number is considerable.   A few are taken for the oil which
they afford; and their skins are applied to various useful purposes, being formed into shoes and thongs, and particularly
into a kind of travelling bag, in which the Icelanders carry
their sour butter, fish, and other little supplies, when passing
from place to place.

Very few of the great northern whales (Balæna Mysticetus)
approach Iceland.   The fin-fish (Balæna Physalus) is more
common.   A species of dolphin, the bottle-nose, (Delphinus
Bidens?) is sometimes driven on shore in very considerable
shoals.   During the winter 1809-10, eleven hundred came
towards the shore in the Hvalfiord, and were captured.

Of the Linnean order Accipitres, we only saw one, the Great Erne, or Cinereous Eagle (Falco Albicilla). According to Pennant, the following other species exist in Iceland:—the white-headed eagle (Falco Leucocephalus); the Iceland falcon (Falco Gryfalco); Falco fulvus; and the Lanner. Of all these, the Erne is, at present, certainly the most frequent, the others being very seldom seen. It is constantly observed hovering over the shores, and is a determined enemy to the Eider-duck; and, as such, of course draws upon itself the hatred of the Icelanders. The Iceland falcon, once so much valued in Denmark for its excellence in falconry, is now suffered to remain unmolested; yet it does not seem to multiply as might be expected; and during our residence in the island, we had not a single opportunity of seeing it, even at a distance.

The raven is very common in Iceland. A pair or more sit near every habitation on the sea-shore, ready to feed on the offal of the fish; and they frequently do great mischief to the fish itself, when split and left on the beach to dry. They build their nests in the cliffs, and sometimes resort for this purpose to rocks a considerable way inland.

The snow-flake, or snow-bunting, (Emberiza nivalis), resides here during the whole year, occurring in pairs, or solitary, during the summer, when it loses much of its snowy plumage; and collecting into flocks in the winter. This is the only bird in Iceland which can truly be said to attempt singing. The song is pleasing, but short, and much resembles the first two or three notes of the robin-redbreast.

The wheat-ear (Motacilla Oenanthe) was not uncommon; and we sometimes saw another small bird, of a brownish colour, in the marshy places, which we had no opportunity of examining.

2 x

The white wagtail (Motacilla Alba) frequents the margins of the pools and rivulets. Very few of the swallow tribe ever arrive in Iceland. Some of our party saw one or two flying about the church at Reikiavik early in the month of July; but to what species they belonged, was not ascertained.

Ptarmigans (Tetrao Lagopus) are generally very abundant in this country; but when we were there, we were told that they were scarce in the neighbourhood of the town, and in some other parts of the country. Towards the latter end of July, we observed a bird of the grouse kind, with a brood of young ones : it was possibly the species which Paulson, in his catalogue of Iceland birds, has called the hazel grouse. It had less white, and in general differed somewhat in its plumage from the common ptarmigan, and appeared to be larger.

Of all the land-birds which are seen in Iceland, none are more common than the golden plover and the curlew. These birds are frequently the only enliveners of dreary plains and extended marshes, where their wild and inharmonious notes accord well with the surrounding scenery. The snipe is likewise common in the same situations; and in some instances seemed to have lost much of that wildness of disposition which it exhibits in this country. Thus we saw it associating, as it were, with the Eider-ducks, and sitting on its eggs within an hundred yards of the house at Vidöe.

The variety of birds which frequent the seashore is very great. The high rocky islets on the south are covered with gannets (Pelicanus Bassanus). The shag (Pelicanus Graculus) and the corvorant (Pelicanus Carbo) sit constantly on the rocks. Innumerable gulls, fulmars, and shearwaters, breed in the cliffs. The black gull (Larus Crepidatus) we saw frequently in the swamps, in considerable numbers. Ducks, mergansers, and divers, in great variety, are at one time seen float-ing on the bays, and at another, suffer themselves to be car-ried along by the rapid streams, or accompany their young broods in the marshes.   Large flocks of auks and guillemots live about the coasts, which, together with the kittiwake, and other species of the gull, present in their eggs and feathers a valuable reward to the Icelanders for the fatigue and labour they undergo in their pursuit.

The tern (Sterna Hirundo) is another bird which is very common, always choosing, for the purpose of breeding, a piece of fresh water situated in a marsh near the sea-shore. The egg of this bird is a very delicate article of food, and frequently formed a principal relish in our homely repasts. We saw the tern, for the first time, on the 27th of May, at Grundivik ; and, as we had not seen it at Reikiavik when we were there only a few days before, this was probably about the time of its arrival in Iceland.   Mr Macwick, in the Lin-nean Transactions, gives as an average of 26 years observa-tions, that the Sterna Hirundo is first seen in England April 1st, and last seen October 8th. He likewise represents the snipe as appearing November 20th, and disappearing March 20th.

The most majestic bird of Iceland is undoubtedly the Swan. It in general seeks the more remote lakes among the moun-tains, resorting at times to the salt marshes about the sea-shore, where forty or fifty are sometimes seen feeding toge-ther. During the breeding season, they retire in pairs to small lakes, where they may be concealed among the reeds, and thus protect themselves from the attacks of the Icelanders, who receive the value of a few shillings for their skins from the Danish merchants. Of the eggs, we once had an oppor-tunity of partaking ; and though somewhat heavy, they were very palatable.

So much has already been said respecting the manners and habits of that most curious and interesting bird, the Eider-duck, that it will be unnecessary to do more than merely mention its name in this place.

Some parts of the coast of Iceland, particularly the bays on the west, abound with varieties of very fine cod ; for which, before the discovery of Newfoundland, a considerable fishery was carried on; so that, in the reign of James I., no less than 150 British vessels were employed in the Iceland fisheries. Great numbers are still taken by the Icelanders, chiefly for the Danish merchants, who dry them, either with or without salt, and export them to Denmark. Some are consumed by the Icelanders themselves ; but their number is comparatively small, as they either prefer haddocks, or are obliged to eat them because the merchants will scarcely take any thing but cod. The best season for fishing is from the beginning of February to the middle of May. In June, the fish become meagre and watery, as this is the month in which they generally cast their spawn.

The haddock is likewise very plentiful, apparently associating with the cod, for they are always taken together. They grow to a size not inferior to the cod, frequently measuring above three feet in length ; and are to the inhabitants of greater importance than any thing with which nature or art has supplied them.

The ling, the skate, and the hollibut, occur in considerable numbers, though not nearly so common as the two last. The hollibut arrives at a great size; and, like the wolf-fish, is cut up and dried for winter use. Flounders abound on the shores ; and herrings are taken in great numbers on the north coast. They come in extensive shoals in the months of June and July, not less than 150 barrels of them having been taken at one draught of a net.   Sharks are taken in great abundance on the north and western coasts.*

Eels are found in the rivers; and we once observed a very fine one in a stream, which was rendered tepid by the admixture of the water arising from a hot spring.   Two or three species of the salmon frequent the rivers and lakes, among which the sea-trout is in great perfection.

The Zeus Opah has been seen in Iceland.   One, of which we saw a tolerable drawing, was taken about two years ago.

Of the insect tribes, we saw nothing very remarkable.   A large Tipula (Plumosa?) began to appear in considerable numbers about the middle of May; and although, as the summer advanced, a few of the most common species of flies and moths were seen,—once only, in a low and marshy situation, in the month of July, did we experience any inconvenience from them.   At that time, the air was thickly peopled by a small yellow-coloured fly, probably a species of Empis.

' The entomological productions of Iceland,' says Mr Hooker, ' are extremely scanty.   A very small collection of ' insects indeed, rewarded my researches in this department ' of natural history; and of these there were none that were ' in the least remarkable for their beauty.   Some of the Le- ' pidopterous species were new to me; among which I think ' I had five or six nondescript Phalenæ.   No Papilio or ' Sphinx has ever been met with in the country.   Of Coleop- ' terous insects, there is scarcely a greater variety; and I

---

* It has been mentioned in the journal, that it is probably the species known by the name of the basking-shark; but the colour is different from that of the Squalus Maximus, being of a pinkish tinge.   From figures I have seen of the white shark, from the general shape of those we saw, and from other circumstances, it appears to be that variety which is so common on the coasts of Iceland, and not the basking-shark.   G. M.

' saw only a single Scarabæus, and a very few Curculiones
' and Carabi, most of which, however, to make me amends,
' were such as I was unacquainted with. I, by mere acci-
' dent, have still preserved a specimen of an undescribed
' species of Coccinella, which I found killed by the steam of
' one of the hot springs at the Geysers; it was the only one
' of the genus that I saw.

Small crabs, of two or three species, are thrown upon the
shore, together with the star-fish and echinus; of which latter, we once observed a great number carried by the birds,
and dispersed along an extensive marsh to a considerable distance from the sea.

Muscles are in great abundance, and also whelks, snails, and
limpets; and the barnade often forms a firm coating to the
rocks.'

BOTANY.

In the variety of soil which Iceland presents, from the
deep bog to the light burnt earth which can scarcely stand
against the storm, even the verge of the arctic circle may afford some objects of interest to the botanist. There is, in
fact, no small variety of plants in Iceland; and what is wanting in the gaiety of the individual blossom, is frequently com-
pensated by its abundant production, and the length of time
that it remains in perfection. It was partly, perhaps, from
the contrast with surrounding objects, and partly from the
effects of association, that the flowers of the Dryas Octopetala have excited in us as much pleasing admiration as the
lively scenes of an English flower-garden. The blossoms of
this plant last but for a short time, but their succession is

rapid and uninterrupted during the greater part of the months of July and August; and at the same time the intervening spaces were filled by a great variety of less conspicuous flowers, among which the wild thyme and the Cerastium latifolium were particularly observable; and in two or three places the Konigia Icelandica, a plant peculiar to the island, grows with great luxuriance. Less striking in itself, but still more abundant than the Dryas, or perhaps than any other flower, may be mentioned the Statice Armeria, which in many places, especially near the sea-side, was thickly and widely distributed among the grass; and in many high vallies, at a distance of fifty miles from the sea, formed the only tufts of vegetation on which our eyes could fix as a relief, when wearied with the constant succession of cinders and lava. Sometimes different species of sorrel, or the yellow poppy, would assist in cheering the dreariness of such places; and occasionally a great variety of moss and lichen grew amidst the cinders. It was in the roughest tracts of lava that the mosses grew with the greatest luxuriance. The Trichostomum Canescens, was the prevailing species, which in dry weather renders the places on which it grows of a greyish colour, but after a little rain it becomes of a light and cheerful green. Many species of Saxifrages are found among the moss; and in some places the Bearberry (Arbutus Uva ursi), the Crowberry (Empetrum Nigrum), and a little heath (Erica vulgaris), afford a more substantial covering to the stones. As you leave this broken scene, you may become insensibly involved in an Icelandic forest, where the most stately birch rises to the height of only ten feet. These trees find it difficult to raise themselves above three or four feet from the ground. They consist of the common and dwarf birch (Betula alba and nana), several varieties of the willow, and a few solitary individuals

of the pyrus domestica. Though in some instances we travelled over four or five miles of such forests, they are by no means
common, and we did not see above five of them. A great part
of the surface of the ground is covered with bogs, where vegetation is more luxuriant, affording considerable variety of
carices and coarse grass, the dark green of which was but
little enlivened with flowers. Such is the general view of vegetation in Iceland. It is, however, very unequally distributed. In some parts, as in the Guldbringè Syssel, the traveller may pass over many miles, and scarcely find a single
spot of earth covered with vegetation; while, in the Borgarfiord and Myrè Syssels, the verdure at times almost equals
that of the pasture districts of England. But were the verdure much greater than it is in the most favoured spot, the
entire want of any thing deserving the name of a tree would
give a wearisome nakedness to the prospect.

It is not consistent with the object of this sketch to enter
into a detailed enumeration of the different species of plants
observable in Iceland, with their localities. None of our
party was skilled in botany; and the collection we made was
chiefly intended to make up, in some degree, the severe
losses which Mr Hooker sustained in the destruction of his
collection. That gentleman has kindly permitted us to make
extracts from his interesting work; and we shall now present
our readers with his general view of the botany of Iceland;
and in the appendix will be found the Icelandic Flora by the
same gentleman.

 ' My inclination, rather than my ability, leads me, in the
' first place, to offer a few remarks on the botany and zoology
' of the country. In these two great kingdoms of nature,
' perhaps it would be difficult to find any spot of land of a
' similar extent, in an equal degree of latitude, which can,

' lay claim to so small a number of species　The arctic re-
' gions of Norway, Lapland, and the Russian empire, are
' comparatively rich in these departments; a circumstance
' most probably to be attributed to their warmer summers,
' and to the undisturbed state of the soil.　In spite of this,
' however, a botanist, coming from the more temperate cli-
' mate of Great Britain, will still meet with many vegetable
' productions that will interest him, such as Azalea procum-
' bens, Cardamine hastulata, of English botany, Rubus sax-
' atilis, Erigeron alpinum, Saxifraga nivalis, rivularis, cernua,
' and oppositifolia, Silene acaulis, Veronica alpina, and fru-
' ticulosa, with many other species, which he has been ac-
' customed to see only on the summits of his loftiest moun-
' tains, but which will here be found growing in the plains
' and vallies, and near the shores of the sea.　Ranunculus
' lapponicus, glacialis, and hyperboreus, Eriophorum capi-
' tatum, Konigia islandica, Gentiana tenella, detonsa (the
' ciliata of Retzius), and aurea, Andromeda hypnoides, Cha-
' mænerium halamifolium, Angelica Archangelica, Lychnis
' alpina, Papaver nudicaule, Draba contorta of Retzius, Or-
' chis hyperborea, Carex Bellardi, Salix Lapponum, and
' other plants peculiar to high northern latitudes, together
' with some, as yet undescribed, will likewise offer themselves
' for his examination, and afford him a pleasure, of which no
' one but a naturalist can form an idea, as well as what is
' happily termed by Dr Smith one of the highest sources of
' gratification attending upon this and similar pursuits, " the
" anticipation of the pleasure he may have to bestow on kin-
" dred minds with his own, in sharing with them his disco-
" veries and his acquisitions." *　But a richer field is open

---

* Preface to the " Introduction to Botany."

' before him in the class Cryptogamia. The Muscologia of
' the country is little known; and I am sure, from what I
' myself found, that many new and rare species would re-
' ward a careful search among this tribe, though, like me, he
' might seek in vain for the magnificent Splachna of the Nor-
' wegian and Lapponian Alps, rubrum and luteum, two plants
' that I had most earnestly reckoned upon gathering. Tor-
' tula tortuosa, Polytrichum sexangulare, and hercynicum,
' the former always barren, as in Scotland, Buxbaumia fo-
' liosa, Dicranum pusillum, Hypnum revolvens, Silesianum,
' and filamentosum, Meesia daalbata, Conostomum boreale,
' Splachnum vasculosum, and urceolatum, Trichostomum
' ellipticum, Fontinalis squamosa, and falcata, both abund-
' antly provided with capsules, and Encalypta alpina, as well
' as many other mosses, which I cannot with any degree of
' certainty now call to my remembrance, are met with upon
' the lava, in the morasses, or in the rapid torrents. Most
' of the known alpine species of Jungermannia are also na-
' tives of Iceland, and some new ones, the loss of which I
' peculiarly regret. Of Lichens, there are comparatively but
' few, as indeed may reasonably be expected, from the ex-
' treme scarcity of trees, to which so many of them are ex-
' clusively attached; and even the rocky species are far from
' abounding; the lava which covers so great a proportion of
' the island, being eminently unfavourable to the growth of
' them. On the primitive mountains I observed the more
' common crustaceous Lecideæ and Parmeliæ, with some
' others unknown to me, which the exceeding severity of the
' weather prevented my examining carefully in their places of
' growth, and the exceeding hardness of the stone prevented
' my getting specimens of. The perennial snow that caps
' the higher hills, forbids any of them to grow on very high

' elevations, as in more temperate climates. In the plains,
' Bæomyces rangiferinus, so useful in Lapland as the food of
' the rein-deer, is found in the greatest profusion and luxu-
' riance; and the singularly elegant Cetraria nivalis, which
' is almost equally abundant, though always barren, makes
' amends by its beauty for the absence of a greater variety
' of species. The shores of the island are too much exposed
' to the most heavy and tempestuous seas, to suffer the more
' delicate species of submersed Algæ to attach themselves to
' the rocks; and the violence of the surf prevents such as
' come from more sheltered spots, from being thrown unin-
' jured upon the beach. Ulvæ I saw none, except U. lactuca
' and umbilicalis; and, among Fuci, F. ramentaceus was the
' only one which came under my observation, that has not a
' place in the British list. With the larger kinds employed
' in the making of kelp, the rocks everywhere abound; and
' I should think, that the advantages resulting from the ma-
' nufacture of this article, which is carried on in Scotland to
' such a great extent, and has proved so enormous a source
' of wealth to many of the Hebrides, might also, with the
' fostering aid of a benevolent and liberal government, be
' extended to the wretched Icelanders, who have so much
' greater need of it. A plant, which has been found in Lap-
' land, and which Dr Wahlenberg, in a letter to Mr Dawson
' Turner, calls Rivularia cylindrica (see page 71 of his MSS.),
' is extremely common in the rivers and fresh-water lakes of
' Iceland, but appears to me to have no nearer an affinity to
' the genus Rivularia, than it has to Conferva; to which lat-
' ter, Dr Roth has lately referred a plant formerly known
' under the name of Ulva lubrica, with which, in its texture
' and the disposition of its seeds, it appears exactly to coin-
' cide. It extends from three inches to as many feet in length,

' unbranched, and, as its name implies, cylindrical, forming
' an uniform tube, of a pale green colour, and thin delicate
' semigelatinous substance, studded all over with darker green
' seeds, that are almost universally placed in fours, standing
' in small squares.   As I have been fortunate enough to save
' specimens of this plant, and a drawing that I made upon
' the spot, I shall probably, at some future time, take an op-
' portunity of making a figure, and more full description of
' it public.   The water of the pools, that have been formed
' in the morasses, by cutting away the turf for fuel, generally
' abounds with our common species of Confervæ, such as
' C. nitida and bipunctata ; and a few of our marine ones are
' found in the basins among the rocks, and upon the sea-
' shores.   But other more interesting species are met with on
' spots of earth and rock that are heated to a great degree,
' either by the steam of the boiling springs, or by the waters
' themselves.   Most of these seem to belong to the Vaucherian
' genus, Oscillatoria.   Of Fungi, the island can boast but few,
' except some Agarici, scattered in such small quantities, that
' they are not used for food, and Lycoperdon Bovista, which
' is found everywhere.'

CHAP. IX.

MINERALOGY.

IN describing, as well as in examining a country, where the operations of subterraneous heat have been very widely diffused, accurate discrimination is indispensable. When new facts are met with, the difficulty of description becomes so great, while the necessity of clearness and precision is so obvious, that even the most expert mineralogists are rendered diffident when called upon to give an account of what they have seen. The contest between the Huttonians and Wernerians, has become so keen, and the economical branch of mineralogy depends so much upon a system of geology, which shall account satisfactorily for every appearance presented in the crust of the earth, that new facts bearing upon any system must not be received, unless they are distinctly unfolded. Being fully aware of the difficulty of the task I have undertaken, I am the more sensible of my inability to do justice to what I have seen. It is much in my favour, however, that I am not under the necessity of appealing to solitary facts, nor of describing any thing which fell only under my own observation. My sole desire is to be faithful, and to render my details as clear as possible without the aid of specimens. I cannot help, however, expressing a wish, that the language of mineralogy were divested of theory. In so far as it relates to his own system, the language of Werner is admirably contrived; and it is most curiously fitted to involve a student, who is not cautious, in a persuasion that what Werner has assumed is absolute fact. But few, I believe, will acknowledge the right of the proposer of a theory to alter the whole language of a science, expressly for the purpose of adapting it to that theory. Neither the disciples of Dr Hutton, nor Werner, have any title to assume such authority; and there ought to be a language, independent of all theory, which might be used in all discussions, as being universally understood. Such a language does not at present exist.

With respect to individual minerals, a settled language is not so indispensable, though certainly very desirable. But in what regards the great rocky masses in which individual minerals occur, and which compose the crust of the earth, without a fixed language, geologists will always be at variance. While the science of geology is thus deficient in terms, I feel myself at a loss when I attempt to describe a country hitherto unknown to mineralogists. The language of Werner having excited a good deal of attention, is perhaps better understood than any other; and though I by no means approve of it, I shall use some of its terms. Were I to use merely a descriptive language, and leave mineralogists to guess at what I meant, my labours would infallibly prove fruitless; and therefore I consider it better, in the present instance, to make use of a faulty language, which can be explained, than one in any respect ambiguous. I beg to be understood, in adopting any of Werner's expressions which may involve theory, that I do not use them in any other than a descriptive sense. Having deposited my collection of minerals in the cabinet of the Royal Society of Edinburgh, I shall annex a catalogue, making use of the marks and numbers affixed to the specimens. This will admit a short and easy mode of reference to any particular rock or substance. And as, on proper application, any person interested in the science of mineralogy may have access to the cabinet, I trust that such an appendix will be found useful. This chapter will therefore be chiefly confined to geological discussion.

I propose, in the first place, to describe the appearances that present themselves in Iceland, which are similar to those observed in other countries, and then to point out such geological facts as distinguish that country from all others hitherto described. This arrangement I consider to be the best; and I presume that the reader is now well acquainted with the map, and the names of different places, from the perusal of our journal.

The external efforts of volcanoes have excited a greater degree of interest than any other natural operations, to the influence of which our globe is subjected. In the infancy of geological science, the minds of philosophers naturally turned towards fire, as an all-powerful agent, demonstrating its tremendous effects by awful appeals to the senses, and displaying its capacity of producing mineral bodies, as well as of destroying those that had been already formed. As soon as chemistry had advanced so far as to assume the character of a science, and the effects of fire upon mineral bodies began to be investigated, it was discovered that a theory of the earth, founded on the external operations of volcanoes, could not be maintained. Recourse was then had to water, as an agent whose power was sufficient to remove all difficulties;

and a certain class of philosophers still continue to insist, that every appearance presented by the crust of the earth can be *demonstrated* to owe its origin to water.   While no analogous phenomena could be found to support that doctrine, it was reserved for Dr Hutton to discover not only a striking analogy between the acknowledged productions of heat, and various rocks which compose the crust of the earth, but a satisfactory reason why differences should subsist between them.   It was not till Mr Playfair removed the obscurity of Dr Hutton's writings by his luminous " Illustrations," that philosophers fully comprehended that the effects of heat on any known substance must be different when the volatile parts are restrained, and kept in union with it, from those when they are allowed to evaporate.   Since the publication of these " Illustrations," many ingenious arguments have been advanced against the Huttonian theory, and in support of the system of Werner ; and matters are now brought to such a state, that a great deal, perhaps every thing, depends on the collection of facts.

In Iceland, the effects of heat constitute not only, as in some other parts of the world, one of the principal geological features, but they seem to embrace the whole mineral masses of the country.   If we take a general view of the continent of Europe, we find that only in a very few spots subterraneous heat has shown itself externally in full activity.   With regard to the extent to which its operations may be traced, geologists differ very widely.   Those who are disposed to ascribe the most extensive influence to fire, even the most decided Huttonians, consider heat as having acted with but little intensity on by far the greatest proportion of rocks.   Thus, wherever stratification occurs, they consider the heat that operated in this instance as having been comparatively moderate, since a strong degree would have reduced the strata to such a state of liquidity as would have obliterated every trace of deposition. Even in those strata which appear to have lost almost the whole of their original structure, the heat supposed to have acted upon them has been confined within certain limits. The instance of tuffa excepted, I saw no trace of stratification in Iceland ; all the rocks having, according to the igneous system, been subjected to a degree of heat sufficient to reduce them to a state of perfect fusion. We can judge of the intensity of subterraneous heat, only by comparing its effects with others which are known ; and, even here, we are greatly limited ; for the degree of liquidity in any mass depends on its fusibility ; and its peculiar characters, when solid, depend on the circumstances under which it has cooled, and are totally independent of the intensity of the heat.

There is, however, one general distinction among the rocks of Iceland. Some bear the marks of having been in a state of fusion in the open air, as those of Vesuvius, &c.; others possess the same characters which distinguish the unstratified rocks about Edinburgh and other places in Scotland, from lava, by containing nodules of calcareous spar and zeolite, substances which, being destructible in open fire, could not exist, or be produced, in the matter ejected from external volcanoes. There is a set of rocks of this description seen in Mount Etna, which Dolomieu regarded as lavas which had been covered by the sea. That some, if not all, of the Icelandic masses which are not the production of external eruptions, are really submarine lavas, various circumstances conspire to prove ; circumstances which could not come under the contemplation of Dolomieu, since they are derived from principles which were unknown to him. In the present state of geology, nothing can be of greater importance than to ascertain with accuracy what are the results of heat acting on bodies under strong compression; since, by means of that knowledge, we are enabled to compare the ordinary deductions from Dr Hutton's principles, with the phenomena of nature, and bring to the test of actual observation the merits of a system which promises fair to put us in possession of a most simple and beautiful view of the mineral kingdom.

Dr Hutton has clearly pointed out the reason why calcareous spar may occur in the productions of subterranean heat, while it could not exist in those of open fire, by ascribing the presence of carbonic acid to the effects of powerful compression. Since Dr Hutton's death, Sir James Hall has confirmed, by actual experiment, the truth of these theoretical deductions. The rocks of Etna, to which Dolomieu ascribed a submarine origin, present a result in the works of nature by which these principles are brought to a test. But that philosopher having no such ideas in contemplation, could derive no such advantage from the facts before him. His natural sagacity, however, induced him, from other circumstances, to conclude that the rocks of Etna, which contained calcareous spar and zeolite, had actually been in a state of fusion, and that their vesicles had been filled by submersion in the sea. Doing every justice to so admirable an observer, we may be allowed, availing ourselves of the elucidation of the subject by Dr Hutton, to derive still greater light from the phenomena which present themselves in Iceland; and by tracing every circumstance peculiar to this intermediate class of bodies, acquire the means of rigorously scrutinising the application of the same principles to the rocks of our own country, in which the influence of pressure seems to have had its full effect. I shall have occasion to resume the consideration of this subject hereafter; and shall now proceed to offer some observations on common lavas, of which a vast quantity is to be seen in Iceland.

All lavas that have flowed in the open air, have assumed a rugged aspect, which cannot be mistaken. Those who have seen lava actually flowing, have described the mode in which this rugged appearance is produced. On its first issuing from the volcano, it flows with the rapidity of a torrent of water: but in its progress it becomes gradually more and more viscid, it proceeds more tardily, the surface cools, and in that process a crust is formed upon it. This crust is burst, and the broken masses are tossed about, by the slow, but powerful current. Occasionally the formation of a crust stops for a time the progress of the lava, which at last forces its way through, and continues its course. This distinguishing character of erupted lava, its rugged aspect, soon becomes familiar. Instances may occur, in which, owing to the current being impeded, the surface may remain comparatively smooth; but it is probable that, on being traced, every stream of lava will be found to possess a greater or less degree of ruggedness in different parts.

The escape of volatile matter renders the upper surface of streams of lava very porous and slaggy. The internal portions are more or less vesicular, and possess characters more or less similar to those of basalt, greenstone, &c. A great many varieties of those substances may be matched by specimens from streams of lava. So striking, indeed, is the resemblance, that the most skilful mineralogists cannot, from hard specimens, distinguish the one from the other. The chemical analyses, by my much lamented friend Dr Kennedy, have proved their perfect similarity in composition; and there now remains no doubt that the materials of which both consist,

are exactly the same. But the supporters of the two great
theories which divide geologists, differ materially in their accounts of the production of lava. The followers of Werner
maintain that lava is melted greenstone; and this supposition
is not grounded upon the result of the chemical analysis of
the two substances, but is a necessary acknowledgment arising out of the Wernerian theory of volcanoes, which assumes
the heat to be occasioned by the burning of beds of coal.
The theory of Hutton is not thus limited; for whatever variety there may be in lavas, the matter composing them is
supposed to have been produced as well from materials in
the bowels of the earth, independently of any rock formation,
as from the destruction of rocks; to which last circumstance
the Wernerians confine themselves. They have given no
satisfactory account of the mode in which beds of coal may
be set on fire, or how the combustion is to be kept up, when
a sufficient mass of fuel is provided. The presence of water
they very properly consider indispensible for producing an
eruption; but a variety of other causes are required to combine and act in regular succession. The Huttonians are not
under the necessity of going in search of any accidental
causes. They infer the existence of an internal source of
heat; but how, or where that heat is produced, or maintained, is not of any importance to the fundamental principles
of their system. Philosophers may speculate respecting the
existence of a central source of heat, the casual effects of
electricity, or the inflammation of the metals of the earths
and alkalies; but though it was absolutely impossible to prove
that any of these causes was the true one, it is surely more
reasonable to infer the existence of internal heat from the
phenomena of volcanoes, than to believe in the Wernerian
rising and falling of the waters of the globe, without the evi-

dence of any analogous fact whatever. It is plain, that, as the Wernerians cannot prove by any analogy that the waters did actually rise and fall, or support their assumption otherwise than by round assertion, they cannot direct their attacks with consistency, against what the Huttonians have assumed from evident and striking analogy. Much ingenuity has been exercised in combating and defending what is really not necessary for the support of the Huttonian theory; and it is by no means fair to interweave detached speculations on the possibility or probability of the existence of a central source of heat, with a theory which has the widely extended phenomena of volcanoes to refer to, in proof of the existence of subterraneous fire. The point, whether lava is, or is not, melted greenstone, is really not worth disputing. In making the assertion that it is, the Wernerians virtually acknowledge the resemblance between the two substances to be remarkable, and that greenstone may be reproduced by heat; thus granting to their antagonists the use of a very powerful argument, though probably not sensible of having done so.

The lava of Mount Hekla cannot be distinguished from some varieties of basalt; and that of the Snæfell Jokul is of the same character. A great many lavas in Iceland contain olivin and felspar, and several of them have in addition to these a large proportion of augit. Their peculiar vesicularity is the only character which would entitle a mineralogist to guess that they were lavas, and not greenstones. Since erupted lavas have this great resemblance, it is impossible to doubt that, had the volatile matter, which caused the vesicularity, been confined, complete greenstone would have been formed. Thus it appears that the lavas of Iceland have the same affinity to that class of rocks, as the lavas of other countries.

There are two minerals which have of late become of very

great importance in geological discussions, but which I think myself entitled to range with volcanic productions. I allude to obsidian and pumice. These had been considered as indubitably volcanic, till, having been discovered in connection with rocks whose origin was not so apparent, they were likely to overset the great system that had been constructed by Werner, who had no other resource but to deny altogether their igneous origin, and to assert that they were of aqueous formation. Obsidian and pumice having been found connected with rocks, supposed by Werner to have been produced by water, is the only proof he has adduced, to render of no avail the testimony of many philosophers, who have asserted from their own observations, that these substances are distinctly among the productions of volcanoes; though they have been observed in countries where, so far as is known, no external volcanoes appear ever to have existed.

So intimately have Werner and his pupils interwoven the minerals in question with other rocks, that, if their origin shall be proved in any one instance to be volcanic, the whole of Werner's theory, as far, at least, as it concerns unstratified flœtz rocks, will be unhinged. The importance of a strict examination of them becomes, therefore, very great, both to the supporters and antagonists of the Wernerian geognosy.

It would not be a difficult task to shew the weakness of the arguments by which the Wernerians imagine they have proved, that the origin of obsidian and pumice is not igneous. They talk much of demonstration; but I must confess my utter inability to discover where any thing like it is to be found, either in the writings of Werner, or those of his pupils.

I shall not, therefore, take up the time of the reader with combating a phantom; but content myself with bringing into view the evidence we possess of the igneous origin of the substances under consideration. The first authority I shall quote is that of Dolomieu.

This celebrated observer of nature, after describing a number of funnel-shaped cavities, and a rent about twenty feet deep, which opened into the crater of the island of Vulcano, proceeds thus,*—' It was from a fissure of this sort that, a ' few years ago, a black vitreous lava flowed, the course of ' which is still seen on the side of the mountain, and I ' walked along it while ascending. This melted glass had ' reached the base of the cone, without entering the valley.'

Of this vitreous lava the stream principally consisted, and he mentions that some of it was of a grey colour. At the top of the new cone he found the most perfect glass adhering to this stream, and thus describes it,—' A black, solid, heavy, ' and very hard glass, giving strong sparks with steel. In ' somewhat thick masses, it is perfectly opaque, and re- ' sembles the bitumen of Judæa; but when its fragments are ' thin, they are semitransparent.' There can be no doubt that the substance here described is obsidian. Dolomieu mentions another black glass not differing from the former, except in being traversed by veins of pumice. Another is of a greyish colour, and has the grain of porcelain, and approaches to pumice, and is stated to be a very common production of the volcano.

Spallanzani mentions the existence of obsidian and pumice in the cone of Vulcano. After describing the obsidian on the mountain of Della Castagna, he observes, that the appearance of streams of it, on the top and sides of the mountain, indicate its having flowed.

---

* Voyage aux Iles de Lipari.

Dr James Home, who travelled through the Lipari islands with Sir James Hall, has often expressed to me his conviction of the volcanic origin of obsidian and pumice, and has gone over with me the specimens which he and Sir James brought from these islands, in order to evince how clearly the specimens themselves point out their origin. He has very kindly permitted me to peruse his journal, and to extract from it whatever I thought worthy of notice in regard to the minerals in question. Dr Home's observations amply confirm those of Dolomieu, in respect to the stream of obsidian that issued from Vulcano in the year 1775; and he notices its passing into pumice, and being very like the lava in the northern part of Lipari. In that island, Sir James Hall and Dr Home visited a mountain that had not been observed by Dolomieu, whom they afterwards met. Arriving at a stream of what they took at a distance for common lava, they were surprised to find it entirely composed of obsidian and pumice, which passed into each other. The former was split into large masses, and white specks were disseminated through some of it. The pumice, of which there were varieties, had evidently flowed along with the obsidian, and formed the upper surface of the stream. The travellers mounted along this lava, and observed that it had burst by different mouths from the great crater. The greatest breadth of the stream was about two miles and a half, and it had flowed about three miles; and seemed to have been produced by the last effort of the volcano.

After such descriptions, coming from observers of acknowledged acuteness and industry, it appears to me surprising that the least doubt should be entertained respecting the origin of obsidian and pumice.

What I observed in Iceland will be found to agree well

with the descriptions of the naturalists I have quoted. After having gone round Mount Hekla to the northward, and then travelled in an easterly direction about twenty miles, winding through valleys covered with volcanic productions, and surrounded by hills of tuffa, we descended by a steep bank of loose earth, into a small valley closed on all sides, and having a small lake in one corner. Opposite was what both my friends and I took to be a stream of common rough lava, presenting its usual rude and fantastic appearance. On seeing this, we were greatly disappointed, as we had been on horseback during ten hours, and had seen neither obsidian nor pumice, excepting some detached masses scattered on the ground. On reaching this supposed stream of lava, we were most agreeably surprised to find it consist of obsidian, pumice, and slags. Those who have seen a stream of rough lava, filling up a valley to the depth of thirty feet, will have a just conception of the appearance which was before us. Having scrambled to the top, we saw this stream extending about two miles, its limits being concealed by hills, behind which it disappeared. At some distance from the place where we had ascended to the upper surface of the stream, we observed a separation, as if the matter had divided, while flowing, into two branches. We went to examine the further branch, as it had an aspect somewhat different from that on which we stood. We found the difference to be owing to a greater quantity of pumice being on the surface; when that was removed, obsidian was found beneath. The great extent of this stream, and several untoward circumstances, prevented our ascertaining its limits. We traversed it in various directions, and selected specimens, which bear every mark of the action of heat, and of which a description will be found in the catalogue.

No doubt can be entertained of heat having affected this mass of obsidian and pumice, in some way or other. This being the fact, it is fair to infer from its general aspect, without any reference to what has been seen in the Lipari islands, that what we saw was no other than a stream of lava of a particular species.

There is another very remarkable fact in the history of the volcanoes of Iceland, which seems to have favoured the idea of the igneous origin of pumice. About the end of January in the year 1783, flames were observed rising from the sea, about thirty miles off Cape Reikianes. Several small islands also appeared, as if they had emerged from the sea; but no new ones were afterwards found (though it has been asserted that there was one); but a reef of sunk rock now exists in the direction in which the flames were seen, terminating in what is called the Blind Rock, over which the sea breaks. The flames lasted several months, during which time vast quantities of pumice and light slags were washed on shore, along the southern coasts of the Guldbringè and Snæfell Syssels, and along different other parts of the shores of the Faxèfiord. In the beginning of June, earthquakes shook the whole of Iceland; the flames in the sea disappeared; and the dreadful eruption commenced from the Skaptaa Jokul, which is nearly two hundred miles distant from the spot where the marine eruption took place. This remarkable fact seems to indicate a communication having subsisted between those two places from which the fire burst out; and, when we consider that Hekla is almost in the direct line between them, we may conjecture that the depth of the source from whence both proceeded was very great.

The connection of obsidian and pumice is so very intimate, that the origin of the one must also be the origin of the other;

and the evidence we already possess, seems to be perfectly sufficient to establish their igneous origin. The connection of obsidian with pitchstone, &c. is of very great importance. In composition they are similar; heat affects them in the same manner; and from both, pumice may be produced. Pumice generally occurs above obsidian; and I consider it as bearing the same relation to the latter, as the common slags of a stream of lava bear to the body of lava.

It appears to me, from the great pains they have taken to shew the connection of obsidian and pumice with other rocks, that the Wernerians have really staked the stability of their system on the origin of these substances.

Numerous cones are to be found in Iceland, composed entirely of loose volcanic substances. The whole of Mount Hekla is a collection of these matters. But the most remarkable appearance is that of hills of tuffa, which invariably accompany lava. This substance forms whole ranges of mountains in the Guldbringè Syssel; and the hills round Hekla, and wherever eruptions have happened, are composed of it. The tuffa of Iceland closely resembles that of Italy and Sicily; and the mode of its formation is a curious subject of inquiry; but I shall defer entering upon it, till I have described some other appearances with which it is connected.

The general appearance of the sulphur mountains of Krisuvik have been already described.* The rocks above the highest banks of sulphur are a species of tuffa, (a considerable portion of which appears to be pearlstone), decomposed greenstone, and what has been called porphyry slate.

The rocks of the portion of Iceland which I examined, that have no external marks of the effects of heat, are of that

* P. 115.

class known by the general appellation, trap. They are
chiefly amygdaloidal, and contain all the varieties of zeolite,
calcedony, calcareous spar, &c. Greenstone and basalt occur in the island of Vidöe, and on the opposite shore. They
are seen massive, in beds and veins, and often assume the
columnar form. Some of the columns in Vidöe are composed of tables several inches in thickness, and from three to
five feet in diameter. Others have the tables composed of
oval masses, rendered visible by the decomposition of the
rock. There is a peculiarity in all the veins I examined in
various parts of the country, which I first observed in the
island of Vidöe. It is a black vitreous coating on their sides,
which varies in thickness in different instances, and gradually
blends with the substance of the veins. This is accounted for
by the Huttonian hypothesis, by supposing that, when the matter of the vein was forced into the fissure which it fills, the portion next the walls was suddenly cooled, and reduced to the
state of glass, while the internal part, by cooling more slowly, became gradually more crystalline in its texture. Werner
supposes that, when a solution entered the fissure, the first
deposition was so rapid as to have little or none of the crystalline appearance, and, as it became more gradual, the matter
shewed more tendency to crystallize. I shall have occasion
to recur to the subject of veins.

The structure of the mountains is, in general, finely displayed. The beds of greenstone and amygdaloid are of great
thickness, often from twenty to thirty feet. We frequently
observed beds of tuffa, alternating with the greenstone and
other rocks. The most remarkable rock to which the Wernerian term *tuff* can be applied, occurs in the island of Vidöe,
and on the road from Reikiavik to Havnefiord. It is composed chiefly of masses of black pearlstone imbedded in a dull

blackish green matrix. It contains also masses of amygdaloid, and I found a few nodules of pyrites in it, coated with pitch-coal. Professor Jameson found a rock very like this in Dumfries-shire.

Pitchstone and pearlstone occur in Iceland. Near Houls, on the Hvalfiord, I found the former among the debris of a precipice, where it appeared in a bed along with amygdaloid and claystone, and perhaps also porphyry slate and tuffa. These last I did not see *in situ*, but found them among the masses that had fallen from the rocks in the neighbourhood. I found pitchstone in veins traversing greenstone, on the western side of the mountain of Baula. This mountain greatly resembles, in external appearance, that called Drapuhlid, to be immediately described. Near one of the veins, where the rocks were much broken and contorted, was a bed of what appears to be a variety of porphyry slate.

The situation in which pearlstone occurs is somewhat curious. The mountain of Drapuhlid is remarkably distinguished from those in its neighbourhood. Viewed from Stikkesholm, it is conical. It is almost entirely destitute of vegetation, being covered by loose fragments of stone, giving it a reddish-yellow colour.

It was with great difficulty, and considerable hazard, that we succeeded in reaching a precipice, where the rocks of this mountain were visible *in situ*. During our ascent we found some small fragments of pearlstone, which the Icelanders mistook for obsidian, which they call Iceland agate. We also met with masses of wood, mineralised in a manner different, I believe, from any hitherto observed. In external appearance, it is similar to a mass of charcoal; but on taking it up, one is surprised to find it so heavy. It was found to contain calcedony, which occurs in transverse fissures. When exposed to heat, it burns without flame; and when the carbonaceous matter is consumed, (which is done with difficulty on account of the interposition of the mineral matter), the texture of the substance remains unaltered, and the weight is scarcely affected. Its colour and general aspect is now precisely that of fresh wood, but its other properties are very different. From the minute division of its fibres, it strongly resembles asbestos; these fibres are transparent, but very brittle.

This substance has not yet been chemically examined. It is quite different from the surturbrand of Iceland, found in the north-west quarter of the island, which we had not time to visit. I procured, however, a very good specimen of it, which was used as a table in the farm-house of Snoksdalr, and I have preserved it with the frame as I got it. The surturbrand burns with flame. I could procure no satisfactory account of the situation in which it occurs. It was said by some that it was found in rocks; and by others, in alluvial soil. From Olafson and Paulson's account, it would appear that it occurs in both, and also in connection with volcanic productions; but their description cannot be relied on; and we may look on this substance as one of the interesting objects that remain to be investigated in this remarkable country.

The wood seems to be oak. I recollect having seen, on board of a Russian frigate that was in Leith roads, many years ago, a chest of drawers made of black oak taken out of the Neva. This wood was not, however, of so deep a black as the surturbrand. This last substance can easily be cut and shaped for use; but it does not cut into shavings, being somewhat brittle. In this respect it resembles very much masses of oak that I have sometimes taken out of lakes and rivers in the highlands of Scotland, and which, at first, are soft, but afterwards, on drying, become remarkably hard.

To find the fossil wood of the mountain of Drapuhlid in its place, was an object of considerable importance. After some fruitless attempts, we at last succeeded in discovering a track on which we could fix our feet with tolerable firmness, and reached the lowest bed that was visible. This was at an elevation of about 700 feet, and proved to be a bed of tuffa, of a loose texture, containing pieces of lava and slags, in which my friend Mr Bright discovered some of the wood. Though the mouldering of this bed, which was in a decomposed state, accounted for our having found masses of the wood below, it did not expose the origin of this singular substance. Immediately above the tuffa was a bed of ash grey pearlstone, which passed into greenish black. From the manner in which this stone disappeared among the other rocks, becoming gradually mixed and blended with them, and having some appearance of veins, I was led to think that the bed itself was an interposed branch, or the trunk of a vein. Resting on the pearlstone was a bed of rock, resembling some varieties of pitchstone, which, in different places, appeared to approach to pearlstone. It contains reniform, botruoidal, and globular masses, the external aspect of which is smooth and somewhat polished; and varying from a bluish green to a brownish grey colour. These are in some instances mere specks, in others half an inch in diameter, and all of them appear radiated from the centres. In some instances, the masses are seen imbedded in a yellowish white substance, having a smooth shining fracture; it may be cut with a knife, and appears to be steatite. Except insomuch as regards the steatite, this rock greatly resembles specimens I have seen, from one of the pitchstone veins in the island of Arran.

Above this is a rock greatly resembling some we observed near Houls, in connection with pitchstone. Its colour is dark bluish grey, with reddish white round specks.   The fracture is rough, uneven, and earthy, and the structure is somewhat slaty.   The bed above this appears to be a rock of the same nature; but it is entirely slaty in its structure, and the specks are hardly visible. These two are also similar to fossils which accompany the veins of pitchstone in Arran, and are probably varieties of claystone.   All these beds were apparently horizontal; and those above, to the top of the mountain, appeared to be precisely of the same nature.

In all that part of Iceland which we had an opportunity of examining, the only rocks that did not bear external marks of heat, were greenstone, basalt, amygdaloid, and some varieties of claystone and porphyry slate.   There was nothing resembling sandstone, except some of the tuffas; and no trace of transition rocks was seen.   But we were so fortunate as to discover a class of rocks, which appear to me to be of very great consequence in a geological point of view, as I believe they will be found to form an important link in the great chain, which is supposed to connect the external operation of volcanoes, with that modified condition of heat; which is the foundation of the Huttonian theory of the earth, and to give certainty to the conjecture of Dolomieu respecting certain rocks of Etna.   The rocks in question extended over a very large tract of country.   For particular description, I shall select a very fine example in the mountain of Akkrefell.

This mountain stands perfectly detached, being separated from the mountains of Esian by the Hvalfiord; and from those towards the north, by a flat swampy country extending several miles.   It is bounded on the south and west partly by the

Submarine Lava & Tuffa on the coast near Krisuvik

Tuffa

Tuffa

Tuffa

Tuffa

Diagram illustrative of the Structure of the Mountain of Akki-sdal

Submarine Lava and Tuffa

Part of a vein of Basalt

Varieties of Amygdaloid and Tuffa

Debris

Swampy Flat

Sea

Sketched by Sir G. M.

E. Mitchell sc.

Published by A. Constable & C.º Edinburgh

Faxèfiord, and partly by the Borgarfiord. On the southern side, the structure of the mountain is exposed almost from top to bottom in a precipice about 2000 feet high. From Indreholm the beds appear horizontal; but on going round the mountain, we found that they dipped towards the north-east, forming an angle with the horizon of about 12 degrees. From the sea at Indreholm to the foot of the mountain, there is a flat swamp extending more than a mile.

To the height of about 800 feet, the mountain of Akkre-fell is composed of beds, generally from ten to twenty, and sometimes extending to forty feet thick of varieties of amygdaloid and tuffa. The latter sometimes occurred not more than a foot in thickness, and was interposed between the beds of amygdaloid. Where it was so thin, it much resembled red sandstone. While scrambling among the loose stones, we had met with quantities of slags, for which we were at a loss to account, as we had been told that nothing like lava existed on, or near this mountain. Its appearance indicated nothing volcanic, and my surprise on finding any undoubted productions of fire in such a place was increased, when at the height above mentioned we saw the under part of a bed completely slaggy, and bearing the most unequivocal marks of no slight operation of fire; and, on continuing to ascend, we found every bed, excepting those of tuffa, one of which was at least forty feet thick, presenting the same appearances, and many of them having an amygdaloidal character. My astonishment was not lessened on discovering a vein of greenstone, about four feet thick, cutting these beds, and having a vitreous coating on its sides, which seems to be common to all the veins of the country.

Similar beds occur in the mountains of Esian, where the most important fact connected with them was found in a mass of slag, which contained calcareous spar.* Forcibly struck with these facts, I soon endeavoured to account for their existence. That heat had affected these rocks in some way or other, was too apparent to require any further consideration; but under what circumstances that had happened, it is of some consequence to inquire.

Having long meditated on this subject, the earliest conjecture that I formed appeared to be the most rational. It was, that the rocks in question were really lavas that had flowed, but under circumstances very different from those which have determined the aspect of common lavas. We all agreed that they must have flowed under the sea. The existence of submarine volcanoes seems to be put beyond a doubt, by the extraordinary phenomena that I have mentioned as having appeared off the south-west point of Iceland in the year 1783, and by many other facts which might be cited.

In order that the theory I have formed, for explaining the singular appearances described above, may be fully understood, I must refer to the experiments made by my friend Sir James Hall, on the effects of heat modified by compression, which are recorded in the Transactions of the Royal Society of Edinburgh. After a long and unwearied application to the object he had in view, Sir James was not content with merely ascertaining the fact, that carbonate of lime could be melted and artificially brought to a state of crystallization; but with that perseverance which distinguishes him, and which is almost always sure of meeting at length with its due reward, he estimated the amount of the pressure required to retain carbonic acid in union with lime at a high temperature. From a table he has given, it appears ' that, under a

---

* See No⁵. 88 and 89 in the catalogue.

' sea no deeper than 1708 feet, near one-third of a mile, a
' limestone would be formed by proper heat; and that, in a
' depth of little more than one mile, it would enter into
' entire fusion. Now, the common soundings of mariners
' extend to 200 fathoms, or 1200 feet. Lord Mulgrave found
' bottom at 4680 feet, or nearly nine-tenths of a mile; and
' Captain Ellis let down a sea-gage to the depth of 5346 feet.
' It thus appears that, at the bottom of a sea which would
' be sounded by a line much less than double of the usual
' length, and less than half the depth of that sounded by
' Lord Mulgrave, limestone might be formed by heat; and
' that, at the depth reached by Captain Ellis, the entire
' fusion would be accomplished, if the bed of shells were
' touched by a lava at the extremity of its course, when its
' heat was lowest. Were the heat of the lava greater, a
' greater depth of sea would of course be requisite to con-
' strain the carbonic acid effectually; and future experiments
' may determine what depth is required to co-operate with
' any given temperature. It is enough for our present pur-
' pose to have shewn, that the result is possible in any case,
' and to have circumscribed the necessary force of these
' agents within moderate limits. At the same time, it must
' be observed, that we have been far from stretching the
' known facts; for when we compare the small extent of sea
' in which any soundings can be found, with that of the vast
' unfathomed ocean, it is obvious, that in assuming a depth
' of a mile or two, we fall very short of the medium. M. de
' la Place, reasoning from the phenomena of the tides, states
' it as highly probable that this medium is not less than eleven
' English miles.'

In making these remarks, Sir James Hall had not submarine volcanoes in view; but let us consider what would be

the effect of a submarine eruption. Let us imagine one to take place at a depth sufficient to restrain carbonic acid. The bottom of a mass of fluid by heat, flowing over a cold and wet surface, will certainly become vesicular and slaggy. In its progress the matter may have a slaggy crust formed upon its front, which may occasionally be burst, propelled, and turned downwards.

We must believe that, in a case of this sort, steam will be produced abundantly ; and although in such circumstances it must also be rapidly condensed, still there will be a constant supply, and a stratum will always be present separating the hot mass and the water, in the same manner as a drop of water is kept detached from a plate of red hot iron. Thus no water could enter the substance of the lava from above, for, when in the form of steam, it always has a tendency upwards. In its progress we may suppose the lava to displace the water ; but still a sufficient quantity of moisture may also be supposed to remain on the bottom over which it flows, to produce the appearances for which I am now endeavouring to account. This moisture at the bottom when converted into steam, operates differently from that above the lava. It must, from its tendency upwards, act upon the fluid lava, and produce that great porosity and slagginess exemplified by the rocks in question, and render the lava more or less vesicular, according to its degree of fluidity. When the lava is very hot, and consequently very liquid, the steam will have less difficulty in penetrating it, than when it is viscid. We may conceive cases in which the lava burst forth in such a high state of liquidity, as to permit the whole of the moisture to pass through it in the form of steam ; in such a state of viscidity as to admit of its escaping very slowly, so that the lava may become solid, and by confining the steam, more or less vesicular;* and lastly, so tough that the exertions of the elastic vapour shall be confined entirely to the lower surface of the lava. In the first case a compact mass of stone would·be formed, having no appearance of the action of heat; in the second, on account of the pressure of the superincumbent water being sufficient to prevent the escape of carbonic acid and other volatile ingredients, a vesicular, and amygdaloidal mass would be produced; and from the last would result a mass entirely compact, excepting at the under surface. It may occur to some, that the lava under such circumstances as those I have stated, would be so rapidly cooled as to have none of the stoney character. But a moment's consideration of the high temperature to which the superincumbent water, to a certain extent next the hot mass, must have been raised, will shew that the cooling of the lava could not have been rapid. Sir James Hall, when he had recourse to steam to assist the compression in his experiments, did on a small scale, what I suppose Nature to have done on a large one, in producing what I shall now distinguish by the name of submarine lavas. On the mountain of Akkrefell, and in various other parts of Iceland, the three cases I have stated are illustrated. Some of the beds in the upper part of Akkrefell are very compact, others are vesicular, and the greatest number, when closely examined, appear to have more or less of the amygdaloidal character. On the coast near Krisuvik is a series of beds, very slaggy at the bot-

---

* In such a case, it is possible that the steam when condensed, would in some instances remain confined in the stone in the form of water; and thus the fact of water being found in the vesicles of basalt and other rocks may be accounted for. It must be observed, that the steam, in such circumstances, must have been very much condensed, so much indeed as to be almost in the state of water greatly heated, much more so than in the familiar experiment made with Papin's digester.

tom, and so compact above as not to differ from some varieties of what is called porphyry slate. On the banks of the Thiorsaa, near Eyalstadir, we saw, on the west side, a bed of compact basalt the under surface of which was slaggy; and on the east side a bed of amygdaloid in the same circumstances, resting on tuffa full of slags. Near the pass called Bulands-höfdè, on the Breidèfiord, is a bed of amygdaloid columnar throughout, and having the same appearance on its lower surface. The extensive and beautiful ranges of lofty columns at Stappen present the same fact, and have slaggy masses included in them. The same thing occurred in the columns of greenstone, between Olafsvik and Bulandshöfdè, but the under surface in this instance, instead of having a marked slaggy appearance, was waved, and covered with the same vitreous coating I have mentioned as peculiar to the veins of Iceland.

Such is the way in which, in my opinion, the curious fact of beds of rock being found, having all the characters of common lava in one part of their substance, and not at all in others, is satisfactorily explained, and the truth of Sir James Hall's conclusions amply confirmed. But there yet remains to be considered a fact connected with these, which adds greatly to their importance, and, if the theory I have offered be correct, gives confirmation to the doctrines of Hutton, and places them almost beyond dispute. I allude to the veins of greenstone and basalt which traverse these rocks.

Along the west side of Esian, on the eastern shore of the Hvalfiord, a vein is exposed to a great height, and to a considerable distance. Its side is seen sometimes to a great extent, and where the rocks have been ruptured the section appears. This vein cuts the rocks from top to bottom, as well those that have no slaggy matter on their under surface,

as those that have. The same is the case of the vein we saw
on Akkrefell, though we could not trace it so far. These veins
are all split horizontally into fragments of a columnar form.
It is hardly necessary to observe, that the matter of a vein
must be of newer formation than the rocks which it traverses.
Here then, we have veins of greenstone and basalt cutting
submarine lavas. I do not suppose that even the keenest or
most prejudiced Wernerian, granting the existence of submarine lavas, will insist that rocks of aqueous origin were ever
deposited above them, and that at the same time the fissures
were filled and veins formed; for, an alternation of igneous
and Neptunian rocks can never be admitted by either Huttonians or Wernerians. The only mode of reconciling both
parties to the appearance of these veins, is to suppose that
they have been formed of lava which has been forced up from
below, and that the matter filling the rents has become greenstone by slow cooling. On the coast between Havnefiord and
Bessestad, below Gardè, veins of greenstone are seen inter-
secting tuffa, and their course is marked by confusion, and
evidences of the action of heat. Veins of greenstone, which
has more of the characters of erupted lava, by being much
mingled with slaggy matter, are seen to the westward of Olafsvik, cutting through the bank of loose gravel, which there
forms the beach, and appears to be an unconsolidated tuffa.

The veins which appear in the mountain of Somma, and
were taken notice of by Sir James Hall in both his memoirs,
have been described by Breislak.* ' The southern face of
' Mount Somma cut into a peak, is intersected by many veins
' of lava, some vertical, others inclined at different angles,
' or zig zag, which in some places are hardly three feet wide.

* Voyages Physiques, et Lithologiques dans la Campanie. Paris 1801.

' They form the support of the mountain ; for the spaces
' which they separate are only heaps of loose matter. Their
' substance is the same, a lava full of crystals of pyroxene,
' and containing a great quantity of leucites ; *it is often pris-*
' *matic, the prisms having their axes lying across the direction of*
' *the vein.* Some of these prisms being detached by decom-
' position, they form by that means steps which recal the
' memory of trap.' Our author goes on to state the theory
of these veins, which have been considered, by every one who
has seen them, as lava that had filled up the rents during an
eruption. These veins of Somma appear to bear a strong
resemblance to most of the veins in Iceland, so far as respects
the prismatic or columnar form, a tendency to which is very
common in the veins of that country. Here then is another
very powerful analogical argument in favour of the Huttonian
theory. Breislak supposes the prismatic form to have been
occasioned by sudden refrigeration ; that form seems, however, to be owing to some other cause. The submarine lavas
are sometimes columnar and sometimes not. I did not see
any common lava in Iceland that had assumed the columnar
form, though in other volcanic countries it is said to be frequent.
But it appears in rocks of igneous origin, and is remarkable
in one which I shall soon have occasion to describe.

There is no way left for Werner and his followers, to evade
the striking application of the facts I have described to the
Huttonian theory, but to deny their existence. I can hardly
believe that any geologist who has paid due attention to the
specimens I have brought from Iceland, will shrink from the
conviction of sense. I am well aware that decomposed amygdaloidal greenstone has been frequently mistaken for slag.
But whoever has been accustomed to compare the undoubted
productions of fire with such stones, can very easily distin-

guish the one from the other. The latter are always in a state of decomposition, whereas true slag is capable of resisting the effects of the weather, from which those I have described are entirely free.*

I proceed to make some remarks on the tuffa of Iceland. When we find it alternating with the submarine lavas, the two substances mutually assist in proving the origin of each other. In this instance the tuffa invariably includes masses of lava, and slags of various sizes, many of very great bulk, and all of them are more or less rounded by the action of water, and the beds are sometimes not less than forty feet in thickness. That the materials composing these beds have been ejected by a submarine volcano, seems to be very probable, both from their nature and their connection with the lavas. On the coast near Krisuvik this connection appears very remarkable, and is best illustrated by the sketch. The quantity of black slags that appear in the tuffa at this place, leave no doubt of the source whence it has been derived.

Tuffa is found alternating with the beds of amygdaloid which form the lower part of the mountain of Akkrefell, but they present nothing indicative of the action of heat. We must therefore conclude, that these beds are also submarine lavas that have been erupted at a much greater depth than those above; at a depth sufficient to prevent water taking

---

* I have some doubts of what has been very often stated, and is at this time generally believed, that rich soil has been formed by the decomposition of lava. I saw much lava in Iceland which had flowed long before that country was inhabited, but had not been affected in the slightest degree by exposure to the air for about a thousand years. I state this for the purpose of inducing travellers to examine lavas strictly, and to distinguish its varieties, some of which may possibly be really liable to decomposition. But there are other volcanic substances which are capable of being speedily reduced to the state of soil, and these may perhaps have been mistaken for decomposed lava.

the elastic form of steam; unless we suppose that they were placed in the situation described in the first of the three cases I have already stated, viz. that the lava was so hot and so fluid, as to admit of the escape of all the moisture on the surface over which it flowed.

Tuffa is found in another form, composing whole mountains, and even long ranges of them. In these there is no appearance of regularity, but all is heaped up in confusion. One of the best examples of this is the mountain of Helgafell, a few miles distant from Havnefiord. This mountain is about 1000 feet above the level of the sea, and stands quite insulated. It is a very remarkable fact, that wherever the operations of volcanic fire are seen, hills of tuffa are found. Lava, and tuffa, are sure indications of each other. The mountains of the Guldbringè Syssel are almost entirely composed of tuffa. The hills round Mount Hekla, near the Snæfell Jokul, and in the neighbourhood of every lava I met with, consist of that substance. Whether these hills and mountains have been formed under the sea, or have been produced by the ejections of the volcanoes which they surround, is a question of some interest to a geologist. We are not yet sufficiently acquainted with the tuffas of Italy, which are known to have been formed by external volcanoes, to enable us to compare their situation and appearance, with those of other countries. I do not know that the tuffa which overwhelmed Herculaneum has been ever fully explored so as to determine whether it proceeded from the volcano in a stream, or whether the materials composing it fell down in an ejected shower.* While

* Something which, from the description of Mr Stephenson, appears to be tuffa, covered part of the country after the eruption of 1755 from Kattlagiau Jokul. He says,—' The sand which fell, afterwards united, and covered the ' meadows with a yellow and copper-coloured crust, quite compact, and in many ' places exhibiting a perfect lava.' Yellow is the prevailing colour of the paste of the tuffa.

this is the state of our knowledge respecting one of the most curious and interesting events in the history of volcanic eruptions, it is needless for me to enter into much discussion.

I am disposed, however, to attach very great importance to the investigation of the tuffa of Iceland. It has been already seen interbedded with trap rocks, and with submarine lavas placed above them; and is a connecting link in that strong analogical chain, which seems to bind all together to a common origin. The paste of almost all the tuffas appears to be the same, and they only differ in the nature of the included masses. Those interbedded with trap rocks contain masses of greenstone, basalt, and amygdaloid; those connected with submarine or external lavas, contain lava and slags. The submarine lavas are always uppermost; and the fundamental principles of the Huttonian theory, confirmed by experiment, teach us to believe, that the greater the depth at which an eruption took place, the formation of rock would be the more perfect. Whether tuffa contains perfect rock, or lava and slags, there can be no doubt of its mode of formation being the same in both cases.

The rounded appearance of the masses indicates the operation of water, and leads us irresistibly to the bottom of the sea, as the scene of the great labours of nature. The indubitable proofs of the action of heat presented by submarine lavas; the striking connection between the operations of external volcanoes and tuffa; and the presence of that substance among rocks which bear no evident marks of having been acted upon by heat, connect the rocks together, in an origin which the enlightened Hutton first conceived, and which, since that profound philosopher's death, has been almost demonstrated.

Nor must the fact of veins of greenstone cutting tuffa be

omitted in the consideration of this interesting subject. Some remarkable facts of this sort we observed to the westward of Havnefiord, below Gardè, on the shore. There we saw the veins fairly cutting tuffa, and their sides in contact with it, presented the same vitreous appearance, which is so remarkable a feature of the veins in Iceland. In some places, the irruption of the matter of the veins has caused the utmost confusion; and the marks of heat, which the specimens I have brought from the place in question present,* seem to me unequivocal.

Nothing, in my opinion, bears more strongly against the Neptunian system than the alternation of what, in its language, are called mechanical and chemical deposits. It is nothing uncommon in this country to see highly crystallised greenstone resting on sandstone; and, in Iceland, we see it placed above tuffa, which has been formed by mechanical exertions, far more violent than those required to form sandstone. To suppose (which is necessary in such cases) that, when the waters of the globe rose at one time, they were fit only to act mechanically; while, afterwards, they were in every respect adapted to act the part of a chemical solvent; appears to me to be a violent trespass on the steady, simple laws of nature. We are left to conjecture, whence the waters came; for the philosopher who has attempted to make the world out of water, has scouted the idea of caverns, to which the waters might retire, and from which they might be brought forth at his great command.

There yet remain to be mentioned some effects of subterraneous heat, which seem to be peculiar to Iceland; at least, nothing similar has been described as existing in other volcanic countries, so far as I know.

* See 47 to 52 in the catalogue.

The first rocks which we examined in Iceland were those in the vicinity of Reikiavik. Along the coast, we traced two beds of rock, both of which bore strong marks of having been affected by heat. Both of them assumed the columnar form in many places; but this was most remarkable in the lowest, which was entirely columnar, the columns varying from a few inches to several feet in diameter. From descriptions and specimens, I had formed a very different idea of the appearance of lava from that now before me; and I concluded, rather hastily perhaps, that though the rocks in question had been subjected to heat, they had never flowed from a crater. On examining the elevations in the neighbourhood, we found them composed of the same substance, which appeared as if it had been raised up, and burst in many places. * This circumstance seemed to confirm my conjecture; for I could not conceive how a stream of lava should flow over, and cover the hills, while nothing appeared to prevent it from going round on all sides. On the way to Havnefiord, we found the hills to consist of the same substances. When about two miles from that place, we saw a real stream of lava, one that accorded fully with the ideas I had formed; and I was instantly struck with the importance of attending particularly to the distinction between two lavas, one of which seemed to have been formed in a manner hitherto unnoticed. We observed in many places that the stream which had passed by Havnefiord had flowed over the other lava. Having travelled a few miles along the lava, and passed across a ridge covered with slags, we descended into a valley where I had an opportunity of ascertaining that there really were two distinct formations of lava, one of which had every

* For a description of these rocks, see catalogue A. 1. to A. 5.

appearance of not having flowed. It was heaved up into large bubbles or blisters, some of which were round, and from a few feet to forty or fifty in diameter; others were long, some straight, and some waved. A great many of these bubbles had burst, and displayed caverns of considerable depth. On account of this peculiar character, I shall distinguish it by the name of *cavernous lava.* This was stretched as far as the eye could reach, over a widely extended plain. In some parts of it, we observed numerous little craters, from which flames and ejected matter had issued, but no lava. They were generally covered at the top with domes of slaggy matter, and the flames seemed to have got vent from the sides. This lava occurred in some places much broken, but presenting a different aspect from streams of lava. Deep rents traversed it, and shewed that the rock below was more compact than that above.

Having now ascertained the characters of the cavernous lava, and guarded myself against deception, I carefully noticed every appearance of it, that occurred during our travels, and I had soon reason to be surprised at its great extent. We did not see any thing in the common streams of lava like a defined approach to the columnar form; in the cavernous lava, nothing can be more common, or more characteristic. The external characters of this substance vary a little, some of it being pretty compact; but, for the most part, it is very vesicular, and generally contains abundance of felspar.

We traced this lava to a great distance. It forms large valleys; is often covered by more recent lava, sometimes with alluvial sand, and very generally with soil. Sometimes it is difficult to distinguish cavernous lava from such as has flowed over it. It is very distinct at the very base of Hekla, and has, no

doubt, been always taken for lava erupted from that mountain. On comparison with those great streams which may be
traced from the volcano, the difference between the two lavas
becomes at once apparent. The whole of the great plain
below Hekla is composed of the cavernous lava, which is
more elevated and more broken in some places than in others.
On the map this species of lava, and such districts as have
undergone the action of fire, are marked by light blue.* The
greatest extent of it which we traced was from the Guldbringè
Syssel, where it reaches from Cape Reikianes, towards Thingvalla in a north-easterly direction, and eastward towards the
Markarfliot. The country towards the Geysers I believe to
be chiefly submarine lava.

At Thingvalla there are two great parallel rents, about a
mile distant from each other, eighty or ninety feet deep, one
of which exposed the structure of the cavernous lava. The
other we did not examine. The rock exhibited the remains
of beds or strata, the whole bearing the marks of fire in a
greater or less degree, the uppermost one being heaved up
into large bubbles. The formation of these rents has been
owing to the ground sinking; at the time when that happened, it is probable that the lake of Thingvalla was thus
formed.

From all these facts, I was led to conclude that the cavernous lava had been formed by subterraneous heat having affected a vast extent of surface, and softened, and indeed
melted the rocks *in situ*. But perhaps it may belong to the
class of submarine lavas, and has been erupted at no great
depth. The circumstance of alluvial sand covering it to a
considerable thickness in many places, particularly near

* Lava that seems to have flowed in streams, is shaded darker. The submarine lavas are not distinguished.

Mount Hekla, and the appearance of a tract of gravel upon it in the Guldbringè Syssel, seem to be sufficient evidence of the sea having once been above it. This seems also probable from the appearance of the little craters mentioned above. The water may have been the cause of the domes being formed, as it would prevent the flames from darting straight up, and would cool the matter thrown out.

With little interruption the cavernous lava is seen dipping into the sea from Cape Reikianes to Krisuvik, and from Eyarback to the Markarfliot. But had it flowed, even under the sea, it is probable that it would have assumed a different appearance, and the bubbles would not have preserved so much regularity in figure, where they do not happen to be burst. I am disposed, therefore, to think, that internal fire has attacked the surface, and penetrated the rocks so as to soften and even fuse them. I have thrown out these conjectures chiefly with the view of inducing travellers to ascertain whether any thing similar is to be found in other volcanic countries; and those who may visit Iceland, to explore so remarkable an appearance.

Volcanic countries, though their aspect has been described by the most eminent naturalists, have never been minutely examined, since it has been in the power of geologists to make themselves understood by one another; and to view the materials which compose the crust of the earth, as connected together in a system, disclosing the various revolutions to which the globe has been subjected. Whether the views I have taken of the rocks of Iceland will hereafter be found to accord with the structure of other volcanic regions, I dare not positively assert; but it is very probable that, on close investigation, Mount Etna will be found to contain geological facts, far more interesting and important than any which have yet been discovered in the vicinity of that celebrated volcano. I consider it probable that the active volcanoes of the present day, are merely the prolonged exertions of subterraneous fire, which first broke out in the depths of the ocean, forming the beds and veins of the trap genus at the greatest depth, then those which I have denominated submarine lavas, until the solid materials reached the surface, and formed dry land. The subsequent eruptions have produced accumulations of matter differing from the above mentioned substances in their structure, but not in their component parts; differing, because they were no longer exposed to the important influence of the pressure of the superincumbent water.

Beside the filling up of the ocean, and the formation of land by such repeated operations of internal heat, this powerful agent seems to have raised vast masses of rock out of the sea, to a great elevation. But it would lead me into a discussion foreign to the objects of this chapter, to treat, generally, of the elevation of land above the surface of the sea. I may just observe, that the aspect of the rocks of Iceland exhibits striking evidence of violent disruption, such as seems to have been the universal cause of the present uneven appearance of the surface of the globe. To consider the slow operations of the atmosphere, or rivers, as sufficient for shaping out huge mountains, and forming stupendous precipices, which are known to defy the most violent external attacks the destructive agents of nature can make, is a poor resource, either for those philosophers who can raise and sink the waters of the ocean as fancy may prompt, or for those who have seen the effects of the earthquake and the volcano, and can appreciate the power of subterraneous heat.

The celebrated transparent calcareous spar of Iceland is

found chiefly on the eastern coast, from whence it is obtained
in very large crystals, some of which have been found two
feet thick.   There can be no doubt that the rocks of that
part of Iceland are of the same nature with those which constitute the country through which we travelled.   At the same
time, I have been led to think that the situation of the spar
will be discovered to be somewhat curious.   A great deal of
it is picked up from the rubbish brought down by the rivers;
and I doubt whether a specimen was ever otherwise procured.
When I considered how very destructible the spar is, this
fact certainly appeared to me surprising.   It could not long
exist in the bed of a torrent; and it therefore follows, that
those beautiful masses which enrich various cabinets, could
not have travelled far from the rock in which they were contained.   Were the entire crystals fixed in solid rock, in the
manner we usually find calcareous spar, we know from experience that the spar would be destroyed by the weather
before the rock.   This circumstance leads us to conclude, that
there is some probability of the spar being included in a rock
of a peculiar character, and either naturally soft, or liable in
a more than ordinary degree, to be destroyed by atmospherical
agents.   On examining a beautiful specimen of the wine-yellow
variety which I procured from Berufiord, and which is now
in the cabinet of my friend Mr Allan, masses of stone, the
same as the lava A 4. in the catalogue, are seen adhering to it.
Be what it may, the appearance of its connection with the spar is
such, that I am inclined to think that the latter will be found in
a situation which, though it would not surprise me, must, till
fully ascertained, be deemed very doubtful.   The situation I allude to, is in beds of some sort of tuffa.   I have hazarded this
conjecture for the reasons above stated, which seem to be such
as to justify a belief that this spar is to be found in some peculiar

position. The surturbrand, I suspect, will also be discovered in tuffa. Many specimens of the calcareous spar are studded with foliated zeolite. This is an indication of trap rocks, with which tuffa is also found connected, and certainly is a strong fact in opposition to my conjecture. It is known that, in Fife-shire, very large masses of spar have been seen by Professor Jameson; still I do not decline to state what appears to me somewhat probable, from the very remarkable fact exhibited by Mr Allan's specimen. So many wonderful things are presented in Iceland, that I shall not take any merit to myself, should my idea of the position of the spar be confirmed; nor shall 1 have any cause to repent my boldness, should it be proved to be groundless. I wish, however, to adhere chiefly to the wine-yellow variety found at Berufiord.

Such, then, is the result of my mineralogical researches in Iceland, where, I am confident, a very wide field is yet open for geological discovery. If the details I have given shall have afforded any amusement to geologists; and if they shall induce them to pay more accurate attention to volcanic countries than has hitherto been bestowed upon them, I shall recollect only with pleasure, the difficulties and dangers to which I subjected myself, in exploring a country, which every where presents objects to fill the scientific mind with astonishment and delight.

MINERAL WATERS.

The most remarkable mineral water in the world, is that of the most wonderful fountain, the Great Geyser. The deposition of siliceous earth was a fact at which chemists were greatly surprised, and which announced a property of water unexpected and important. The celebrated Dr Black, in his elegant and accurate analyses of the Geyser water, and that of a spring at Reikum, has pointed out the cause of these waters containing silica in solution; and has given it as his opinion, that the uncombined alkali contained in them, assisted by the great heat to which the water is subject in its subterraneous caverns, is the agent which enables the water to dissolve that refractory earth. It is to be regretted that Dr Black did not analyse the matter which is deposited by these waters; but I hope that the public will soon be favoured with an examination of this, and other fossils from Iceland. The appearance of the ancient depositions is various, and presents some singular forms; and it will not be an uninteresting inquiry to ascertain whether, at former periods, the water has held different substances in solution.

The contents of the water of the Geyser, as ascertained by Dr Black, are as follow:

In gr. 10,000 of Geyser water—

Soda - - - - - -	0.95 gr.
Alumina - - - - -	0.48
Silica - - - - -	5.40
Muriate of soda - - -	2.46
Dry sulphate of soda -	1.46
Total - - -	10.75

In gr. 10,000 of Reikum water—

Soda - - - - - - 0.51
Alumina - - - - - 0.05
Silica - - - - - - 3.73
Muriate of soda - - - 2.90
Dry sulphate of soda - 1.28

Total - - - 8.47

Dr Black has also given the proportions contained in an English gallon of the waters;

Geyser.

Soda - - - - - - 5.56 gr.
Alumina - - - - - 2.80
Silica - - - - - - 31.58
Muriate of soda - - - 14.42
Dry sulphate of soda - 8.57

Reikum.

Soda - - - - - - 3. gr.
Alumina - - - - - 0.29
Silica - - - - - - 21.83
Muriate of soda - - - 16.96
Dry sulphate of soda - 7.53

There are some hot springs at Reikum which deposit carbonate of lime in considerable quantities. The muddy springs near the Geyser have not yet been examined; nor the waters which escape from the sulphur mountains.*

The first cold mineral water which we procured was from

* Owing to the loss of a box, in which these waters were packed to be conveyed to Dr Davy, their analyses cannot be inserted here. I have some hope, however, that I may be enabled to give an account of their composition in the Appendix, as Mr Holland has fortunately preserved a quantity sufficient for the purposes of analysis.

Stadarhraun. The minister of the place informed us, that he had discovered the spring not very long ago; and that, on drinking some of it, he felt considerable uneasiness in his bowels. When fresh from the spring, from which it issues in a small quantity, the water tastes acid and astringent, and is not perfectly transparent. It rises at the distance of a few yards from the bank of a small river, seemingly from amongst lava, of which the valley is full. This water was put into the hands of Dr Thomas Thomson, who favoured me with the following remarks.

1. This water is transparent and colourless; its taste is acid, and rather unpleasant. Its specific gravity, at the temperature of 60°, is 1.0025. This, according to Mr Kirwan's formula, gives us 3.5 as the quantity of saline matter in 100 parts of the water.

2. Ten cubic inches of this water, which are equivalent to 2533.52 grains, were found to contain $2\frac{1}{2}$ cubic inches of carbonic acid gas, or $\frac{1}{4}$th of their bulk. No other gas could be detected in the liquid.

3. From 10 cubic inches of the water, $3\frac{1}{2}$ grains of carbonate of lime were obtained, partly by boiling, partly by evaporation.

4. The only other constituent detected in the water was muriate of soda. The quantity was not determined; but it was not much greater than what is found united to water in most cases.

5. This water, then, owed its peculiarities entirely to the portion of carbonate of lime (held in solution by the carbonic acid gas) which it contained.

The water of the spring called Ölkilda, or the ale-well, was contained in a hollow near the banks of a small stream, about two miles from Roudemelr. The temperature was 45°.

The water was in constant and violent agitation, owing to the escape of a great quantity of carbonic acid gas, with which it was very highly impregnated. Unluckily the water which I succeeded in bringing home, froze and burst the bottle in which it was contained, so that its other gaseous contents could not be ascertained. When thawed, the water had nothing peculiar in taste or smell. According to Dr Thomson, its specific gravity, at 60°, was 1.0001, which indicates a water of great purity. Accordingly, nothing could be detected by re-agents, except traces of muriate of soda.

The spring at Lysiehouls was of the temperature 96°, and was considerably agitated by the escape of a gaseous fluid; probably carbonic acid gas. The following is Dr Thomson's account of it.

1. This water was transparent and colourless. Its smell was peculiar, and probably proceeded from a minute portion of sulphurated hydrogen gas which it contained, as was indicated by the brown colour of the precipitate when nitrate of silver was dropped into it. Its taste was acid and disagreeable. Its specific gravity, at 60°, was 1.0020.

2. The constituents detected in 10 cubic inches of this water were as follows:

> 3.7 cubic inches of carbonic acid gas, probably mixed with a minute portion of sulphurated hydrogen gas;
> 0.9 grain of carbonate of lime;
> 1.7 grain of crystallised subcarbonate of soda;
> 1 grain of muriate of soda; and
> Traces of a sulphate, probably sulphate of lime.

The next mineral water we found not far from Buderstad. It was in a hole, in a flat piece of ground, close to a swamp.

There was a slight agitation from the escape of a small quantity of gas, and the water appeared slightly turbid. Its taste was acid and disagreeable, and very like the former. Dr Thomson found,

1. Its specific gravity, at 60°, to be 1.00217.

2. Ten cubic inches were found to contain the following constituents :

> 3.8 cubic inches of carbonic acid gas mixed, probably with some sulphurated hydrogen gas. For nitrate of silver was thrown down, of a dark brown colour, by the water; and this took place even after the liquid had been boiled for an hour in an open flask;
>
> 4 grains of crystallised subcarbonate of soda;
>
> 0.8 of carbonate of lime;
>
> 0.33 of a white powder, which does not fuse with soda, but dissolves without effervescence in hot sulphuric acid. It is therefore alumina.
>
> A minute portion of muriate of soda.

Thus the examination of the mineral waters of Iceland has shewn the possibility of two of the earths, not before suspected to be soluble in water, being held in solution by means of a small portion of soda. Dr Black does not seem to have attributed any influence to the neutral salts; though it is not unlikely that the presence of alumina may be owing to the action of the sulphate of soda. Sulphate of alumina will probably be found to constitute a principal ingredient in the water of the muddy springs. The source from which the water derives the substances it holds in solution, must be involved in mystery, till some convulsion of nature shall lay open its hidden laboratories. Dr Black has supposed that the muriate and sulphate of soda, conveyed by sea water, has been applied, under the influence of violent heat, to the strata which contain siliceous and argillaceous earth. The analyses by Dr Kennedy have shewn, that silica, alumina, soda, and muriatic acid, exist in lava and in greenstone; so that the conveyance of the saline matter, by means of sea water, need not be had recourse to, when we can suppose a more probable source of supply. The springs that are turbid, seem really to have acted on rocks liable to decomposition; and pyrites being found in considerable quantity among the clay which they throw out, is sufficient evidence of the fact. But the remarkable transparency of the water of the Geyser, while it is subject to the most violent agitation, renders the supposition of decomposed rocks being the source of the ingredients found in the water improbable, unless we suppose that the greatly heated water corrodes them; for, otherwise, we might expect it to be turbid, especially when we recollect how great a proportion of iron is contained in trap rocks. The iron, no doubt, may be left by the water in its progress to the place from whence it is thrown out. Some may be disposed to think, that both the heat and the solution, are caused by water meeting with the metals of the earths, and that of the alkali; but, notwithstanding such conjectures being interesting and reasonable, they do not carry us beyond bare possibility, and leave the inquiring mind still dissatisfied.

———————

APPENDIX.

APPENDIX.

No. I.

ON THE DISEASES OF THE ICELANDERS.

THE poverty of the Icelanders, and the dispersion of their small community over so vast an extent of country, render it almost impossible that medical practitioners should obtain an independent subsistence in the island. To obviate, as far as possible, this evil, a small medical establishment is provided at the public expence; consisting of a superintendant physician, who has the title of Landphysicus, an apothecary, and five subordinate medical men, who are stationed in different parts of the island. The physician and apothecary are settled in the vicinity of Reikiavik; where a house, somewhat superior in size and accommodation to the common class of Icelandic habitations, is provided for their reception. Independently of this provision, and the use of some land annexed to the house, the Landphysicus has an annual salary of 600 rix-dollars, with the liberty to avail himself of the profits of any practice which his situation may afford. The present possessor of the office is Dr Klog, a native of Iceland, but educated at Copenhagen. Of the country practitioners, one is stationed on the southern coast of the island ; another on the eastern coast ; a third on the northern ; and two in the western province. The reader will readily conceive how entirely destitute of medical assistance many parts of the country must be, when it is mentioned that some of these districts, subject to the care of a single individual, extend nearly 200 miles along the coast, with a breadth varying from ten to thirty miles. We had the opportunity, while in Iceland, of seeing two of the country practitioners ; both very respectable men, and well informed in their profession. One of them, Mr Paulson, has already been noticed, as possessing a more extensive knowledge of natural history than any of his countrymen.

With the exception of three hospitals, in which a few incurable lepers receive gratuitous support, no medical institution exists on the island. These hospitals are maintained at the public expence ; and in a method worthy of being noticed from its singularity. On a certain specified day, at that time of the year, when the fishery on the coast is most abundant and successful, every fishing boat in the island is required to contribute one man's share of the capture that has been made. A provision is added to the law, that if the number of fish taken by any boat on this day does not afford a share of five to each fisherman, the contribution to the hospitals shall be delayed, until the next time when the produce of a day's fishing equals or exceeds this amount.

In speaking of the diseases of Iceland, it will be necessary to allude only to those, which furnish any facts peculiar and interesting ; or which are more especially connected with the climate and mode of living among the inhabitants.

The diet of the Icelanders consists almost solely of animal food; of which, fish, either fresh or dried, forms by far the largest proportion. During the summer they have milk and butter in considerable abundance; but of bread, and every other vegetable food, there is the utmost scarcity, and, among the lower classes, an almost entire privation. The want of cleanliness in the personal and domestic habits of the people has been frequently alluded to in the preceding pages. It is an evil incident to their situation; the removal of which could probably only be accomplished by the sacrifice of other habits, still more essential to their comfortable existence. As an effect of these circumstances in the mode of life of the Icelanders, cutaneous diseases, arising from a cachectic state of the body, are exceedingly frequent among them, and appear under some of their worst forms. Scurvy and leprosy are common in the island; occurring especially in the districts of Guldbringè and Snæfell Syssels, and on other parts of the western coast, where the inhabitants depend chiefly upon fishing, and where the pastures are inferior in extent and produce. The scurvy *(kreppusott)*, as it appears in Iceland, presents no remarkable peculiarity of symptoms. The disease is observed to occur with greatest frequency at those periods, when there has been a deficiency of food among the inhabitants, or when the snow and frost of the winter succeed immediately to a wet autumnal season. For its cure, a vegetable diet is employed, in as far as the circumstances of the Icelanders will allow of such means. Fruits of every kind are altogether wanting to them; but some advantage is derived from the employment of the cochlearia (Officinalis et Darica), of the trefoil (Trifolium repens), of the berries and tops of the juniper (Juniperus communis), and of the Sedum-acre; plants which are all indigenous in the island.

The leprosy of the Icelanders, *(Likthra, Holdsveike,* or *Spitelska,)* exhibits in many instances all the essential characters of the genuine Elephantiasis, or Lepra Arabum;* and is a disease of the most formidable and distressing kind. Indolent tumours of the face and limbs are generally among the first symptoms of the complaint, attended by swellings of the salivary, inguinal, and axillary glands. The nostrils, ears, and lips are progressively affected with swelling and deformity. The skin over the whole, or different parts of the body, becomes thick and hard ; sometimes exhibiting a shining or unctuous surface, sometimes one rough and scabrous, which at a more advanced period of the disease, displays numerous cracks or fissures. The senses are usually much enfeebled ; and anæsthesia of the extremities generally occurs. The voice assumes a peculiar hoarseness and nasal tone, frequently with swelling of the tonsils, but without any hindrance of deglutition, until the disease has made great progress in the habit of the patient : the breath and perspired matter are extremely fetid ; and the hairs and nails frequently fall off. The tumours in different parts of the body gradually pass into malignant ulcers, which discharge an acrid, unhealthy matter. In this state the patient often lingers during a long time ; or where the disease has a more speedy termination, all the symptoms are rapidly aggravated, and he is carried off in a state of extreme debility and wretchedness.†

When it is considered how frequently unsuccessful the treatment of this disease is in more auspicious regions, it will not excite surprise that in Iceland the attempt at cure should

---

* The Elephantiasis of Cullen: the Elephantiasis Legitima of Sauvage.

† The description given by Dr T. Heberden of the Elephantiasis occurring in the island of Madeira, accords very exactly with the appearance of the disease as it exists in Iceland. See London Medical Transactions ; Vol. 1.

generally be unavailing. Where, from the situation of the patient, medical assistance can be obtained, laxatives, diaphoretics, and issues, or sometimes even venesection, are employed in the earlier stages, or with a prophylactic view. The indigenous plants which the natives employ as remedies, are the Juniper, the Vaccinium Myrtillus, the Rhodiola-Rosea, and the Dryas Octopetala; the latter of which particularly grows in great abundance on the island. These remedies, however, appear to be of little avail in relieving any of the urgent symptoms of the disease.

It does not appear that any distinct record exists in Iceland of the first appearance of the leprosy in this country. The Chevalier Bach, in his letter to Dr Van Troil on the subject, thinks it probable that the disease was brought into Iceland from Asia or the South of Europe, at the time of the crusades; in which he asserts that these islanders bore a part with the other nations of Europe.* From the Ecclesiastical History of Iceland, it appears that the latter statement is not well founded : but though not participating in the holy wars, the Icelanders had at this period an intimate connection with the European continent; and the disease, of which we are speaking, when once introduced, would readily be kept up, partly by its contagious character, principally perhaps by the food and personal habits of the people. In the rest of Europe, it has gradually disappeared, in consequence of the progressive improvement in the modes of living among every class of society.

The ravages committed by the small pox in Iceland have been such as to render this disease important even in the political history of the island. Introduced from the conti-

---

* See Van Troil's Letters on Iceland.

nent at different periods, and these in general distant from each other, it has spread rapidly, and under its most virulent form ; producing effects almost unexampled in the history of this dreadful disease. The most remarkable instance of this kind occurred in 1707 ; during which year the mortality amounted, according to the most accurate estimate, to about 16,000 souls ; more than a fourth part of the whole population of the country at that period. Several similar instances are recorded in the history of Iceland, though none attended with effects so extensively disastrous. A few years ago, the vaccine matter was introduced into the island from Denmark ; but owing to the smallness of the population, and its dispersion over so wide a surface, this was soon lost again ; and at the time of our arrival, we found the practice of inoculation entirely suspended. In the contemplation of this circumstance, we had taken out with us a few vaccine crusts, with the design of recommending the method lately proposed by Mr Bryce. Almost immediately on our arrival, we inoculated several children at Reikiavik, and afterwards in other parts of the country ; and having a communication with the Landphysicus on the subject, we had the satisfaction of knowing, before we returned to Britain, that the vaccine crust had found its way into every part of the island. The adoption of the plan of inoculating from the crust will doubtless secure to the inhabitants a permanent continuance of this blessing.

The Icelanders have occasionally suffered much from the measles, as well as from the small pox ; the epidemic being attended with inflammatory affections of the chest. In 1797, six hundred people were carried off by this disease.

Syphilis cannot be said to exist in Iceland. Single cases have sometimes occurred from communication with foreigners; but the disease has always been intercepted before it made any progress in the country.

Psora is an almost universal complaint in Iceland, appearing indiscriminately among all classes of the inhabitants. No discredit is attached to it, nor does it seem that any means of cure are attempted, though the most efficacious remedy exists in so much abundance in the country.

Inflammatory visceral affections are very common among the Icelanders. The variable nature of the climate, and the constant exposure to wet and cold which is incurred in the occupation of fishing, give a strong tendency to pulmonary complaints; and out of the annual number of deaths in the island, a very large proportion are referable to this cause. This fact was ascertained from the examination of certain statistical registers, which are annually drawn up by the priests of the several parishes, and transmitted to the Bishop at Reikiavik. In these pulmonary affections, and especially in cases of Pthisis, the Lichen Islandicus is much employed by the natives; and possesses a reputation among them, which the experience of its effects in other countries would scarcely seem to warrant. As a demulcent remedy, however, it probably in some degree alleviates the symptoms, and, as an article of diet in such cases its use may certainly be advantageous.

Inflammatory affections of the abdominal viscera are likewise very common among the Icelanders; chiefly perhaps in consequence of the peculiar nature of the diet to which they are accustomed. It is possible also that a disposition may be given to these complaints by the treatment of the children in their early infancy. A mother in Iceland seldom suckles her child; but nourishes it from the time of its birth, with cow's or sheep's milk, which the infant sucks from a piece of moistened rag, or a sponge. Where from extreme poverty, or

3 F 2

other circumstances, milk cannot be obtained, a little fish or flesh meat, rolled up in cloth and linen, and put into the infant's mouth, is the substitute most commonly employed. The diet of the Icelanders likewise gives much disposition to worms; and the ascarides are observed to be particularly frequent.

The climate and the occupations of the people, particularly that of fishing, render rheumatic affections very common. It is said that gout also occasionally occurs; but it may be doubted whether it is not some modification of rheumatism which obtains this name.

Hypochondriasis is a frequent complaint among the natives of Iceland; induced probably by the physical circumstances of their situation, and the long confinement to their habitations, which is necessary during the winter season. Yet the general temperament of the Icelanders does not appear to be a melancholic one, and the vivacity of their manner frequently forms a striking contrast to the wretchedness which their external condition displays.

Besides the diseases which have already been noticed, I had the opportunity, while in Iceland, of seeing cases of Epilepsy, Hysteria, Amenorrhœa, Menorrhagia, Asthma, Icterus, &c. No case of idiopathic fever, either intermittent or continued, occurred to my observation. With respect to intermittents, however, I was informed that they occasionally appear among the inhabitants under a well marked form; an effect no doubt of the vast extent of bogs and marshy ground, which are found even in the most populous districts of the island.

A singular complaint remains to be noticed, the effects of which, though limited to a small spot, are eminently disastrous as far as they extend. This is the disease, called

Ginklofe by the Icelanders ; the Tetanus or Trismus Neonato-
rum of medical writers ; which invades children at a very early
age, and almost invariably proves fatal in its event. It occurs very rarely, if at all, on the mainland of Iceland ; but is
confined principally to the group of islands, called West-
mann-Eyar, situated on the southern coast, of which a description has already been given in the Journal, p. 258. The
population of Heimaey, which is the only one of these islands
that is inhabited, does not amount at present to 200 souls,
and is almost entirely supported by migration from the mainland ; scarcely a single instance having been known, during
the last twenty years, of a child surviving the period of infancy.
During a great part of the year, the island is wholly inaccessible in consequence of storms, currents, and the nature of
the coast. The inhabitants are therefore left almost solely to
their own resources. Their chief article of food is the sea-
fowl, called the Fulmar, which they procure in vast abundance ; using the eggs and flesh of the bird, and salting the
latter for their winter food. The destructive effect upon the
fishery around these islands, by the great volcanic eruptions
in 1783, has before been mentioned. Of vegetable food the
inhabitants have none, and there are only a few cows and
sheep on the island.

The distressing consequences of this disease led the Danish
government to give an official direction to the Landphysicus
of Iceland, to visit the Westmann Islands, for the purpose of
investigating its nature and causes. This gentleman went over
to the islands during the summer of 1810, and remained three
weeks on the spot. Though he did not himself see a case of the
disease, he obtained all the principal facts connected with it
from the priests, and those of the inhabitants who had had
children. The symptoms of the complaint are briefly these.

Very soon after birth, strabismus and rolling of the eyes are observed; subsultus tendinum occurs; and the muscles of the back are often drawn together and stiffened, evidently by incipient spasm. These appearances infallibly denote the approach and event of the disease. Having continued during a period varying from one to seven days after birth, trismus generally comes on, sometimes attended by Opisthotonos, which is strictly called the *Ginklofe;* occasionally with Emprosthotonos, to which the name of *Klums* is given by the natives. The trismus present impedes deglutition, and the paroxysms becoming more violent, the infant is speedily carried off. When the rare event of a favourable termination occurs, it is portended by a critical diarrhœa, or by an exanthematous eruption, with the evacuation of the meconium.

The following Table, which includes a period of twenty-five years, shews the mortality consequent upon this disease in the Westmann Islands; and exhibits also the days upon which death has happened.

Children.		Days.	Children.		Days.
1	lived	2	18	lived	9
3		3	10		10
14		4	2		11
16		5	1		12
22		6	1		13
75		7	5		14
16		8	1		21

It will be seen from this Table, that the number of deaths on the 7th day greatly exceed those on any other; and also that they are more frequent on the 14th day, than on the days immediately preceding or succeeding it. From the propor-

tion which these cases of fatal event bear to the whole population of the island, it is probable that few, if any, instances
of recovery have occurred, during the period included in the
Table. No methods of cure have hitherto been resorted to by
the inhabitants.

This disease is well known to prevail in other parts of the
world; and has been particularly described as it appears in
the West Indies, and in the island of Minorca.* It exists also
in Switzerland, and in some northern districts of Scotland;
especially in the island of St Kilda, the inhabitants of which, in
their diet and mode of life, much resemble the natives of the
Westmann Islands. The exciting causes are involved in much
obscurity. It may be presumed, however, that they must
vary considerably, when the disease appears in countries, so
widely different with respect to climate, and the situation of
the inhabitants. Its occurrence in the Westmann Islands
may reasonably be supposed to have some connection with
the extraordinary diet of the natives; and this is the more probable, as it appears that the complaint has been much more
frequent, since their fishery was destroyed by the volcanic
eruptions in 1783. Independently of any effect which the
peculiarity of the mother's constitution may have upon her
offspring, the practice of giving to the infant a strong and
oily animal food almost immediately after birth, will necessarily create irritation in the bowels, and dispose to spasmo-
dic affections. Dr Klog, in some remarks he gave me on
this subject, attributes much to the effects of the sea air, and
of a moist atmosphere; but had these causes any considerable influence, we might expect that the disease would be more

---

* See the works of Hillary, Chisholm, and Clarke, on the diseases of the West
Indies; and Cleghorn's Diseases of Minorca.

frequent in different parts of the world, than is actually found to be the case.

The age which the Icelanders usually attain presents nothing very remarkable in either extreme. From the table of population, given in a preceding part of this volume, it appears that in 1801, when the number of inhabitants was 47,207, there were 41 persons between the ages of 90 and 100; 443 between 80 and 90; and 1698 between 70 and 80. The number of females was 25,371: of males, only 21,746. The longevity of the females exceeds considerably that of the males; owing no doubt to their less exposure to the severities of labour, and the hardships of the climate. Of the 41 persons between 90 and 100, 35 were females; of those between 80 and 90, 285 were females, while the number of males was not more than 158. A comparison of facts would probably prove, that the longevity of the Icelanders rather exceeds, than falls short, of the average obtained from the continental nations of Europe.

The Icelanders are in general of a tall stature; arising, however, rather from the length of the spine, than of the limbs: the head is of the middle size: the countenance open: the complexion exceedingly fair, and among the women, often very florid. The hair is almost universally of a light colour, and seldom curled. Corpulency is rarely observed among the natives of the island.

No. II.

LIST OF ICELANDIC PLANTS.*

By W. J. Hooker, Esq.

I. MONANDRIA.

1. *Monogynia.*
Hippuris vulgaris
Zostera marina
2. *Digynia.*
Callitriche aquatica
— — γ autumnalis

II. DIANDRIA.

1. *Monogynia.*
Veronica officinalis
— serpyllifolia
— Beccabunga
— Anagallis

Veronica scutellata
— alpina
— fruticulosa
— marilandica
Pinguicula vulgaris. —" Les Islandais s'en servent quelquefois en guise d'ail." *Voyage en Islande.*
— alpina
2. *Digynia.*
Anthoxanthum odoratum

* This catalogue is principally taken from Zoega's *Flora Islandica,* (attached to the Danish edition of Povelsen and Olafsen's account of Iceland) and Mohr's *Forfög til en Islandsk Naturhistorie,* published at Copenhagen in 1786. The few additional species, which I am enabled to insert by means of Sir George Mackenzie's and Mr Paulsen's collections and my own researches, are distinguished by being printed in Italics.

III. TRIANDRIA.

1. *Monogynia.*

Valeriana officinalis
Schœnus compressus
Scirpus palustris
— lacustris
— cæspitosus
— acicularis
— setaceus
Eriophorum polystachion. Of the *pappus* of this plant the natives make wicks for their lamps.
— vaginatum
— capitatum. *Hoppe.*
— alpinum
Nardus stricta

2. *Digynia.*

Phleum pratense
— nodosum
— alpinum
Alopecurus geniculatus
Milium effusum
Agrostis rubra
— stolonifera
— canina
— vulgaris
— — β pumila
— alba
— arundinacea
— cærulea
Aira cæspitosa

Aira flexuosa
— montana
— subspicata
— alpina
— aquatica
— præcox
Holcus odoratus.—Said to be used by the Icelanders to perfume their apartments and their clothes.
Sesleria cærulea
Poa pratensis
— trivialis
— compressa
— annua
— angustifolia
— alpina
— maritima
— *glauca.*—Both this and the following species are far from uncommon in Iceland.
— *cæsia*
Festuca ovina
— rubra
— elatior
— fluitans
— duriuscula
— *vivipara*
Arundo Phragmites
— Epigejos
— arenaria
Elymus arenarius.—The seeds

are occasionally made in-
to a sort of bread.
Triticum caninum
— repens

3. *Tryginia.*

Montia fontana
Koenigia islandica

IV. TETRANDRIA.

1. *Monogynia.*

Scabiosa succisa.—The Icelandic names for this
plant, *Pukabit* and *Die-
velsbid*, have both the
same signification as our
Devil's bit.
Galium verum
— palustre
— Mollugo
— *pusillum*
— boreale
— Plantago major
— lanceolata
— maritima
— *alpina.*—This I recollect
seeing, in some plenty, at
Thingvalla, and I have
since received specimens
from Sir George Macken-
zie and Mr Paulsen.
— Coronopus
Sanguisorba officinalis
Alchemilla vulgaris

Alchemilla alpina

3. *Tetragynia.*

Potamogeton natans
— marinum
— compressum
— lucens
— crispum
— perfoliatum
— pectinatum
— pusillum
Sagina procumbens
Tillæa aquatica

V. PENTANDRIA.

1. *Monogynia.*

Myosotis scorpioides α and β
Pulmonaria maritima
Echium vulgare
Primula farinosa
Menyanthes trifoliata.—This
plant is important to travellers who are not ac-
quainted with the route
in the morasses; for they
are well aware that where-
soever it grows they may
safely pass; its closely
woven roots making a firm
bed upon the soft sub-
soil. The Icelanders call
it *Reidinga,* and employ
the matted tufts to pre-
vent the saddle or any

load from chafing the horses' backs.

Azalea procumbens

Campanula rotundifolia

— patula

Viola canina

— tricolor

— palustris

Glaux maritima

       2. *Digynia.*

Gentiana campestris

— amarella

— nivalis

— aurea

— detonsa

— bavarica

— tenella

— verna

— rotata

Hydrocotyle vulgaris

Ligusticum scoticum.--To this plant, which Mr Paulsen named by mistake *Imperatoria Ostruthium*, was attached the following observation: ' Hæc (in ' Islandiâ) rarissima her- ' ba, in saxis solùm et ' montibus præruptis ma- ' ritimis reperiunda. De- ' voratis radicibus hìc tra- ' ditur divinos edidisse

' effectus in hydaridibus ' abdominalibus. (isl. *me-* ' *inlæti*).'

Angelica Archangelica.—The Icelanders gather the stems and roots of this plant, which they eat raw, and generally with the addition of fresh butter.

— sylvestris?

Imperatoria Ostruthium

Carum Carui.—Naturalised in Iceland, according to Sir George Mackenzie.

       4. *Tetragynia.*

Parnassia palustris

       5. *Pentagynia.*

Statice Armeria

Linum catharticum

       4. *Hexagynia.*

Drosera rotundifolia

— longifolia

     VI. HEXANDRIA.

       1. *Monogynia.*

Convallaria biflora

Juncus effusus.

— *arcticus.* — Discovered by Mr Bright.

— squarrosus

— trifidus

— articulatus

— bulbosus

Juncus bufonius
— biglumis
— triglumis
— pilosus
— campestris
— spicatus

2. *Trigynia.*

Rumex digynus.—All the species of *Rumex* are boiled and eaten by the Icelanders; though only the young shoots of *acutus* are employed. Of the *Acetosa* a beverage is made by the common people, by steeping the plant in water till all the juice is extracted. This drink is kept some time; but soon becomes bad and putrid in warm weather.

— acutus
— Acetosa
— Acetosella
Triglochin palustre
— maritimum
Tofieldia palustris

VIII. OCTANDRIA.

1. *Monogynia.*

Chamænerium halimifolium.
—From specimens now before me it appears that this species is subject to considerable variation, as well in the proportional breadth of its leaves, as in the size of the flowers. —Mr Paulsen remarks, ' Crescit ferè solum ad ' fluvios montium glacia- ' lium, in argillâ et arenâ ' vulcanicâ.'

— angustifolium
Epilobium montanum
— palustre
— *origanifolium*
— alpinum
— tetragonum
Vaccinium Myrtillus
— Oxycoccos
— uliginosum.

Erica vulgaris.—' Ex ejus magnâ florescentiâ de magnâ nivis hyemalis copiâ augurantur Islandi.' *Paulsen in Epist.*

2. *Trigynia.*

Polygonum viviparum
— Bistorta.—The roots are often eaten raw, and sometimes converted into bread.

Polygonum Hydropiper
— amphibium
— Persicaria
— aviculare

　　　3. *Tetragynia.*
Paris quadrifolia

　　X. DECANDRIA.
　　1. *Monogynia.*
Andromeda hypnoides
Arbutus Uva Ursi
— alpina
Pyrola rotundifolia
— secunda
— *minor*

　　2. *Digynia.*
Saxifraga Cotyledon
— stellaris
— nivalis
— Hirculus*
— *palmata*
— punctata
— oppositifolia
— autumnalis
— aizoides
— bulbifera
— *cernua*
— rivularis
— tridactylites

Saxifraga cæspitosa
— groenlandica.—My speci-
　　mens of this, gathered by
　　Sir George Mackenzie,
　　exactly accord with the
　　figure of this species in
　　the *Flore de Pyrenees.* La
　　Peyrouse has observed it
　　growing at the height of
　　sixteen hundred toises
　　above the level of the sea.
— hypnoides
— *tricuspidata*
— petræa
Scleranthus annuus

　　3. *Trigynia.*
Silene maritima
— acaulis.—Boiled and eaten
　　with butter by the Ice-
　　landers.
Stellaria media
— biflora
— cerastoides
Arenaria peploides.—This is
　　steeped in sour whey,
　　where it ferments; then
　　the liquid is strained off,
　　and fresh water added to

---

* I am informed by Mr Holland, that this beautiful Saxifraga which is abundant in Iceland, is only found on one small spot in Britain, near Knutsford in
Cheshire.　G. M.

the beverage, which is said to taste like olive-oil; whence the name of the plant in Iceland, *Smidiu-kaal.—Voyage en Islande.*

— serpyllifolia

— ciliata

4. *Pentagynia.*

Sedum *saxatile*

— rupestre

— annum

— acre.—'Vulgatum in Islandiâ vomitorium.'—*Paulsen in Epist.*

— villosum

Lychnis Flos-Cuculi

— alpina

— — *var. fl. albo.*

Cerastium viscosum

— vulgatum

— alpinum

— *latifolium*

Spergula arvensis

— nodosa

— saginoides

XII. ICOSANDRIA.

2. *Pentagynia.*

Pyrus *domestica.*—This was found by Sir George Mackenzie, growing eight feet high, in a cleft of lava near Buderstad in Snæfell Syssel. Another plant of the same was brought to Sir George Mackenzie, from Eyafiord, on the north coast.

— aucuparia*

Spiræa Ulmaria

3. *Polygynia.*

Rosa hibernica.—This, the only species of *Rosa* discovered in Iceland, was sent me by Mr Paulsen with the following remark: ' Nulli hìc priùs obvia. ' Crescit in rupe unicâ ' ad villam Seljaland.'

Rubus saxatilis

Fragaria vesca

Potentilla verna

— anserina.—The roots are frequently eaten in the

---

* It is probable that the Pyrus Domestica has been taken for Aucuparia, which, on that account, perhaps, should have no place in the Flora. It was only on close examination, that Mr Hooker and Dr Smith discovered the specimen I found to be Domestica.

southern parts of the island.

— aurea

Tormentilla officinalis

Geum rivale

Dryas octopetala.—Its leaves are gathered, and made into a sort of tea.

Comarum palustre

XIII. POLYANDRIA.

1. *Monogynia.*

Papaver nudicaule

5. *Polygynia.*

Thalictrum alpinum

Ranunculus acris.--Often used for making blisters.

Ranunculus hederaceus

— reptans

— aquatilis

— lapponicus

— repens

— glacialis.—A rare plant in Iceland. I was not so fortunate as to meet with it myself. Sir George Mackenzie has favoured me with the only specimen which he procured: and which he found growing among loose stones on the declivity of a mountain between Stadar-hraun and Kolbein-stadr.

— nivalis

— hyperboreus

— Caltha palustris

XIV. DIDYNAMIA.

1. *Gymnospermia.*

Lamium purpureum

Galeopsis Ladanum

— Tetrahit

Thymus Serpyllum.—An infusion of the leaves is often used to give an aromatic flavor to the sour whey.

Prunella vulgaris

3. *Angiospermia.*

Bartsia alpina

Rhinanthus Crista-Galli

Euphrasia officinalis.—I possess alpine varieties of this plant from Iceland, which (though bearing perfect flowers) scarcely rise a quarter of an inch above the surface of the ground.

Pedicularis sylvatica

— flammea

Limosella aquatica

XV. TETRADYNAMIA.

1. *Siliculosa.*

Subularia aquatica

Draba verna
— muralis
— incana
— — *var. contorta. Retzius.*
Thlaspi Bursa Pastoris
— campestre
Cochlearia officinalis
— danica.—Occasionally eat-
　　en as spinage, and reck-
　　oned of service in the cure
　　of the scurvy, though sel-
　　dom made use of.
Bunias Cakile.

2. *Siliquosa.*

Cardamine pratensis.
— hirsuta.—A singular variety
　　of this plant, if not a dis-
　　tinct species, has been
　　sent me both by Sir George
　　Mackenzie and Mr Paul-
　　sen, having the lower
　　leaflets round, the upper
　　ones linear, and all very
　　entire.
— bellidifolia
Sisymbrium terrestre
Arabis alpina
— hispida
Brassica alpiua.—Sent me by
　　Sir George Mackenzie.

XVI. MONADELPHIA.
5. *Decandria.*
Geranium sylvaticum
— pratense
— montanum
XVIII. DIADELPHIA.
3. *Octandria.*
Polygala vulgaris
4. *Decandria.*
Lathyrus pratensis
Vicia cracca
Pisum maritimum
Lotus corniculatus
Anthyllis vulneraria
Trifolium arvense
— pratense
— repens.—' Les gens de la
　　' campagne, dans la par-
　　' tie Nord et Est de cet-
　　' te ile, en mangent en
　　' légume.' — *Voyaye en*
　　' *Islande.*
XIX. SYNGENESIA.
1. *Polygamia Æqualis.*
Leontodon taraxacum
— autumnale
Hedypnois Taraxaci
Hieracium Pilosella
— Auricula
— alpinum
— præmorsum
— Murorum

Serratula arvensis

Carduus lanceolatus

— heterophyllus

2. *Polygamia Superflua.*

Gnaphalium alpinum

— uliginosum

— *sylvaticum*

— fuscatum.   *Pers.*

Erigeron alpinum

Senecio vulgaris

Pyrethrum inodorum

— *maritimum*

Achillea Millefolium. — The Icelandic appellation, *Vall-Humall* (field-hops) seems to imply that this plant has been used instead of hops in that island, as it is still in some parts of Sweden. At present the natives only make an ointment of its leaves with butter, which they apply to cutaneous and other external sores.

XX. GYNANDRIA.

1. *Diandria.*

Orchis maculata

— Morio

— mascula

— latifolia

— hyperborea

Satyrium viride

Satyrium albidum

— nigrum

Epipactis ovata.—I possess the only specimen of this ever gathered in Ireland ; it was found at a place called *Vik*, by the son of Mr Paulsen.

— *Nidus avis ?*—Either this or a new species of *Epipactis* has been sent me by Sir George Mackenzie. The specimen is destitute of its root, so that I cannot ascertain it with certainty.

Cymbidium corallorhizon.

XXI. MONŒCIA.

1. *Monandria.*

Zostera marina.--This the cattle eat, and the natives gather and dry for their beds.

Chara vulgaris

— hispida

3. *Triandria.*

Sparganium natans

Cobresia scirpina. Willd.-Carex Bellardi of preceding authors.

Carex dioica

— capitata

Carex pulicaris
— arenaria
— uliginosa
— leporina
— vulpina
— muricata
— loliacea
— canescens
— elongata
— flava
— pedata
— montana
— rigida
— limosa
— *atrata*
— pallescens
— capillaris
— Pseudo-cyperus
— acuta
— *ampullacea.*--The specimen sent me by Sir George Mackenzie is a slight variety with branched spikes.
— vesicaria
— hirta

4. *Tetandria.*

Urtica dioica
— *urens.* — This I only saw growing in Mr Savigniac's garden at Reikiavik.

8. *Polyandria.*

Myriophyllum spicatum

Myriophyllum verticillatum
Ceratophyllum demersum
Betula alba
— nana

XXII. DIŒCIA.

2. *Diandria.*

Salix Myrsinites
— arbuscula
— herbacea. — The downy substance from this and other species of Willow is applied by the natives to wounds both of man and beast. The leaves steeped in water are employed in tanning skins. The wood is used in making ink, being steeped in a decoction of the leaves, to which is added some of the earth used in dying; it is then all boiled together until the liquid has acquired a proper consistency.

Salix purpurea
— reticulata
— myrtilloides
— glauca
— lanata
— Lapponum
— arenaria
— fusca

Salix capræa
— pentandia
    3. *Triandria.*
Empetrum nigrum
    8. *Octandria.*
Rhodiola rosea
    13. *Monadelphia.*
Juniperis communis
    XXIII. POLYGAMIA.
    1. *Monœcia.*
Atriplex laciniata
— patula
    XXIV. CRYPTOGAMIA.
    1. *Filices.*
Equisetum sylvaticum.—Various species of Equisetum are given to the cattle in Iceland, where they are said to be excellent food for the saddle horses.
— arvense
— limosum
— palustre
— fluviatile
— hyemale
Osmunda Lunaria
Ophioglossum vulgatum
Lycopodium alpinum
— clavatum

Lycopodium annotinum
— Selago
— selaginoides
— dubium*
Polypodium vulgare
— fontanum
— ilvense
— arvonicum
— Phegopteris
— Dryopteris
Aspidium Lonchitis
— Thelypteris
— Filix mas
— Filix fæmina
— fragile.—I possess a curious and elegant species of Aspidium (Cyathea of Dr Smith) somewhat allied to this, but hitherto undescribed.
Asplenium septentrionale
Isoetes lacustris
    2. *Musci.*
Phascum muticum
Sphagnum obtusifolium.-The same use being made of this moss in Iceland as in Lapland, I shall be readily excused for inserting Lin-

---

* Surculis simplicissimis, erectis, compressis; foliis complicatis, carinitas, acutis, alternis, distiche imbricatis. *König.*

næus' words upon the sub-
ject. ' Feminis *Lapponicis*
' maxime notus est hic
' muscus ; hunc enim, lin-
' teis cùm destituantur, in-
' fantibus, dum cunis suis
' continentur, undique cir-
' cumponunt, qui et pul-
' vinaris et tegmenti vices
' servat, urinam acrem ab-
' sorbet, calorem conser-
' vat, sericisque stragulis
' gratior est tenellis ; mu-
' tatur deinde vesperi et
' mane, dum purus et re-
' cens substituitur in pri-
' oris locum.'—*Fl. Lapp.*
' *p.* 337.

Sphagnum capillifolium
Gymnostomum truncatulum
— *fasciculare*
Tetraphis pellucida
Andræa rupestris
— *Rothii*
Splachnum ampullaceum
— urceolatum
— mnioides.
— rubrum
— vasculosum
Conostomum boreale
Encalypta vulgaris
— *alpina.*

Grimmia apocarpa
— *maritima.*--Not uncommon
on rocks by the sea shores.
Weissia cirrata.
— *lanceolata ?*
Dicranum scoparium
— undulatum
— heteromallum
— purpureum
— flexuosum
— squarrosum
— *pusillum*
— pulvinatum
— taxifolium
Trichostomum fontinalioides
— fasciculare
— canescens
— *ellipticum*
Syntrichia ruralis.
— subulata
Tortula *tortuosa*
— convoluta
Catharinea *hercynica*
Polytrichum commune
— *alpinum*
— *sexangulare*
— urnigerum
— aloides
— *subrotundum*
Orthotrichum striatum
Neckera curtipendula
Bryum androgynum

Bryum argenteum
— *Zierii*
— cæspititium
— *dealbatum*
— hornum
— crudum
— *turbinatum*
— serpyllifolium
— pyriforme
— dendroides
Hypnum sericeum
— abietinum
— *filamentosum*
— prælongum
— velutinum
— *proliferum*
— nitens
— illecebrum
— purum
— filicinum
— aduncum
— *uncinatum*
— *revolvens*
— denticulatum
— triquetrum
— squarrosum
— cuspidatum
— Crista castrensis
— cupressiforme
— scorpioides
— *silesianum*
Bartramia fontana

Bartramia *ithyphylla*
— pomiformis
Fontinalis antipyretica
— *squamosa*
— *falcata*
Funaria hygrometrica
Buxbaumia *foliosa*

3. *Hepaticæ.*

Jungermannia *concinnata*
— *julacea*
— *asplenioides*
— *scalaris*
— *Sphagni*
— *angulosa*
— *byssoides*
— *bicuspidata*
— disticha.　*Mohr.*
— albicans
— nemorosa
— resupinata
— complanata
— dilatata
— ciliaris
— epiphylla
— pinguis
— furcata
Marchantia polymorpha
— hemispherica
— tenella
Targionia hypophylla
Blasia pusilla
Riccia crystallina

Riccia glauca
Anthoceros punctatus

4. *Lichenes.*

Lepraria botryoides
— Jolithos
Lecidea sanguinaria
— fusco-atra
— fusco-lutea.—About Rei-
kiavik.
— atro-virens α and γ
— pustulata
Gyrophora glabra β
— deusta
— erosa
— cylindrica.—Used, in times
of scarcity, as food, but
more frequently for dying
woollen cloth of a brown-
ish green clolour
— *hirsuta.*—‘ Longè optimum
‘ in re cibariâ Lichenis
‘ genus.—Pagina inferior
‘ pilosa. Crescit unicè in
‘ lapidibus magnis discre-
‘ tis, et rupibus alpinis,
‘ imprimis summis cacu-
‘ minibus, ubi Falcones
‘ sæpiùs insident.’—*Paulsen in Epist.*
— vellea
Endocarpon Hedwigii

Endocarpon *tephroides.* About
Reikiavik.
Sphærophoron compressum
Isidium defraudans
Urceolaria calcarea
Parmelia tartarea
— subfusca
— pallescens
— candelaria
— *brunnea.*--About Reikiavik.
— gelida
— stygia
— fahlunensis
— omphalodes
— saxatilis
— stellaris
— parietina
— olivacea
— *scrobiculata*
— nigrescens
— physodes
— furfuracea
— ciliaris
— Prunastri
— fraxinea
— farinacea
— ochroleuca
— *sarmentosa*
— jubata
Peltidea venosa
— resupinata
— canina

Peltidea apthosa
— crocea
— saccata
Cetraria islandica
— nivalis
Cornicularia lanata
— pubescens
Usnea hirta
Stereocaulon paschale
— *globiferum.*--About Reikiavik and other places, not
uncommon.
Bæomyces cocciferus
— digitatus
— deformis
— pyxidatus
— cornutus
— gracilis
— *endivifolius.*-About Reikiavik.
— uncialis
— subulatus
— rangiferinus
— *vermicularis*
— *tauricus*

5. *Algæ Aquaticæ.*

Fucus serratus.—This and various other large species of
Fucus serve occasionally
for food for the cattle and
fuel for the poor natives.
— vesiculosus

Fucus vesiculosus
———— *var.* divaricatus
———— — excisus
———— — inflatus
———— — spiralis
— ceranoides
— canaliculatus
— distichus.   (*Fl. Dan.* 351.)
— nodosus
— siliquosus
— loreus
— aculeatus
— *purpurascens*
— lycopodioides
— ramentaceus
— muscoides
— Filum
— lanosus.   *Mohr.*
— fastigiatus.   (*Fl. Dan.* 393.)
— digitatus
— palmatus.—This, the *Sol* of
the Icelanders, is the most
frequently prepared and
eaten of any of the genus.
— esculentus
— saccharinus
— edulis
— sanguineus
— ciliatus
— crispus
— alatus
— dentatus

Fucus rubens
— plumosus
— cartilagineus
— spermophorus
— gigartinus
— confervoides
— *flagelliformis*
— plicatus
— albus. (*Fl. Dan.* 408.
— corneus
— fungularis. (*Fl. Dan.* 420.)
— clavatus. *Mohr.*
— coronopifolius
— *fœniculaceus.—(Conferva*
   Huds.)
Tremella lichenoides
— verrucosa
— hemispherica
— adnata
— Nostoc
Ulva umbilicalis
— intestinalis
— latissima
— compressa
— pruniformis
— Lactuca
— lanceolata
— Linza
— plicata. *Mohr.*
Rivularia cylindrica.—Wahl.
   MSS.
— *angulosa*

Conferva dichotoma
— *spiralis*
— *bipunctata*
— *nitida*
— *flavescens*
— æruginosa
— *vaginata*
— *limosa*
— littoralis
— scoparia
— cancellata
— polymorpha
— rupestris
— ægagropila
— corallina
Byssus Cryptarum
     6. *Fungi.*
Agaricus campanulatus
— fimetarius
— campestris
Boletus luteus
— bovinus
Helvella atra. (*Fl. Dan.*354.)
— æruginosa. (*Fl. Dan.* 354.)
Peziza lentifera
— scutellata
— cupularis
— zonalis
Clavaria coralloides
— muscoides
Lycoperdon Bovista
Mucor Mucedo

In the collection of Mr Bright, a specimen of a very minute lichen has been lately discovered, which Mr Hooker and Mr Brown think has been mentioned only by Wahlenberg, who found it in Lapland, but has not yet published an account of his travels; at least his work has not reached this country. If it is really the one discovered by Wahlenberg, its name is ' Polytrichum Glabratum.'

No. III.

CATALOGUE OF ICELANDIC MINERALS, BEING CHIEFLY
GEOLOGICAL SPECIMENS.

*Presented to the Royal Society of Edinburgh, and deposited
in their Cabinet.*

A

No. 1 ROCK in the vicinity of Reikiavik. Its colour is ash
grey; it has a rough uneven fracture, and the fragments have blunt irregular edges; it is not very compact, and with difficulty scratches glass: it contains
particles of olivin. There is much resemblance, I am
informed, between this rock, and the clinkstone of
Andernach on the Rhine, where it is said to alternate
with pumice. As some specimens contain a minute
portion of hornblende, and the rock appears to
be principally composed of compact felspar, it might
be considered a variety of greenstone. It is however a lava, (page 389.)

A2,A,A′ On examining these specimens, which are from the
lower part of the same bed of rock, they all bear the
most unequivocal marks of fusion. This is the case
on the whole of the lower surface; hence I conclude that this rock is a species of lava.

A

No 3   In this specimen, which is broken from a columnar mass, the effects of decomposition are apparent, and the olivin seems to have been destroyed.

4, 5, 6 Are from the bed of rock under A 1. It seems to be the same, only it is entirely vesicular ; the vesicles are partly filled with a reddish white decomposed matter, quite soft and friable. This bed may be readily distinguished by its columnar form, as exemplified in the two last specimens. Some of the specimens contain specks of a brilliant golden lustre, which are olivin in an altered state.

7   Wacke lying horizontally above A 1, but visible only for a few yards.

8, 9   The same with A 4, 5, 6, from a place where it appears mixed with clay, slags, and other matter forming a tuffa.

10   Friable white steatite, forming a tumulus near the hot springs in the neighbourhood of Reikiavik. It is

11   also found where the hot water bubbles up. This circumstance, and the tumulus being hollow at the top, make it probable that the latter has been the site of a boiling spring, of a larger size than any which now exist at this place.

12   The steatite occurs reddish brown, and has, in some places, recently deposited matter adhering to it, some

13, 14 of which effervesces with acid, though not all. Here

15   is also found tuffa the same as A 9.

16   Greenstone containing splendent crystals of felspar from the island of Vidöe.

17   Fine grained basalt, having a conchoidal fracture, from a vein cutting the greenstone. It contains small vesicles, some of which had water in them.

A

No. 18 This specimen shows a peculiarity common to all the veins we saw in Iceland; a vitreous coating on the sides which becomes gradually blended with the substance of the stone.

19, 20 Columnar greenstone, from the same place. The latter contains small specks of a black vitreous substance, the fracture of which is conchoidal. The same has been observed by Professor Jameson in the trap rocks of Fifeshire, and by Mr Allan in the western parts of Mid Lothian.

21 From a vein of vesicular greenstone which cuts a mass of trap tuff.*

22 A slaggy mass from the tuff.

23 Mass of vesicular greenstone from the same; many of the vesicles coated with zeolite.

24 This rock may be considered as a non-descript. It is composed principally of the brilliant black substance, small specks of which were observed in A 20.; but here it occurs in larger masses, mixed with a dull blackish green matrix, which circumstance denotes it

25 to be a tuff. It contains masses of amygdaloid; and

A 1 nodules of pyrites, some of which have a small quantity of pitchcoal adhering to them. Professor Jameson informed me, that he observed a rock in Dumfries-shire which resembles this. The black substance seems to be pearlstone; and the rock may therefore be called pearlstone tuff. This rock appears irregularly connected with greenstone. There is in the island of Vidöe, a rock of fragmented amygdaloid, which in some places appears divided into large

---

* I use the term tuff here, because this rock is similar to the trap tuff of Werner.

A

No. 26   masses by a sort of net work of veins; the substance of which is similar to A 24. Columns of vesicular greenstone are seen resting on it.

27   A great portion of the island of Vidöe consists of A 1. In A 27, there are specks of olivin, of a brilliant blue colour, in a state of decomposition. In one place we observed a vein of greenstone about forty feet thick, in such a situation that it must have cut A 27, though we did not see the junction of the rocks; several yards between the beds on each side and the vein, being covered with soil.

28   From columnar greenstone, near the above mentioned vein.

29   Columnar greenstone; the columns being composed of tables from three to six inches thick, and from three to five or six feet in diameter, from one of which, A 29 was broken.

30, 31   From a vein of basaltic tuff cutting greenstone.

32   Broken from the wall of a rent in greenstone which appeared to have been once filled with matter forming a vein. This mass has a curious reniform appearance, and the vitreous coating mentioned as being common to the veins of Iceland.

33   From the same place; tinged green by some metallic substance.

34   Another from the same place, having a vitreous metallic glaze on its surface.

35   The rock from which this specimen was taken, is on the mainland opposite to Vidöe. It has, on the great scale, an external appearance from which one might be led to think that it had a slaty structure; but this seemed to be owing to decomposition. It

A   is a fine grained greenstone, and it passes into the columnar form, the columns being horizontal. Near the place where this was observed, we saw diverging columns of amygdaloid resting on vertical columns of greenstone. We could not discover any connection of A 1 and A 5 with this greenstone.

No.36   Tuffa, found a few miles to the south of Reikiavik, on the sea shore. It contains masses of greenstone, basalt, amygdaloid, small specks of the substance forming so large a proportion of A 24, (black pearlstone?) of A 1, and A 5.

37   Under the tuffa is a bed of wacke, containing shells, in some places four or five feet thick. This was traversed by a vast number of cracks, on each side of which at right angles with them, were innumerable

38   minute columnar masses, of which A 38 is a specimen.

39   From an included mass in the tuffa.

40   Appears to be wacke much indurated.

41   Under these we observed A 1.; and also a rock the same as A 24.

·42   Is a specimen of the tuffa, with the wacke adhering to it.

43   Is part of a large mass of columnar greenstone, contained in a rock similar to A 41, which we saw on the sea shore, in going to Havnefiord.

44, 45   Specimens of the lava near Havnefiord. This lava is of a bluish grey colour, dense and vesicular. It contains crystals of felspar, and has olivin disseminated through it. In several parts of this stream of

46   lava, we observed that the olivin, from decomposi-

A

tion or alteration, presented a beautiful irridescent appearance.

No.47　At the extremity of this lava, towards the west, on the shore of the bay, we found a considerable extent of rock similar to A 1. We did not see the junction of this with any other rock; but we soon came to a tuffa, with a vein of the same rock passing through

48　it. The sides of this vein have the vitreous appearance

49　already mentioned. A little beyond this vein, a large

50, 51　extent of tuffa occurs, with the same rock passing through it in so many directions, that the two seemed as if mixed together. The tuffa has here a paste similar to A 9, inclosing round, black, vitreous masses like obsidian, perhaps black pearlstone.

52　A specimen of this has part of a vein adhering to it, presenting an appearance which many will consider to be the effect of heat, and which strikes me as such. Above these we found A 1. Not far from this place are some hot springs, which are covered by the sea at high water.

53, 54,　Are specimens from the cave, mentioned page 108 of

55, 56,　the Journal. The first three are from the roof; 56 is

57　from the bottom, and 57 is part of a mass which appeared to have been squeezed, while soft, from the side.

58　Is from one of the little craters, mentioned in the same part of the Journal.

59　Is a specimen of tuffa, of which whole ranges of mountains are composed in the Guldbringè Syssel.

60　Soft white clay, from a bank on the side of a mountain on the road to Krisuvik. It has evidently been

A
No. produced in the same manner as the banks on the
sulphur mountains, which are not far distant.

61, 62, Masses from the same place; the first two are de-
63 positions, the last is an altered rock. The specimens of sulphur are deposited in this part of the collection, but are not marked, on account of their delicacy.

64, 65 A rock above the sulphur banks; it appears to be a
tuffa in a state of decomposition, and very friable.

65 Similar to the last, but not so much decomposed;
from the same place. It may perhaps be a variety
of A 24, from the appearance of specks of pearlstone.

66 Porphyry slate, from the same place.

67 We observed a great quantity of the rock A 63, appearing above the surface of the clay and sulphur.
It is difficult to give it a name. It is composed of
soft roundish masses about the size of a walnut, of a
greyish yellow colour, separated by irridescent ferru-
ginous films; and is extremely fragile. It has evidently been altered, and is probably wacke. It has
too, some resemblance to steatite.

The specimens of sulphate of lime which are arranged in this part of the collection, are not marked.
They were taken from different places where masses
occurred irregularly projecting through the clay.
They are very beautiful; chiefly white, tinged with
red; and are confusedly crystallised, some of them
fibrous, and some of them stellated.

68 From the submarine lavas on the coast near Krisuvik. It greatly resembles porphyry slate, and the
specimen marked above B.

A

No. 69 Lava between Krisuvik and Grundevik.

70, 71 Specimens of the pumice and slag, which were washed on shore during the marine eruption of 1783.

72, 73 Specimens of lava from Grundevik, containing felspar and olivin, the latter irridescent.

74, 75, Varieties of A 1. The first was found on the road from
76   Grundevik to Kieblivik, where subterraneous heat
had acted in a tremendous manner; the two last near
Kieblivik, in beds.

77   Part of an amygdaloidal vein, near Brautarholt. It
seems to be a variety of basalt. Its colour is dark
bluish grey. The fracture is imperfect conchoidal
passing into uneven. It is difficultly frangible; and
the fragments have very sharp edges. It may be
scraped by a knife; but it scratches glass easily. It is
very compact; and on the whole greatly resembles
indurated clay. Beside calcareous spar, and common radiated zeolite, it contains nodules for the
most part long and cylindrical, or rather of the shape
of an egg much elongated, and sometimes flattened. These nodules are lined with Laumonite, a variety of zeolite lately described; the crystals of which
are characterized by their extreme brittleness; so
much so, that we could not preserve a single entire
specimen. The outside of the nodules was coated
with green earth.

77′   Part of one of the largest of the nodules.

78   Vesicular slaty clinkstone through which the former
79   passed. It likewise appeared to traverse greenstone,
which also had a slaty structure, and contained much
green earth. The mountain of Esian, and those which
belong to the same range, are composed of varieties

A

of greenstone and amygdaloid, traversed by veins of
basalt, such as have been described, and of jasper

No. 80 of various colours. The veins have the vitreous coat-
81   ing on their sides. The jasper is often mixed with cal-
      careous spar, and passes sometimes into opal jasper.

82   A specimen taken from the centre of a vein; it is
      much less compact than the jasper, and appears to
      have been an included portion of some other rock in
      an altered state. Sometimes the jasper, from decom-

83   position, is vesicular.

84, 85, Are varieties of jasper. The last specimen is inte-
86   resting, in so far as it shows the jasper passing into
      opal jasper, and from that into pitchstone.

87   Amygdaloid containing agate.

88, 89 These two specimens are particularly deserving of
      notice, especially the last. A 88, is an amygdaloid
      containing calcareous spar in elongated vesicles. I
      do not wish to lay any particular stress upon this
      specimen, because it has been unfortunately damaged,
      and because certain appearances which it presents
      may be attributed to the effects of the weather. But
      as it was found along with the next specimen, I may
      state what strikes me in regard to it, in order to in-
      duce future travellers to attend particularly to the
      spot where these were found, which is in the face of
      rock on the shore of the Hval Fiord, before turning
      into the valley on the road to Houls. Several days
      might be well spent in this district.

      The spar in A 88, is not attached closely to the
      sides or bottom of the vesicles, which are lined with
      a number of minute, round, yellowish coloured
      masses, which have left impressions on the spar.

A
No.   These are also seen in the body of the stone, and
must have lined the vesicles before the spar was
formed. If the spar entered in a state of solution,
it ought to have reached the bottom of the vesicles,
and adhered closely to the sides. If any empty space
was left at all, it should have been in the heart of the
spar itself; but A 89, which was found at the same
place, exhibits marks of fusion which cannot be mistaken. I may here mention, that, among the debris of the rocks in this place, great quantities and
varieties of slags were observed; but these did not
at first excite particular attention, as we were at
the time quite ignorant of whence they came. Nor
did I take particular notice of the specimen under
consideration, excepting as a slag, till I was repack-
ing it to be sent home. This remarkable specimen
contains calcareous spar; and is one which, together with others to be soon described, gave rise to the
discussion in the chapter on Mineralogy. At this
place there are fine calcedonies and zeolites.

The rocks near Houls consist of apparently hori-
90, 91 zontal beds of amygdaloid, porphyry slate, and of
92, 93 blackish pitchstone. Beyond this place, masses of
94, 95 porphyritic pitchstone, of a dull black colour, were
96, 97 found; and also a species of tuffa, and a variety of
wacke in a state of decomposition.

On the western side of the Hval Fiord, nothing
particular occurred, all the rocks being greenstone
or amygdaloid, excepting a variety of the former, of
98   an ash grey colour.

The following specimens illustrate the structure of
the mountain of Akkrefell, and probably of almost

B

all the mountains in this part of Iceland. It was with difficulty and hazard that we obtained so complete a series of specimens, which are peculiarly interesting, as proving the existence of a new set of rocks in the structure of the crust of the earth.

No. 1, Tuffa which appears on the shore near Indre-
2, 3   holm.
4, 5   Amygdaloidal greenstone containing fine crystals of chabasie, or cubic zeolite, on the shore above the former.
6   Another tuffa, which formed the lowest visible bed of the mountain. It is similar to what is found in the Guldbringè Syssel; but no lava nor slags were observed in it. It contains cavities lined with minute crystals, unconnected with any included masses.
7, 8, 9 Amygdaloid from three different beds.
10   This fossil is very similar to red sandstone; but is in fact a fine grained tuffa. The mass of it which we saw, was not more than a foot thick, and was irregularly interposed between the beds.
11 to 16 Varieties of amygdaloid from different beds. The next specimen is wanting in the series, on account of the package having got wet, and the number having been lost.
18 to 22 Varieties of amygdaloid follow.
23   Is similar to B 10; and, being coarser, serves to elucidate the nature of that substance. Above,
25   amygdaloid again appears. One variety of it is very vesicular; some of the vesicles being empty, and others filled with chabasie, the crystals of which, in
No.   one instance, assume a stalactitic disposition.
1   After experiencing great difficulty, we arrived at a

bed, the lower part of which was slaggy. Under some

No. 2　of the slags was a substance, apparently indurated lithomarga, of the same red colour as that which forms so prominent a feature in the aspect of the Giant's Causeway, and which abounds in many parts of the county of Antrim.

3　This specimen shows the junction of the slag with the rock.

4　The rock itself, which resembles the lava of Havnefiord, (A 44, 45.)

5　Above this, we found amygdaloid, an unexpected occurrence in this situation; but we afterwards found it again in another part of the same mountain; the amygdaloid being placed between two beds, the lower

6, 7, 8　surface of which were slaggy.

9　From a vein of basalt, which cut the beds nearly in a perpendicular direction.

10　Part of the edge of the vein, vitreous at the sides.

Above this, to the top of the mountain, all the beds, except those of tuffa, were slaggy on the lower surface.

11, 12,　Some were amygdaloidal as 11; and others compact

13　as 12; and some were vesicular and scorified looking throughout, as 13. One of the beds of tuffa was very large, not less than fifty feet thick, and contained slags and lava. Many of the included masses were several feet diameter. The average thickness of the beds composing this remarkable mountain, I suppose to be about 20 feet. Above this great bed of tuffa, were several beds slaggy underneath. The uppermost re-

15　sembles the Havnefiord lava.

This singular assemblage of rocks, which I have en-

B

deavoured to show to be a series of lavas erupted at the bottom of the sea, I believe will be found to extend over the whole of Iceland ; and it is very probable that the future researches of geologists will prove that the whole island has been produced by the agency of heat; the power and efficacy of which seems to be vastly underrated by many philosophers who have not seen or sufficiently considered its effects.

No. 26 Deposition of the hot springs near Leira.

27   Conglomerate, formed by the deposition of the same springs.

28, 29, Siliceous petrifactions, apparently of peat, contain-
30, 31 ing roots; from the same place. These have been formed by more ancient springs, which no longer exist.

In passing over the eastern Skardsheidè, the same rocks, we had already observed, occurred ; and among them pitchstone. In the vast precipices, which were everywhere exposed to view, we saw the finest possible display of the structure of the mountains ; and recognised the tuffa, so often mentioned, at a great elevation. Zeolites and calcedonies were scattered about in abundance, but we did not see any that were remarkably fine.

The rocks of the Western Skardsheidè continued amygdaloidal, till we met with lava of the same na-
32   ture as that at Havnefiord. Hills of tuffa were on every side. Leaving the defile which was filled with la-va, nothing particular occurred till we met with a green-stone very highly crystallised, partly amygdaloidal,
33, 34 and partly porphyritic. The crystals of felspar occur-red more than half an inch thick. Beyond the valley

B

of Stadarhraun, which was full of lava, the mountains consisted of the same materials as those we had left behind; and contained great plenty of zeolites of every description. I found one remarkably fine specimen, half of which I placed in the cabinet of Mr Allan, and the remainder in that of the college of this city. It was found entire, among the debris, and was afterwards broken, when it displayed a most beautiful cavity.

No. 35 A specimen of the amygdaloidal rock containing stilbite.

36   Lava from Roudimelr. It contains a great quantity of augit, and altered olivin.

37   A specimen of the range of columns near that place. This rock does not differ, except in its being more compact, from the lava of Havnefiord and other places.

38   Part of a rolled mass of sienitic greenstone, the felspar white.

39, 40, Depositions from the spring at Lysiehouls, chiefly
41   calcareous. Not far from this spring are large quan-
42, 43, tities, several acres, of petrifactions, that have been
44   formed by some ancient springs which held silica in solution.

45   Lava of Buderstad, which differs from that at Havnefiord in containing augit.

46   From the columns at Stappen. Here there is still a resemblance to lava, only this contains less olivin than the lavas we had met with before.

47   From the lower end of a column. Wherever we saw
48, 49, the lower ends they were slaggy. Slags were found in
50   the heart of some, and lining every cavity we observ-

B
No.   ed.   The specimens can hardly leave a doubt of the
        action of heat.

51, 52, From a stream of lava that has flowed from Snæfell
53, 54 Jokul.   The more compact specimens are exactly
        similar to black basalt.   In several parts of this
        stream, we saw masses very different from the lava in
55   general.   They contain a few minute vesicles, some
        very small crystals of felspar, and specks of augit.
        The general colour of the stone is ash grey, spotted
        with white, and it appears to have a slaty texture.

56   A specimen of pumice, picked up from among many
        that still remain of those heaps which were washed on
        shore during the eruption in 1783.   Masses of pumice
        exactly similar to this, have been frequently found
        on the north coast of Ireland ; and, in all probabi-
        lity, were derived from the same source, having float-
        ed on the surface of the ocean from the place where
        the marine eruption took place.

57, 58, Slags, pumice and obsidian, from the Snæfell Jokul.
59, 60, These were picked up by my friends, from a bank
61   composed of them, and which was free from snow.

62, 63, Specimens from a bed on the mountain between
64   Stappen and Olafsvik.   The upper part of this bed
        (B 62) is a perfect greenstone, containing small specks
        of olivin, augit, and felspar.   The middle part of the
        bed (63) has a coarse and scorified appearance ; and
        the lower part (64) is completely slaggy.

65   From another bed on the same road.   It contains
        the largest masses of augit we had observed. I should
        have remarked that we did not see any augit distinct-
        ly crystallized.

66   Part of a rock which is heaved up into blisters, like

B

those near Reikiavik. It is very like A 1, but is more generally vesicular.

No. 67 Tuffa on which a grand range of columns, on the road to Olafsvik, rested.

68 Is a part of one of the columns, which differ from those at Stappen in being more compact.

69 From a large rolled mass in the river near the columns. It is a highly crystallized greenstone, containing augit and large crystals of felspar.

The rocks about Olafsvik are amygdaloidal; and, in several places, beds of tuffa present themselves. The most curious appearance in this neighbourhood, is a vein of slaggy matter, passing through the bank of gravel which forms the beach.

On a point of land several miles to the eastward of Olafsvik, are some fine ranges of columns overhanging the sea. Some of these appear as if they had been twisted. At the only place where they were accessible, they presented an undulated appearance on the lower ends, at the separation

71 of the columns from a bed of amygdaloid on which they rested; but of this it was difficult to obtain specimens. When broken, the fracture exhibits

72 the vitreous appearance so often observed on the

73 sides of veins. In some places the same slaggy ap-

74, 75 pearance is seen on the sides, and also in the very heart of the columns.

76 Specimen of greenstone, from a mass that had fallen from a precipice not far from Bulandshöfdè.

77 Another mass from the same place. It is of an ash grey colour, vesicular, the vesicles being irregularly shaped, and lined with minute transparent crystals.

B

Some of the vesicles contained minute diverging crystals of calcareous spar.   Near a cascade, mentioned
No. 78, in the Journal, p. 186, we observed numerous veins of
79, 80 greenstone passing through rock, of the same substance, tuffa, and amygdaloid, all in the greatest confusion.

In this part of the country, tuffa frequently occurs ; and when it forms the tops of mountains, it is easily recognised by the rugged and fantastic peaks which they present, similar to those in the view of the sulphur mountains.   At Stikkesholm, I observed a vein of greenstone, standing erect like a wall, about the height of ten feet, the beds on each side having been worn away.   The sides, as usual,
82   were vitreous.   It contained nodules of obsidian.
83   Shows both sides of a vein, containing small nodules of the same substance.   A vein of calcareous spar traverses the rock in various directions ; from
84   one of which I took a specimen of semi-opal.
85   Is a specimen of highly crystallized greenstone, which is disseminated through part of the rock near this place.
86   Mineralized wood from Drapuhlid.
87   Ash grey pearlstone from the same mountain.
88, 89 Greenish black pearlstone from the same.
90, 91 This rock was immediately above the preceding.
92, 93 Above the last.   This rock greatly resembles some we observed near Houls, where it was connected with pitchstone.   The colour is dark bluish grey, with round reddish white specks.   The fracture is uneven and earthy, and is somewhat slaty in the texture.

B

No. 94 This appears to be the same rock, entirely slaty, with the specks hardly visible; the beds were horizontal. These two rocks are similar to the fossils which accompany some of the pitchstone veins in the island of Arran.

95 From a vein of pearlstone at the base of the mountain of Baula, on the west side. The colour is greenish grey. This has much of the character of pearlstone.

96 Dark green pitchstone, from the same place.

97 Pitchstone porphyry, from the same place.

98 This is from a rock connected with the pitchstone veins of Baula, probably a variety of porphyry slate.

99, 100 Small masses of coaly matter, which were given to me as having been found on the mountain of Baula. Both have a strong resemblance to wood, but are different from that of Drapuhlid. They contain a small quantity of pyrites, and burn with flame.

C

No. 1. This tuffa was found at Eyalstadir, and is the same with B 10, from the mountain of Akkrefell. It is here connected in the same manner with submarine lava.

2 The under surface of a bed of amygdaloid resting on tuffa, which has the peculiar characters of a slag. The bottom of this bed is not exposed in many places, being concealed by debris. It is probable, that, in the course of the river Thiorsaa, in places higher up, some interesting examples of submarine lava may be seen.

3 The upper part of the rock, containing analcime.

4 The same, with green steatite.

C

No. 5　Black obsidian.　This occurs only in detached masses at the place where this specimen was found.　It exists in great quantity in the neighbourhood of Mount Krabla ; from whence, I was informed, all the specimens of Icelandic agate, in the European cabinets, were brought.　This is not so perfectly vitreous as specimens I have seen from the north of Iceland.

6　The most common variety in the great stream which we saw.　It is vesicular, with white crystals of felspar scattered through the mass.

7　In this the vesicles are elongated, and flattened so much, that, when viewed in one direction, it seems as if composed of plates.　The other fractures show distinctly the vesicular structure.　The crystals of felspar are nearly disengaged.　I have seen many similar specimens from Lipari.

8　Contains more felspar ; the vesicles are minute ; and it approaches to pumice.

9, 10　Show the whole gradation from compact obsidian, to the most perfect pumice.

11　This contains felspar, and is blackish grey.　It wants the vitreous lustre, and its fracture is uneven.　It is dense, and somewhat vesicular.　This and the following have been called compact pearlstone by Mr Jameson.

12　The vesicles of this are studded with minute globular, white, and hard masses.　It is of an ash grey colour,

13　and passes into obsidian.　One variety has a peculiar aspect ; appearing, when fresh broken, as if dusted over with a purplish grey powder.

14,　These are seen, in different specimens, passing into
15, 16　obsidian, which appears in layers.　One specimen

(C 14) exhibits small globular masses of a reddish grey colour, dispersed through the obsidian.

No. 17 In this specimen, all these are seen connected, as well as the gradation into pumice.

18, 19 These specimens have masses of slag attached to them. Pumice occurs above the obsidian; and, from the motion of the stream when flowing, has been sometimes included in it.

20 This is a remarkable and beautiful specimen, the last of the series from the stream of obsidian. It is a mass of slag, in a cavity of which some fusible matter has been included, and reduced to the state of glass. The cavity is lined by it in stalactitic masses; and some of the matter has been drawn out to the fineness of hair. No operation of water could possibly produce these appearances.

21 to 27 Are specimens of lava from Mount Hekla, which are very like those from the Snæfell Jokul, B 51, &c.

The remaining numbers, to C 40, inclusive, are varieties of slags from Hekla; C 33, 34, being from the very summit of that celebrated mountain.

39 Is the only mass we found having the appearance of an ejected stone: it is little altered, and is probably sienite.

40 Is a specimen from the hills of tuffa which surround Hekla.

The specimens from the Geysers are marked from D 1, to D 41.

As the productions of the hot springs in Iceland, are of a nature entirely different from those of any other springs in the known world, it is probably not presuming too much when I propose to mineralogists, to form a separate class of those minerals, which have been deposited from chemical solution in water, under the general name of *Hydrolite*, and to arrange the stony depositions of water under the heads of calcareous and siliceous Hydrolite. This is perhaps a more precise denomination than *Sinter*, the word used by Werner, and at once conveys the *known* mode of the formation of such substances. I propose to draw up a minute description of the depositions of the hot springs of Iceland, and to obtain the assistance of chemical analysis, that it may be known whether composition has any influence in the variety of external appearance. In the mean time, I throw out this hint for a new arrangement of the depositions of water, that I may discover whether it is likely to obtain the approbation of mineralogists. I shall therefore, at present, only point out the specimens in a very general manner.

D

No.1 to 5　The outer part of the mount, 1, 2, being the surface, and much resembling the heads of cauliflowers.

6, 7　From the inside of the bason. This takes a tolerable polish, and is very pretty.

8　Is a mass of old incrustation, coated over with recently deposited matter. It was taken from a hollow on the mount in which the water was retained.

9　Is part of the depositions of the New Geyser, formed, apparently, when that fountain presented phenomena different from what it now does.

10, 11　From the beautiful cavity described p. 214. The specimens resemble the capital of a Gothic column.

13　Is from the same place.

14　A mass of turf, on which the water, after having cooled, was depositing its contents while the grass was yet growing.

D

15 to 18 Masses of petrified leaves, &c.

19   This was picked up among the old incrustations. The opaline matter is arranged in waved lines, which are separated by layers of an open texture, resembling the tables of the skull separated by cellular bone.

20   Contains leaves and rushes, and is discoloured by iron.

21, 22   Masses of petrified peat, containing rushes and branches.

23   Clay from the muddy springs.

24   Shows the opaline matter in layers.

25   Appears to have been produced by the deposition having taken place upon a conferva.

To D 37   Varieties of depositions and petrifactions.

38, 39   From the spring to the northward of the Geysers. It bears a very striking resemblance to opal.

40   A very curious specimen picked up on the clay bank above the Great Geyser. It is in a state of decomposition, and is a good model of a rocky promontory. I am told that it is not unlike that of Fairhead in Ireland.

No. IV.

MISCELLANEOUS ARTICLES, CONNECTED WITH HISTORY, LITERATURE, &C.

The Berserkine Superstition.

A SHORT account of this superstition has been given in a note to the Preliminary Dissertation, p. 39. The Berserkir, or magical wrestlers, are thus described in the Hist. Eccles. Island. tom. 1. p. 45, note :—' Tales athletæ, antiquâ linguâ
' vocantur *Berserkir*, id est, nudi et sine loricâ in cædes et
' pugnas ruentes. Erant viri robusti, sed facinorosi, et ut
' plurimum incantatores, qui cutem arte diabolicâ indurave-
' rant, ne iis ferrum ignisve nocere posset. His furore per-
' citis, ita ad tempus intendebantur vires, ut postea debiles
' et languidi fierent.' In the treatise *De Berserkis*, annexed
to the Kristni Saga, the following account is given of the paroxysms to which these men were subject :—' Effectus furo-
' ris berserkici ex veterum traditionibus præcipui fuerunt,
' quòd eo occupati, ferocitate canum luporumve æmula, ore
' torvum infremerent, clypeosque morsibus non modo attrec-
' tarent, sed et ex parte consumerent ; porro quòd robore
' tauros ursosve æquante augerentur, ferro impenetrabiles

' evaderent, incendia et flammas nudis etiam pedibus percur-
' rerent et penetrarent; denique et torridas prunas igneosve
' carbones deglutirent.' p. 159.   By Snorro Sturleson and
others, the origin of this superstition is ascribed to Odin
himself, who was supposed, in those times, not only to have
instructed the original Berserkir in the magical arts, upon
which their powers depended, but also to exercise an immediate influence upon the mind in every instance where this
furor was present.   With respect to the real history of the
superstition, it is probable that some of the Berserkir were
men of weak judgment and a depraved imagination, who became almost involuntary agents in these absurdities : others,
doubtless, were merely impostors, who assumed this strange
character that they might the better work upon the prejudices and terrors of those around them.

For some centuries past, it has been customary among the,
Icelanders, during the period of any great volcanic eruption,
to appoint a day of general prayer and supplication.   This
was first done in the northern parts of the island in 1477,
when a great assembly was convened of the inhabitants of
the district.   The following is the preamble to the record
preserved of the vows and other religious ceremonies of this
meeting.

' In nomine Domini, amen! Die Martis primo mensis hye-
' mis ultimi, anno P. C. 1477, in Grund in Eyafiord, conve-
' nere clerici et laici inter Vargaa et Gleraa habitantes, et
' locuti sunt de terribilibus prodigiis, quæ tunc premebant;
' eruptio nempe ignis, dispersio decidentis arenæ atque cine-
' ris, tenebræ et horrendi stridores.   Horum prodigiorum
' causâ pecora alimentis destituebantur, terra licet nive va-

' cua esset. Convenit inter eos hoc profectò accidisse in
' peccatorum et prævaricationum humanarum pœnam : hinc
' omnes in id consenserunt, ut gratiam et misericordiam ibi,
' ubi abundant, scilicet apud Deum ipsum quærerent, ut
' ille suæ iræ vindictum a nobis avertere vellet.　Non igitur
' ulterius divinas castigationes et poenas in hoc mundo de-
' precabimur; ne autem regnum cœlorum nobis occludat,
' elegimus, ut Deus, qui omnia scit atque potest, et ea quæ
' optima sunt, vult, nobis omnibus et in præsens et in futu-
' rum consulat.　In primis votum vovebamus omnipotenti
' Deo, illi qui est fons omnium bonorum : illustri deinde re-
' ginæ Virgini sanctæ Mariæ ; sancto Michaeli Archangelo,
' et omnibus Dei angelis ; sancto Johanni Baptistæ,' &c.

The structure of the ancient Icelandic or Scandinavian
verse has been briefly described in the Preliminary Dissertation, p. 22.　The following verses will furnish a specimen of
the modes of alliteration which were employed in the poetry
of these times, and upon which its harmony was considered
so much to depend.　They form the beginning of a sacred
poem, called the Lilium, which was composed during the
14th century, by Eystein, an Icelandic monk, who possessed
much reputation at this period.

Almattigr Gud, allra stetta	All powerful God, who presidest over
Yferbiodandinn, engla og thioda,	all orders of beings, both angels and
Ei thurfandi stadi ne standir,	mortals; who, independent of place and
Stad halldandi i kyrrleiks valldi ;	time, continuest undisturbed in thy so-
Senn verandi uti og inni,	vereign power; who at once art with-
Uppi og nidri og thar i midiu,	out and within, above and below, and in
Lof se ther um alldr og æfi	the midst ; praise be unto thee for ever
Eining sonn i thrennum greinum !	and ever, the true unity in trinity!

Æski ec thin en mikla miskunn	I ask of thee, that in thy great mercy
Mer veitist, er ec epterleita	thou wouldst grant me what I implore
Af klockum hug; thui ec ynnist ecki	with a submissive soul; for I desire no
Annad gott, nema af ther, Drottinn:	other good than what comes from thee,
Hreinsa briost, oc leid med listum	Lord: Cleanse my breast, and dispose
Loflig ord i studla skordum,	suitable words with elegance into poeti-
Stefnlig giord svo visan verdi	cal numbers, that a song of graceful
Vunnin ydur af thessum munni.	structure may be offered up unto thee
	from my mouth.

The alliteration employed here is of two kinds; one correspondence appearing in the initial letters or syllables, another in certain letters which occur in the middle of words. Where the alliteration is initial, the same letters generally occur twice in the first line of each distich, and are once repeated in the second. Thus, in the first distich of the second verse, we find *mikla, miskunn, mer;* in the third we have *leid, listum, loflig;* in the fourth, *visan, verdi, vunnin.* The rythm in the middle of the words is also of two kinds. In the first line, the same consonants twice occur, with different vowels preceding them; in the second, the alliteration is complete, both as to the consonants and vowels. Thus, in the first verse, we have of the former kind, alm*attigr,* st*etta,*— thurf*and*i, st*und*ir,—s*enn,* i*nn*i,—*lof, æf*i. Of the second kind, we have, yferbi*ód*andinn, thi*ód*a,—ha*lld*andi, va*lld*i,—n*i*dri, m*id*iu,—*ein*ing, gr*ein*um.

This poem, which consists of a hundred stanzas, was, at the time it was written, in much repute among the Icelanders for the elegance and accuracy of its style. There is an old saying in the language, *Oll skálld villdu Liliu kuedit hafa;* or, ' Every poet would wish to have composed the Lilium.'

The following verses are part of a poem composed by Finnur Magnuson of Reikiavik, in commemoration of the events which occurred in Iceland in 1809. The first nine verses are descriptive of the earlier condition of Iceland. Those succeeding, which are taken from another part of the poem, describe the usurpation of Jorgensen, in language which certainly assumes some licence in the poetic embellishment of facts. In these verses, it will be observed that the alliteration is very frequent, in imitation of the ancient Icelandic poetry.

Fieck theim frelsi
Frædi skopud,
Rit og mal
Retti nadu.

Liberty brought forth the elegant arts : writing and speech obtained their proper privilege.

Gall thar greppa
Gullinn harpa ;
Fedra saung
Fræg threkvirki.

The golden harp of the bard gave forth its sound : it sung the heroic deeds of our forefathers.

Sin og kunnra
Sam lifenda
Orlog og verk
Adrir skradu.

The fates, and the deeds, of themselves and of their contemporaries, were written by others.

Skalda log,
Skrifs og mælsken,
Lista og ydna
Litu ritinn.

The laws of poetry, of composition, of oratory, of the arts, and of workmanship, were committed to writings.

Undrast enn
Europear
Frodir visindi
Fedra vorra.

The European nations even now admired the learning and wisdom of our forefathers.

Breyttist fron,
Breyttust landar ;
Ærdust their auds
Og æru-syki.

The land was changed, the people were changed : the seeking of wealth and honours became a madness among them.

Mottu hvarfveggia	They preferred these things
Mentum fremur :	to the pursuit of knowledge :
Kings og Klerka	kings and priests obtained the
Kugun hrepptu.	mastery over them.

Soadi meingi	The *Black-death** devoured the
Svartur daudi :	people : its ancient glory de-
Hvarf ur landi	parted from the land.
Hrodur forni.	

Forust lystir	The arts, wealth, and happi-
Fie og sæla :	ness, perished : the ground,
Vesladist fold	diminished in fertility, lost its
Og fegurd tyndi.	pleasantness.

Lietst hann Engla	He pretended that he served
Lofdung thiona	the English king ; that he de-
Hermagtar hanns	pended on the protection of
Hafa fylgi.	his armies.

Vopnadiz brodir	He armed brothers against
Bormum moti :	each other : terror seized the
Enn otti greip	remainder of the people ;
Adra lydi.	

Hofdu ei sied	Who had never before beheld
Sverd nie dreira,	the sword or blood, and unwil-
Lagaleysi	lingly submitted to the inso-
Lutu naudgir.	lent yoke.

Sa hinn oblgari	He more powerful raised for-
Ebldi virki	tifications ; and erected his
Og heldocku	standard, black as hell.
Hreikti mirki.	

* The plague which devastated Iceland at the beginning of the 14th century, is emphatically called in the writings of the country, *Svartur Daudi*, or the *Black Death.*

Tok hann tignar	He took a lordly title; having
Titil jarla,	dared to assume possession of
Vogandi mildings	the supreme power.
Magt eigna.	
Lietst at thiod vorri	He pretended that our people
Thar til kiorim,	wished for these things; and
Ad hun uppreitar	that they all demanded these
Oll svo krefdi.	tumults.

The following verses are an extract from the translation of Pope's Essay on Man by Jonas Thorlakson, the priest of Backa. They include the first twelve lines of the fourth epistle, beginning,

　　" O Happiness! our being's end and aim !"

O ! farsæld ! thu vort einka hnoss,
Astefnda lifs og veru mid !
Lust, gledi, rosemd, alnægd oss !
Oeld hvad hellst nafn thig brufar vid !
Eitthvad sem girnast allir menn
Ævarandi med stundun frekt,
Sem gioerir lifid yndællt, en
Andlat daudlegum bærilegt ;
Er svo ei nalæg shnist oeld
Samt firrist iafnan hendur manns ;
Imist ei siest, eda' ert tvoefold
Augum ens visa' og heimskingians.
Seig, himin-sædis himna blom
Hia oss ef bygd ther gefin var,
Hvoer hreppti joerd, than heilla dom,
Ad hæf se til, thu varir thar ?
Maske hofgoerdum megir a
Med allt thit finnast glansa skraut ?
Ertu med demant hoeddum hia
Hulin malm-æda rikt vid skaut ?

Maske Parnassum byggir blid
Blandin larberia kransa vid?
Hoendla thig their, sem heyia strid,
Og hioervi brytia sigrad lid?

Translation of Anacreon's 34th Ode, by the Assessor Benedict Grondal.

Μη με φυγης, ορῶσα.

Fly thu mig ecki, fagra mey!
Tho mer af hærum herdar vidur
Hrynie fannhvitir lockar nidur,
Fyrilit mina elsku ei,
Medan ther æsku endiz blom,
Thui ofurvel a milli rosa
Miallhuitar ser, at minum dom,
I mai-kroensum liliur hrosa,

The following is a short extract from the *Minnisverd Tidindi*, or Icelandic Historical Register, for the year intervening between the summers of 1801 and 1802. It is a part of the account of the debates which took place in the British Houses of Parliament upon the peace of Amiens. The celebrated speech of Mr Windham on this occasion, is given at considerable length.

' Merkileg var fyrsta seta ens enska Parlaments eptir Fri-
' dar-gioerdina, tha Hertoginn af Bolton, eptir nockra bid,
' —thui thenari hans kom orseint med ræduna,—baud Yfir-
' husinu ad samglediast med ser yfir stridsins luckulegu en-
' dingu, og quad tilhlydilegt, ad thetta Rikisins haa rad leti
' thess vegna Konunginum skriflega thackargioerd i tie, hvoer
' ogsvo af oellum Radherrunum i einu hliodi alyktadist ad
' ske skyldi. I Undir-husinu framsetti Lord Lovaine lika
' osk, og gafu flestir mali hans godann ordrom. Medal

' theirra reis thar upp fyrstur fals-vinurinn Fox, og nærst
' eptir hann hinn nafnkunni Pitt, er badir urdu a einu mali,
' hvad thessi sagdi gleddi sitt hiarta, thar their Fox og hann,
' aldrei fyrr hefdu ordid, samthyckir a æfi sinni, en—moti
' theim reis upp gamall vinur Pitts, og fyrrverandi Stiornar-
' herra Windham, er sagdist engannveginn geta samsiunt
' thessu thacklæti, thar ser vyrdtist fridur vid Frankariki,
' i thessum kringumstædum, vera mioeg htryggur, og jafnvel
' skammarlegur fyrir-hina ensku thiod, og, eins og hann tok
' til ords, *Thungadur med eymd og olucku*—" Eg verd " sagdi
' hann, " ad vera syrgiandi i gledinnar hop," &c. &c.

The following is a catalogue of a few of the papers, which
appeared in the Transactions of the first Literary Society of
Iceland. They are taken indiscriminately from the different
volumes of this work.

List of Icelandic plants, fishes and birds, with their Linnæan names. By
Olaf Olafson.

Treatise on the catching of whales. By John Ericson.

On the mines and merchandize of sulphur in Iceland. By Bishop Finnson.

On the maintenance of orphans and crippled persons. By Sysselman Ketilson.

On the catching of sea-fowl at Skaga-fiord. By Olaf Olafson.

On the sea and river-fisheries of Iceland. By Olaf Stephenson.

On the cultivation of trees in Iceland. By Skule Magnuson.

Review and correction of the Icelandic version of certain passages in the Pro-
phecies. A series of treatises by John Olafson.

On the advantages of horned cattle. By Olaf Stephenson.

On the catching of foxes. By Thord Thorkelson, farmer in Eyafiord Syssel.

On faithfulness and affection in servants ; and how these qualities may best be
produced and cultivated. By Sysselman Einarson.

Treatises on the wheel and axle, inclined plane, screw, &c. By Stephen Biornson.

On the alkaline salt from sea-ware. Translated from the Danish by Assessor
Benedict Grondal.

A short commentary on the flowing back of the waters of the Red Sea for the passage of the Israelites. By Stephen Thorarenson.

A key to meteorological changes, of the sun, moon, stars, air, winds, &c. By Stephen Biornson.

On the causes of the diseases prevailing in Iceland. By John Peterson.

On the building of habitations in Iceland. By Provost Sweinson.

Dr James Home's Essay on the Scurvy.   Translated into Icelandic by Land-physicus Sweinson.

On the cookery of fish, flesh, meat, and milk, in Iceland. By Olaf Olafson.

Some words on the free trade of Iceland. By Olaf Olafson.

The alphabet of the Icelandic language presents no striking peculiarity, except in the letter called *Thorn*, Þ, which has been transferred from the Runic to the modern alphabet. This letter has a double sound. At the beginning of a word, it appears to be equivalent to the Hebrew *Thau ;* and has a sound intermediate between *Th* and *Tsh*, such as is not unknown in the English language.   At the end of a word, or after a vowel in the same syllable, it is pronounced like *d*, as in *mathur*, a man, which is pronounced *madur*.

The following are the cardinal numbers in the Icelandic language; which, to the number *four*, are declinable : the remainder, up to the *hundred*, are indeclinable words.

Eyrn,	One.	Threttan,	Thirteen.
Tveir,	Two.	Fioortan,	Fourteen.
Thryr,	Three.	Tuttugu,	Twenty.
Fioorer,	Four.	Thriatyu,	Thirty.
Fimm,	Five.	Fiorutyu,	Forty.
Sex,	Six.	Fimmtyu,	Fifty.
Sioe,	Seven.	Hundrad,	Hundred.
Aatta,	Eight.	Tvo-hundrad,	Two hundred.
Nyu,	Nine.	Thriu-hundrad,	Three hundred.
Tyu,	Ten.	Thusund,	Thousand.
Ellefu,	Eleven.	Eyrn thusund,	One thousand, &c.
Twolf,	Twelve.		

The following short catalogue of Icelandic words will show the similarity between this language and the English, derived from their origin in a common source.   Numerous other examples of the same kind might have been obtained.

Aska,	Ashes.	Graata,	To weep—Scotch, *to greet.*
Back,	The back.		
Bane,	Bane.	Grey,	A dog — *grey-hund,* hunter's dog.
Barn,	A child—Scotch, *bairn.*		
Bed,	A bed.	Greip,	A gripe.
Bende,	To bend.	Gulur,	Yellow.
Ber,	Bare.	Hæna,	A hen,
Blad,	A leaf or blade.	Hagl,	Hail.
Bladra,	A bladder.	Hil, Pret. Hulde,	To cover—a hull.
Blek,	Black.	Hlaatur,	Laughter.
Bloma,	Flower or blossom.	Hlaup,	A leaping—Scotch, *to laup.*
Dey,	To die.		
Dyn,	To make a noise.	Ida,	An eddy.
Dyr,	Dear.	Illa,	Ill—bad.
Domur,	Doom—judgment.	Klyufa,	To cleave.
Draumur,	A dream.	Kref,	To crave—to beg.
Dregg,	The dregs.	Kioosa,	To choose.
Duyn,	To fall off — Scotch, *to dwine.*	Magur,	Meagre.
		Molld,	Soil—mould.
Erende,	Errand,	Nakenn,	Naked.
Eymnpria,	Embers.	Naut,	An ox—Scotch, *nowt.*
Fader,	Father.	Ol,	Ale.
Fæde,	Food.	Oop,	A weeping.
Fære,	I move my place—(*wayfaring man.*)	Poke,	A bag or poke.
		Rettuys,	Righteous.
Fel,	I conceal—Scotch, *to feal.*	Rif,	A rib.
Fie,	Money,—a fee.	Spade,	Spade.
Fingur,	The finger.	Stam,	A stammering.
Fiskur,	A fish.	Thif,	A thief.
Foolk,	People.	Tidinde,	Tidings.
Frys,	To freeze.	Torff,	Turf.
Gabl,	A limit—(*gable end.*)	Tuinne,	Thread—twine.
Giell,	To shout, or yell.	Vellde,	Power—*to wield a sceptre.*
Giora,	To make—Scotch, *to gar.*		

Translation of a letter of recommendation, addressed by the Landfoged Frydensberg to the inhabitants of Guldbringè Syssel.

' Sir George Mackenzie is come from Scotland to Iceland,
' for the purpose of examining the natural curiosities here ;
' the mountains, lava, and hot-water springs. He, and the
' other gentlemen who are travelling with him, will first make
' a journey through Guldbringè Syssel ; and as they are
' strangers in the places where they are going, it is my ear-
' nest request that all good and respectable people will pay
' every attention to them, and shew kindness to these stran-
' gers, as good and hospitable Icelanders are wont to do to
' travellers ; assisting their journey in every way they are
' able, guiding them in the road from place to place, and
' providing food for their horses. For this purpose, I give
' this letter open into the hands of the Baronet to shew to
' the people.

' FRYDENSBERG.

' *Reikiavik, 21st May,* 1810.'

SPECIMENS
of the
Ancient Sacred Music
of
Iceland

Pl. 15.

No. V.

MUSIC.

IT has been mentioned that, in modern times, the Icelanders have shewn neither genius nor taste for music. It is said, however, to have been formerly taught in the island; and from the merit of some ancient sacred tunes, which we copied from a manuscript in the possession of Mr Stephenson of Indreholm, it appears that music was known as a science. We were assured by Mr Stephenson and others, that these tunes were very old, and really native compositions suited to various hymns. Having no reason to doubt these facts, I think that it may gratify the curiosity of some of our readers, to present them with a selection from the tunes alluded to, transposed from the soprano clef, in which the originals are written, to suit the piano fortè.

No. VI.

REGISTER OF THE WEATHER.[*]

Days of the Month.	Barometer and Thermometer at Reikiavik.		Atmosphere.	Thermometer at different Places of the Journey, and Miscellaneous Observations.
	Barom.	Th.		
May 14.	......	28°	NE. Breeze. Clear.	
15.	......	34°	N. Gale. Do.	
16.	......	30°	Do. Do. Do.	
17.	......	32°	Do. Do. Do.	
18.	......	29°	Do. Do. Do.	Ice within doors early this morning; ¼ inch thick.
19.	......	38°	NW. Moderate. Cloudy.	Snow in the afternoon. Full moon. High water of the second tide at V½, 14 feet.
20.	......	43°	N. Do. Clear.	
21.	......	40°	SW. Do. Clear morning.	Went to Havnefiord. Snow in the afternoon. Thermometer 30°.
22.	......	...	W. Calm. Clear.	
23.	......	...	Do. Breeze. Do.	Evening cloudy. Wind N. Thermometer 32° at Kaldaa.
24.	......	...	SE. Gale. Rain.	Th. 45°. Wind, evening, NNE. Fair.
25.	......	...	S. Breeze. Clear.	— 47°. — E. At Krisuvik.
26.	......	...	SE. Do. Do. Evening cloudy.	— 50°. — — At do.
27.	......	...	Do. Moderate. Cloudy.	— 46°. To Grundevik. Wind freshening.
28.	......	...	Do. Gale. Cloudy. Showers.	— 50°. To Kieblivik.
29.	......	...	Do. Do. Rain.	— 50°. At do.
30.	......	...	Do. Do. Showery.	— 56°. At do.
31.	......	...	Do. Heavy gale. Rain.	— 56°. At do.
June 1.	......	...	SSW. Moderate. Showers.	— 52°. Evening calm. To Havnefiord.
2.	......	...	Do. Breeze. Clear.	— 50°. To Reikiavik. Evening, Wind S.
3.	......	53°	S. Do. Cloudy.	— 52°. Do.
4.	30.311	58°	SW. Morn. clear. Even. cloudy.	

[*] The observations were generally made between 8 and 10 o'clock A. M., unless when otherwise mentioned.

Days of the Month.	Barometer and Thermometer at Reikiavik.		Atmosphere.	Thermometer at different Places during the Journey, and Miscellaneous Observations.
	Barom.	Th.		
June 5.	30.01	58°	Do. Do. Rain.	
6.	29.718	47°	Do. Do. Do. Fog.	
7.	29.87	48°	Do. Do. Do.	Afternoon clear. Breeze from NE. Th. 35°.
8.	30.14	46°	Do. Do. Clear.	
9.	29.851	50°	SE. Do. Do.	Evening. Wind NE. Th. 47°.
10.	29.117	49°	NE. Calm. Do.	Fair afternoon. Th. at 12 o'clock, in the sun 72°, shade 61°.
11.	29.513	50°	Do. Moderate. Showers.	
12.	29.619	55°	NNE. Do. Do.	
13.	29.7	46°	SE. Do. A few clouds.	Showers towards evening.
14.	29.65	48°	Do. Do. Rain.	
15.	29.82	51°	Do. Gale. Rain.	Left Reikiavik.
16.	29.8	51°	Do. Squally. Showers.	Th. at Brauturholt, 53°.
17.	29 8	60°	E. & NE. Breeze. Few clouds.	— at Houls, 56°.
18.	29.83	59°	SW. Do. Clear.	— at Saurbar, 56°; at 2 P. M. 65°; in the sun, 86°.
19.	29.83	62°	Do. Calm. Do.	At Indreholm. Th. 56°. In the sun, at 2 P. M. 85°.
20.	29.92	56°	SW. & S. Moderate. Cloudy.	At do. 49°. Fog in the morning.
21.	29.84	54°	SW. Do. Do.	— — 49°. At 12 at night, 50° at Leira.
22.	29.91	56°	E. Do. Do. Rain, evening.	At Leira, 53°.
23.	29.83	56°	SE. Do. Do. Do.	At Huaneyre, 60°. Heavy rain, evening. At Svignaskard.
24.	29.51	54°	E. Do. Thick rain.	At Svignaskard, 56°. Rain. At 4 P. M. 58°; at 8, 55°.
25.	30.1	48°	Do. Do. Morning foggy, with showers.	Evening, at Stadarhraun, NNW. At 11 A. M. 49°; evening, 39°.
26.	30.32	50°	SSW Calm till evening.	Th. at 6 A. M. 43°; at 2 P. M. 55° at Roudemelr.
27.	30.33	54°	SW. Moderate. Thick fog.	— 49°; 10 P. M. 48°, at Miklaholt.
28.	30.13	53°	Do. Calm. Clear.	— 52°, at 4 P. M. 61°, at 11 P. M. 52°, at Stadarstad.
29.	30.14	60°	Do. Breeze. Morning foggy.	— 52°. To Buderstad.
30.	30.21	61°	W. Calm. Clear.	Wind, evening, N. Th. 56°, at Stappen.
July 1.	30.01	64°	Do. Do. Cloudy.	Th. 57°, at 10 P. M. 56°, at do.
2.	30.02	62°	NNW. Do. Clear.	— 54°; at 2 P. M. in the sun, 82°; shade, 61°; at Olafsvik.
3.	29.94	56°	W. Do. Few clouds.	— 56°.
4.	29.74	58°	NW. Calm. Cloudy.	— 55°. Some very light showers at Olafsvik.
5	29.82	59°	Do. Do. Morning foggy. Evening clear.	— 51°; at midnight, 48°. Sun 2 h. 35 m. under the horizon, and rose 55° E. of N. At Olafsvik.

Days of the Month.	Barometer and Thermometer at Reikiavik.		Atmosphere.	Thermometer at different Places during the Journey, and Miscellaneous Observations.
	Barom.	Th.		
July 6.	29.92	63°	NW. Calm. Cloudy.	Th. 60°; 10 P.M. 51°, at Grunnefiord.
7.	29.9	50°	N. Breeze. Clear.	
8.	29.84	63°	Do. Calm. Do.	— 50°. At do.
9.	29.84	58°	SE. Breeze. A few clouds.	— 50°. At do.
10.	29.84	58°	NE. Gale. Clear.	— at 6 A.M. 45°. At Narfeyre.
11.	29.65	48°	Do. Do. Do.	— 42°. Rain towards evening.
12.	29.62	...	Do. Moderate. Clear.	— at 4 A.M. 42°. At Snoksdalr.
13.	29.71	...	Do. Do. Showery.	— at 5 A.M. 41°. At Hvam.
14.	29.68	...	N. Do. A few clouds.	— 45°. At Huaneyre.
15.	29.72	...	Do. Do. Clear.	— 52°. At do.
16.	29.75	60°	NW. Calm. Clear.	At Reikiavik.
17.	29.91	60°	W. Breeze. A few clouds.	
18.	30.015	60°	Do. Calm. Clear.	Th. in the sun, at 1 P.M. 80°; shade 68°.
19.	30.26	58°	Do. Breeze. Cloudy.	
20.	30.22	57°	Do. Do. Clear.	
21.	30.18	61°	NW. Moderate. Clear.	Evening. Wind E. Rain.
22.	29.77	60°	W. Do. Showery.	
23.	29.69	54°	Do. Do. Cloudy.	Evening. Wind E.
24.	29.23	52°	E. Gale. Heavy rain.	
25.	29.045	54°	Do. Moderate. Showery.	
26.	29.13	54°	SE. Do. Do.	Evening. At Thingvalla. Th. 52°.
27.	29.2	53°	E. Do. Do.	Evening. At the Geysers.
28.	29.2	52°	Do. Breeze. Cloudy.	Th. 55°.
29.	29.72	53°	Do. Do. Showery.	— 56°.
30.	29.91	55°	SE. Do. Do.	— 56°. At Skalholt. Evg. at Kalfholt.
31.	29.94	54°	Do. Do. Do.	— 58°. Do. Evg. at Storuvellir, 52°.
Aug. 1.	29.93	54°	SW. Moderate. Rainy even.	— at Storuvellir 57°.
2.	29.94	60°	SE. to SW. Do. Do.	
3.	29.94	50°	SW. Breeze. Clear.	Fog on M. Hekla. Th. at 10 A.M. 59°.
4.	29.91	57°	S. Do. Do.	Th. 60°.
5.	29.84	50°	S. Do. A few clouds.	— 50° at Hliderende.
6.	29.94	46°	NW. Do. Clear.	— 55°.
7.	29.93	49°	Do. Do. Do.	— 59° at Odde.
8.	29.71	60°	S. Do. Cloudy.	
9.	29.61	56°	SE. Do. Showers.	
10.	29.582	62°	NW. Do. Clear.	
11.	29.61	62°	E. Do. Cloudy. Rain aftern.	
12.	29.723	61°	N. Do. Do.	
13.	29.86	58°	Do. Do. Clear.	
14.	30.18	57°	NW. Do. Cloudy.	
15.	30.21	57°	S. Calm. Clear.	
16.	29.798	55°	SE. Gale. Rain.	
17.	29.06	55°	Do. Do. Do.	
18.	29.08	50°	W. Heavy gale. Cloudy.	

Days of the Month.	Barometer and Thermometer at Reikiavik.		Atmosphere.	Miscellaneous Observations.
	Barom.	Th.		
Aug. 19.	......	...	....	Sailed for Britain.
20.	29.33	54°	SE. Rain.	The observations were now made by
21.	29.21	52°	SW. Do.	Mr Fell, a barometer having been
22.	29.3	54°	SW. Clear.	left with him.
23.	29.31	...	N. Frosty night.	
24.	29.31	50°	NNE. Rain.	
25.	29.53	...	N. Clear.	
26.	29.54	56°	N. Heavy gale. Frosty night.	
27.	29.7	46°	N. Moderate. Clear.	The people observed that frost was about to set in, and dug up their potatoes.
28.	29.84	52°	Do. Calm. Do.	
29.	30.01	46°	NW. Do. Do.	Frost at night.
30.	30.1	49°	Do. do. do.	Do.
31.	30.14	52°	Do. do. do.	Do.
Sept. 1.	30.10	60°	N. Do. do.	Remarks made by the people that so fine a summer was never known, and that a severe winter might be expected.
2.	29.81	50°	E. Breeze. Rain.	
3.	29.92	52°	N. Do. Clear. At night rain.	
4.	29.54	47°	Do. Gale. Cloudy.	Before next morning it blew a heavy gale.
5.	29.41	40°	NE. Do. Rain.	The mountains covered with snow.
6.	29.9	52°	S. Moderate. Rain.	
7.	29.52	50°	Do. do. do.	
8.	29.3	48°	SE. Do. Heavy rain.	
9.	29.2	50°	Do. do. do.	
10.	29.42	48°	NW. Breeze. Cloudy.	
11.	29.5	48°	N. Do. Clear.	
12.	29.2	39°	NE. Very heavy gale. Rain.	
13.	29.6	48°	N. Calm. Clear.	
14.	29.6	50°	SE. Moderate. Rain and fog.	
15.	29.83	51°	NE. Do. do. Gale at night.	
16.	29.8	53°	Do. do. do. Showery.	
17.	29.33	48°	SSE. Do. Rain and fog.	
18.	29.34	48°	S. Squally. Rain.	
19.	29.73	48°	Do. do. do.	
20.	29.91	49°	N. do. do. Gale at night.	
21.	29.42	42°	Do. Calm. Clear.	
22.	29.4	47°	SE. Squally. Rain.	
23.	29.51	47°	SW. Do. do.	
24.	29.54	45°	Do. do. do. Clear afternoon.	
25.	29.4	51°	Do. do. do. At night a heavy gale.	

Days of the Month.	Barometer and Thermometer at Reikiavik.		Atmosphere.	Miscellaneous Observations.
	Barom.	Th.		
Sept. 26.	29.8	53°	NE. Gale. Rain till afternoon.	
27.	29.6	49°	SW. do. do.	
28.	29.64	49°	Do. Moderate. Foggy. Showers	Thunder was heard to-day.
29.	29.71	50°	Do. do. do.	
30.	29.7	50°	Do. do. Showery.	
Oct. 1.	29.43	51°	SE. do. Heavy rain.	
2.	29.61	54°	SW. Breeze. Rain.	
3.	29.4	54°	SE. Squally. do.	
4.	29.64	46°	S. Moderate. do.	
5.	30.03	50°	Do. do. Clear till evening.	
6.	29.81	50°	Do. do. Rain.	
7.	29.9	55°	Do. do. Cloudy.	
8.	29.91	48°	Do. do. Rain afternoon.	
9.	29.8	50°	SE. Gale. Rain.	
10.	30.12	48°	NW. Moderate. Cloudy.	
11.	30.01	49°	S. do. Rain.	
12.	29.82	42°	SW. A heavy gale. Cloudy.	
13.	30.11	41°	Do. do. do.	
14.	29.63	46°	SSE. Gale. Thin rain.	
15.	29.53	45°	S. Moderate. Rain.	
16.	29.7	44°	N. Breeze. Clear. Showers at 2 P. M.	
17.	29.51	44°	NE. Gale. Cloudy.	
18.	29.02	43°	SE. do. Thin rain.	
19.	28.91	44°	N. Breeze. Cloudy.	
20.	29.04	42°	Do. Calm. Clear.	
21.	29.22	41°	ENE. do. do.	
22.	29.41	38°	SE. Breeze. Cloudy.	
23.	29.6	38°	S. do. Showery.	

The following register of the weather, kept by Mr Fell since the above date, has fortunately reached me in time to be presented to the public. It exhibits a dismal picture of an Icelandic winter; and rouses the most lively feelings of compassion for the condition of the inhabitants of so desolate a region.

Days of the Month.	Barometer and Thermometer at Reikiavik.		Miscellaneous Observations.
	Barom.	Th.	
Oct. 24.	29.90	43°	SE. Very dark thick weather. Between 7 and 8 A. M. an earthquake was felt in Reikiavik.
25.	29.54	48°	SSE. Blew very hard, and rained all day.
26.	29.34	41°	SW. Blew heavy, with showers of snow, hail, and rain, the whole day. Frost at night.
27.	29.32	32°	NW Blew very hard all day. Do.
28.	29.81	24°	NE. Blew a gale of wind all day, with showers of snow, hail, and rain.
29.	29.72	40°	SE. More moderate weather, and rain.
30.	29.5	45°	Do. Fine, but rain at times; frost at night.
31.	29.80	24°	N. Clear.
Nov. 1.	30.24	29°	NE. A fine day; thaw before night.
2.	30.02	45°	SW. Blew very hard, and rained all day.
3.	29.92	39°	SSW. Blew very hard all day; with hail, rain, and snow, at night.
4.	29.72	34°	SW. Blew hard all day, with hail and snow.
5.	29.3	24°	N. Blew extremely hard all day, with showers of snow and hail. During the night came on a most tremendous gale of wind. Dark at 2 P. M.
6.	29.84	14°	N. A most terrible and awful gale of wind during the whole 24 hours. Boats on the beach were taken up into the air, and dashed to pieces. The thermometer in my warmest room, though the stove was red hot, could only be brought to 30°. The weather extremely dark; and the country for *several miles round,* was covered with salt water driven from the sea in the form of rain, which destroyed what little vegetation was left.
7.	30.21	24°	N. Blew hard, with clear frost. Towards night the wind increased; and early next morning blew a hurricane.
8.	30.23	28°	N. A fine clear day.
9.	30.10	28°	N. A clear hard frost; the day passed without a cloud being seen. Saw the Snæfell Jokul at 12 P. M. The northern lights were beautiful.
10.	29.90	32°	NE. A day without a cloud. Thaw at night.
11.	29.73	32°	Do. Snow and wind the whole day.
12.	29.84	34°	Do. Fine day; frost at night.
13.	30.14	35°	ENE. Clear weather, but frost at night.
14.	29.71	30°	Do. A fine clear day.
15.	29.12	39°	Do. A dark day.
16.	29.11	40°	NE. A fine day, but frost at night.
17.	29.84	28°	Do. A fine day.
18.	30.13	26°	S. Remarkably fine weather, and hard frost.
19.	29.83	38°	NE. Cloudy, with small rain. A heavy gale, and rain at night.

Days of the Month.	Barometer and Thermometer at Reikiavik.		Miscellaneous Observations.
	Barom.	Th.	
Nov. 20.	29.50	38°	ENE. Rain the whole 24 hours.
21.	29.50	35°	NE. Frost and thaw by turns the whole day.
22.	29.62	34°	N. Fine day; a few showers.
23.	29.80	32°	Do. Fine frosty day, but cloudy at night.
24.	29.81	34°	E. Fine day.
25.	29.60	38°	ENE. Fine weather.
26.	29.53	34°	NE. Fine day.
27.	29.11	39°	Do. Dark day; blew hard at night.
28.	29.64	34°	Do. Do. Do.
29.	29.72	24°	N. Blew hard all day, and very hard at night.
30.	30.02	11°	Do. Blew hard all day.
Dec. 1.	29.74	22°	NE. Dark day, heavy snow, and wind at night.
2.	29.81	30°	SW. Dark weather, and snow at night.
3.	29.22	26°	Do. Dark day, and heavy snow at night. Saw lightning at 11 P. M. but heard no thunder.
4.	29.51	24°	S. Clear.
5.	29.12	24°	NE. Do.
6.	29.41	23°	Do. Blew a heavy gale of wind the whole day.
7.	29.94	18°	Do. Clear weather, and the gale continued till 12 P. M.
8.	30.11	4°	ENE. Clear weather. Thermometer 2° at 2 P. M.; water froze under the stove.
9.	29.84	16°	NE. Clear.
10.	29.72	15°	Do. Do.
11.	29.61	24°	ENE. Dark weather.
12.	29.34	26°	NE. Blew hard all day; but fine.
13.	29.32	32°	Do. Dark weather.
14.	29.43	18°	Do. Clear weather.
15.	29.84	22°	ENE. Very fine day.
16.	29.64	35°	S. Rain the whole day, and very thick.
17.	29.80	27°	NE. Dark weather.
18.	29.73	28°	Do. Dark; rained a little towards night, and froze again.
19.	29.61	27°	N. Dark weather; towards evening began to blow hard, and at night a gale of wind.
20.	29.82	32°	Do. Dark thick weather; during the day a heavy fall of snow.
21.	29.22	27°	Do. Hard frost; at 4 P. M. a very heavy fall of snow, and during the night a very heavy gale of wind.
22.	28.42	29°	SW. A heavy gale of wind the whole day, which drifted the snow to the tops of the houses. At night the lightning (or snow lights) was incessant for a few hours.
23.	28.61	32°	NE. Fine weather.
24.	29.01	22°	Do. Do.
25.	29.10	23°	Do. Do.
26.	29.51	17°	Do. Do.

Days of the Month.	Barometer and Thermometer at Reikiavik.		Miscellaneous Observations.
	Barom.	Th.	
Dec. 27.	29.71	12°	NE. Very fine day.
28.	30.12	32°	Do. Blew hard all day, which increased towards night to a heavy gale. Gentle thaw.
29.	29.94	38°	SE. The gale of wind continued the whole 24 hours, with heavy rain at night.
30.	29.74	43°	Do. The gale still continued blowing very hard, with rain. The whole place inundated.
31.	29.93	43°	Do. The gale still continued blowing very hard, with heavy showers of rain.
1811. Jan. 1.	30.14	44°	SSE. The gale continued the whole day blowing hard, with showers of rain.
2.	30.20	43°	Do. Blew hard all day; but calm towards night.
3.	30.42	37°	SSW. Fine mild day, with rain towards night.
4.	30.50	36°	ENE. Mild weather, with a little rain; frost at night.
5.	30.60	33°	Do. Fine mild weather.
6.	30.22	37°	Do. Fine day; a few showers, which, in some places, froze into ice directly; this is not uncommon.
7.	29.90	34°	S. Fine day, and a little rain, which, in some places, became ice.
8.	29.62	36°	SE. Blew hard all day, with rain; and during the night came on a gale of wind.
9.	28.90	37°	Do. A heavy gale of wind the whole day, with showers of rain; before morning, a hurricane.
10.	28.81	35°	SSE. Heavy squalls of wind and rain all day; frost at night.
11.	28.60	33°	N. Blew hard most of the day; calm at night, with rain, and frost.
12.	28.33	36°	Do. Mild weather, but frost at night.
13.	28.61	32°	Do. Fine day; the whole face of the country like glass.
14.	28.61	28°	NE. Dark day; snow in the evening.
15.	28.80	26°	N. Fine weather.
16.	29.14	21°	Do. Fine day; snow at night.
17.	29.01	22°	NE. Do. do.
18.	29.21	18°	Do. Fine day, and frost at night; when it blew hard with hail, snow, thunder, and lightning: the latter not uncommon in the winter.
19.	28.90	27°	SSW Heavy fall of snow the whole day.
20.	29.12	24°	NE. Heavy snow the whole day, and very dark.
21.	28.83	23°	Do. Fine weather, but heavy snow at night.
22.	28.90	28°	N. Fine day.
23.	29.91	33°	SE. Mild weather, and a little rain; at night, blew very hard, with hail and rain.
24.	29.34	35°	SW. Blew a very hard gale of wind the whole 24 hours, with showers of hail.
25.	29.70	27°	Do. Blew hard in the morning, but was moderate at night.

Days of the Month.	Barometer and Thermometer at Reikiavik.		Miscellaneous Observations.
	Barom.	Th.	
Jan. 26.	30.00	8°	NW. Fine weather; blowing fresh, which increased to a gale of wind.
27.	30.32	10°	N. Blew a tremendous heavy gale of wind the whole 24 hours.
28.	30.30	8°	Do. The gale still blowing.
29.	30.12	10°	Do. Do. Sea frozen from the land out to the Islands (about a quarter of a mile), and strong enough to bear a horse.
30.	30.12	16°	Do. Moderate weather.
31.	29.64	16°	ENE. Blew very hard all day, and towards night a gale of wind.
Feb. 1.	29.74	5°	N. A heavy gale of wind the whole day; Th. at night, 6° below 0°.
2.	29.51	4°	Do. Towards night blew hard.
3.	29.42	29°	NNE. Dark weather, blowing hard all day.
4.	29.54	19°	Do. Blew hard all day, and fine.
5.	29.43	16°	Do. Snow, with a gale of wind the whole day.
6.	29.40	15°	Do. Very fine weather.
7.	29.00	27°	Do. A very heavy gale of wind the whole 24 hours.
8.	28.91	22°	Do. The gale continued the whole day, with heavy showers of snow.
9.	29.60	26°	ENE. Blew hard most of the day, but was moderate towards night.
10.	29 30	30°	WSW. Remarkably fine day.
11.	29.11	12°	ENE. Fine day, but blew hard towards night.
12.	28.83	10°	NNE. Blew the whole day one of the hardest and heaviest gales of wind we have yet had, the whole country to a distance of many miles was covered with salt water snow from the sea.
13.	29.70	16°	NE. Fine weather.
14.	29.41	15°	ENE. Blew hard, with showers of snow, the whole 24 hours.
15.	29.43	8°	NNE. Blew a gale of wind all day; Th. at night, 4° below 0.
16.	30.10	1°	ENE. Fine day, but blew hard towards night.
17.	29.50	26°	Do. Blew a gale of wind the whole day; and before 12 P. M. a thaw.
18.	28.50	38°	SSW. Blew a heavy gale of wind all day, from all points in the compass, with showers of rain and snow.
19.	28.63	33°	ENE. Blew a tremendous gale of wind, with showers of rain and snow; but towards evening it was terrible, many people were blown clean off their legs, and the whole place was inundated.
20.	28.62	36°	SSE. Blew very hard all day, with rain and snow.
21.	29.70	27°	N. Blew hard in the early part of the day, but became fine and moderate.

Days of the Month.	Barometer and Thermometer at Reikiavik.		Miscellaneous Observations.
	Barom.	Th.	
Feb. 22.	29.13	22°	NE. Blew hard, with fine weather; at night the northern lights were brilliant.
23.	29.10	30°	ENE. Fine weather, and calm.
24.	29.02	30°	N. Very fine.
25.	28.94	26°	Do. do.
26.	28.72	36°	ENE. Very fine day.
27.	28.81	38°	Do. Fine weather, but rained a little towards night; and a heavy fall of snow.
28.	29.94	28°	Do. Fine weather, and clear.
March 1.	29.14	25°	SW. Fine day, and sunshine.
2.	28.61	28°	ENE. A dark day, and blew hard.
3.	29.24	19°	N. Fine weather.
4.	29.03	23°	ENE. Fine day, and clear; heavy snow at night.
5.	28.83	23°	Do. Very thick, and snow the whole day.
6.	28.81	22°	N. Very thick, and snow till the afternoon.
7.	29.01	12°	Do. Fine morning; but before noon began to blow hard, with snow, and continued all day.
8.	29.33	24°	SW. Very thick, with storms of snow, from all points in the compass, the whole day.
9.	28.92	33°	NE. Mild weather, and fine.
10.	29.34	24°	SW. A continual succession of terribly heavy storms of snow the whole day and night; and very thick. From 3 A. M. to 6, was an eclipse of the Moon, when the wind thundered like heavy pieces of ordnance. At night came on a heavy gale of wind.
11.	29.54	23°	SSW. One of the most terrible days we have yet had; blew a most tremendous gale of wind all day, with snow.
12.	30.02	26°	ENE. Dark weather; at night blew hard.
13.	29.73	40°	SE. Rain the whole day, which, melting the snow, deluged the whole country; frost at night, with snow.
14.	30.10	33°	SW. A little snow.
15.	29.60	35°	SSW. Blew hard all day; rain at night.
16.	29.31	41°	ESE. A very heavy gale of wind the whole day, with rain; frost at night.
17.	29.02	35°	E. Fine weather, with a little rain; sharp frost at night.
18.	29.03	27°	ENE. A tolerably fine day, but heavy snow towards night.
19.	29.42	26°	SW A fine day, but blew fresh towards night; Ther. 8° 10 P. M.
20.	29.31	22°	ENE. A most terrible storm of snow and wind nearly the whole day, and so thick that one could not see twenty yards.
21.	29.24	33°	SW. A continual succession of heavy storms of snow and wind the whole day, and very dark.
22.	29.73	32°	S. Blew hard all day, with snow; at night a thaw.

Days of the Month.	Barometer and Thermometer at Reikiavik.		Miscellaneous Observations.
	Barom.	Th.	
Mar. 23.	29.43	42°	SW. Blew very hard all day, with small rain; and before night a heavy gale of wind.
24.	29.30	36°	Do. Heavy squalls of wind; snow, hail, and rain, most of the day. By the Icelandic almanack, an eclipse of the Sun, but not visible here.
25.	28.92	39°	SE. Blew hard all day, with rain; frost and snow at night.
26.	29.31	34°	SSW. Blew hard all day, with squals of snow and rain.
27.	29.73	35°	Do. Hail, rain, snow, and sunshine, by turns, the whole day; blowing hard.
28.	30.23	43°	E. Dark weather, blowing fresh, with rain all day.
29.	30.20	43°	S. Dark, calm weather, and rain.
30.	30.20	36°	W. A tolerably pleasant day; little wind.
31.	29.84	32°	NE. Snow with wind almost all day.
April 1.	29.74	35°	N. A fine morning, but wind and snow towards evening.
2.	29.74	31°	NNE. Blew hard all day, with a little snow; sharp frost at night.
3.	29.92	36°	W. A fine morning; but rained towards night.
4.	29.82	35°	NW. Excessively thick, and heavy rain the whole day; but cleared up at night, with a heavy gale of wind, and hard frost.
5.	30.11	16°	Do. Blew an extremely heavy gale of wind all day, accompanied by sea-water rain. Th. 8° at 10 P. M.
6.	30.51	22°	N. Remarkably fine day.
7.	30.23	24°	Do. Blew fresh all day, and severe frost. Th. 8° at 9 P. M.
8.	29.92	15°	Do. Heavy squalls of wind all day, but fine.
9.	30.11	15°	Do. Fine clear weather, with heavy squalls of wind all day. Th. 6° at 9 P. M.
10.	30.11	18°	ENE. Fine clear weather, and squalls of wind; blew a gale before night.
11.	30.01	22°	NE. Fine clear weather.
12.	29.51	34°	ENE. Dark thick weather.
13.	29.21	38°	NE. Dark weather, with small rain; blowing hard all day.
14.	29.84	50°	E. Dark weather, and rain most of the day.
15.	29.10	45°	ENE. Do.
16.	28.84	46°	SE. Do.
17.	29.24	38°	S. Very thick, with hail and rain all day.
18.	29.21	44°	ENE. Blew a gale of wind all day, with showers of rain.
19.	29.54	44°	Do. Dark weather, blowing hard all day.
20.	29.60	40°	E. Fine weather, blowing fresh all day.
21.	29.41	43°	ENE. Dark weather, blowing fresh all day.
22.	29.73	44°	E. Fine weather; blowing fresh, with a little rain.
23.	30.00	50°	Do. Remarkably fine clear weather.
24.	30.03	50°	N. A fine warm day.
25.	30.51	52°	Do. Fine clear weather; frost at night. The first day of summer by the Iceland almanack.

Days of the Month.	Barometer and Thermometer at Reikiavik.		Miscellaneous Observations.
	Barom.	Th.	
April 26.	30.53	28°	N. Fine weather, and hard frost.
27.	30.41	34°	ENE. Cold dark weather; blew hard towards night, and before morning a heavy gale of wind.
28.	30.30	24°	N. Blew a heavy gale of wind all day, and very cold.
29.	30.22	24°	Do. Fine weather, and hard frost; blew hard at night.
30.	30.00	26°	Do. Blew a heavy gale of wind all day, and hard frost.
May 1.	30.00	26°	Do. Fine day, blowing fresh all day.
2.	39.91	26°	Do. do. do.
3.	29.73	28°	Do. Fine weather.
4.	29.80	34°	Do. Fine weather, blowing fresh all day.
5.	29.84	38°	E. Dark weather, blowing very fresh all day.
6.	29.74	45°	ESE. Dark weather, and rain; blowing a gale of wind the whole day.
7.	29.80	44°	Do. Dark weather, blowing a gale of wind all day.
8.	29.93	49°	ENE. Dark weather; but cleared up at night, and froze.
9.	29.84	46°	W. Fine clear weather.
10.	29.91	50°	Do. Fine pleasant day.
11.	30.00	49°	N. Very fine weather.
12.	29.72	46°	E. Dark weather and sunshine by turns.
13.	29.72	50°	W. Dark and sunshine at intervals; rain at night.
14.	29.82	50°	Do. Fine; frost at night.

The Greenland ice grounded in the northern parts of the country some time in February, and continued to increase daily till it inclosed nearly two thirds of the island. In the month of June, the sea was not visible from the tops of the highest mountains; so completely was the land beset with these tremendous Icebergs. Numbers of the polar bears made their appearance. From the accounts I have been able to gather, the winter has been a very severe one; and the oldest people do not recollect such a succession of gales of wind. In the north country, and in those places where the ice grounded, the inhabitants suffered greatly from the want of provisions, not being able to go to sea for fish; they lost also a great number of cattle, sheep, and horses, which died for want of food.

No. VII.

REMARKS OCCASIONED BY THE HONOURABLE MR BENNET'S SKETCH OF THE GEOLOGY OF MADEIRA.

SINCE the preceding sheets were printed, I have read a paper in the Transactions of the Geological Society of London, by the Honourable Mr Bennet, descriptive of some facts he observed in the island of Madeira. The examination which Mr Bennet made, though imperfect, from the shortness of the time he could allot for it, has been productive of much satisfaction to me, as it shews the importance I have ventured to attach to a more accurate examination of volcanic countries; and as the facts which he records appear to have a striking resemblance to those which we observed in Iceland. It is to be lamented that Mr Bennet has indiscriminately called every rock lava, which is a term requiring some limitation in the present advanced state of Geological science. His paper, however, is to me perfectly intelligible; and I now entertain the most sanguine hopes, that future researches in volcanic countries, will lead to a full developement of the progress of subterraneous heat, from the bowels of the earth, through the depths of the ocean, till at length its operations become visible on the surface. To the many remarkable instances already recorded, of volcanic eruptions appearing in the sea, we have now to add the recent formation of the island Sabrina, near the Azores. The great pressure of water, 240 feet, which the volcanic heat quickly overcame in this striking instance, while it conveys a proper idea of the magnitude of the power exerted, enables us to conceive what may be effected at greater depths; and to appreciate properly, the genius of Hutton, and the valuable experiments of Sir James Hall.

No. VIII.

As announced in the Preface, my sentiments respecting the conduct of the individuals concerned in the Icelandic Revolution were printed; but, on receiving notice that Mr Phelps was engaged in a law-suit, with the Underwriters who had insured his ship Margaret and Ann, and fearing lest any remarks of mine, (although they made nothing appear but the truth, and placed the affair in its proper light), might be considered as intended to influence the jury or the court, I have, with regret, cancelled the sheets containing them.

BREIDE   FIORD

FAXE

FIORD

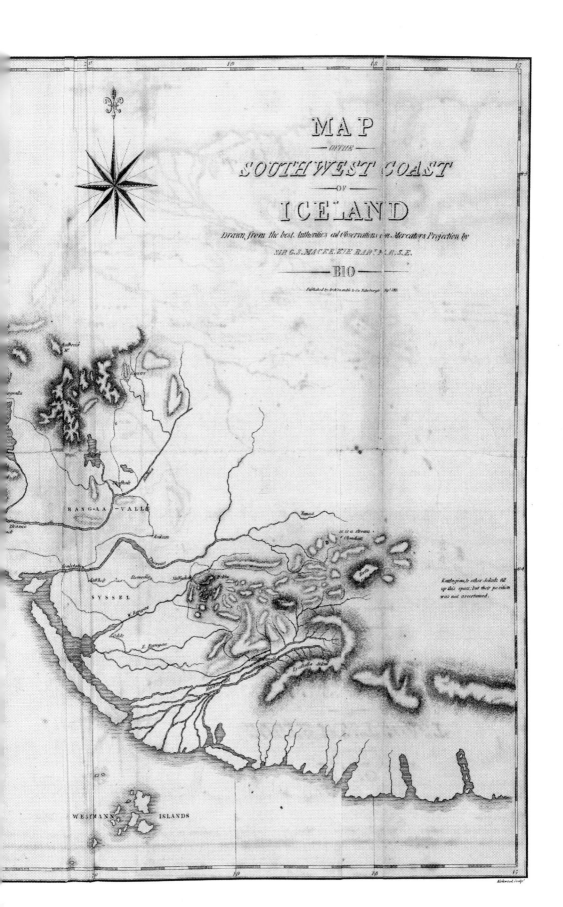

MAP
ON THE
SOUTH WEST COAST
OF
ICELAND

Drawn, from the best Authorities and Observations on Mercators Projection by

SIR G. S. MACKENZIE BART. F.R.S.E.

— B10 —

Published by Archibald & Co Edinburgh Sep.r 1811.

RANG·AA·VALLE

SYSSEL

It is a stream of Ökuldara

Kartogious & other details fill up this space, but their position was not ascertained.

WESTMANN ISLANDS

INDEX.

DIRECTIONS TO THE BINDER.

T. Allan & Co.
Printers, Edinburgh.